# MARINE LIFE

Photographs compiled and collated by
Gillian Lythgoe of Seaphot

**Main Photographic Sources**

Richard Chesher     William Gladfelter
Dick Clarke     Karl Kleemann
Neville Coleman     Larry Madin
Peter David     Christian Petron
Walter Deas     Peter Scoones
David George     Armin Svoboda

**Additional photographs from**

Anthony Baverstock     John Lythgoe
Peter Bertorelli     Michael Mastaller
British Museum (Natural History)     Maria Mizzaro
Peri Coelho     Pat Morgan
Mike Coltman     John Place
William Cross     Howard Platt
Jean Deas     Rod Salm
David Guiterman     Howard Sanders
John Harvey     Helmut Schuhmacher
Keith Hiscock     Alan Southward
Gertraud Krapp-Schickel     Heinz Splechtna
Michael Laverack     Neil Swanberg
Jan Lenart     Peter Vine
Roger Lincoln     Roman Vishniac

Line Drawings by
Stephanie Harrison

# MARINE LIFE

## An Illustrated Encyclopedia of Invertebrates in the Sea

by

J. David George, B.Sc., Ph.D., A.R.P.S.,
and
Jennifer J. George, B.Sc., M.Sc.

with a Foreword by
Sir Eric Smith, F.R.S.

A Wiley-Interscience Publication
**JOHN WILEY & SONS**
*New York*

Marine Life is published simultaneously by

Australia: Rigby Limited
Canada: Douglas & McIntyre Limited
France: Maloine Editeur
Netherlands: Zuidgroep BV Uitgevers
Spain: Ediciones Universidad de Navarra S.A.
United Kingdom: George G. Harrap & Co. Limited
United States of America: John Wiley & Sons

**Library of Congress Cataloging in Publication Data**

George, John David.
Marine life.

"A Wiley-Interscience publication."
Bibliography: p.
Includes index
I. Marine invertebrates.   1. George, Jennifer J.,
joint author.   II. Title.
QL121.G4     592′.09′2     79–10976
ISBN 0–471–05675–8

This book was created and prepared by
Lionel Leventhal Limited,
2–6 Hampstead High Street,
London NW3 1QQ.

Design and production in association with
Book Production Consultants,
Cambridge, England.

Photoset in 'Monophoto' Times by
Northampton Phototypesetters Ltd and printed and
bound in Great Britain by the Fakenham Press,
Fakenham, England.

# Acknowledgements

We would not have undertaken the daunting task of
writing such a book were it not for the encouragement
provided by Gillian Lythgoe to whom we owe a special
vote of thanks. We also thank her for assisting in the
selection of the photographs and for undertaking the
onerous task of compiling the indexes.

The considerable contribution made to this book by the
photographers cannot be overemphasised and we are
extremely grateful for their co-operation and patience
during its gestation. We are grateful also to the illustrator,
Stephanie Harrison, whose line drawings have done
much to enrich the text.

It is a pleasure to acknowledge the invaluable help
given by Dr A. Svoboda whose wide knowledge of the
ecology of marine invertebrates has resulted from over
20 years of diving observation. P. David, Dr K. Kleeman,
Dr G. Krapp-Schickel, Professor M. Mizzaro and Dr H.
Splechtna were also kind enough to provide us with
certain ecological information.

We are grateful to the following specialists, who gave
generously of their time to read and criticise various
chapters of the book, or who have responded to our
cries for help on nomenclatural and phylogenetic
matters:

Porifera: Professor P. R. Bergquist and Miss S. M. K.
  Stone
Coelenterata: Dr P. Cornelius, Dr K. W. Petersen and
  Dr B. Rosen
Platyhelminthes: S. Prudhoe, O.B.E.
Nemertea: Dr R. Gibson
Nematoda: Dr H. Platt and J. Coles
Priapulida, Sipuncula, Echiura: R. W. Sims
Crustacea: Dr G. Boxshall, Dr A. Fincham, Dr R. Ingle
  and Dr R. Lincoln
Chelicerata: D. McFarlane
Tardigrada: Dr C. I. Morgan
Mollusca: Dr A. Scheltema and Dr J. Taylor
Bryozoa: Miss P. Cook
Brachiopoda: Dr H. Brunton and E. Owen
Echinodermata: Miss A. M. Clark
Hemichordata and Chordata: Dr R. P. S. Jefferies and
  G. Patterson
We shall always be indebted to Bob and Dennis Wright,
who patiently guided our first faltering excursions
beneath the water surface. Alec McAulay edited the
typescript and prepared it for the printer. Finally, special
thanks are due to Janet Devenport and Fanoulla
Perendes, who struggled mightily to transform the hand-
written manuscript into a typescript for the publisher.

*David and Jennifer George*

# Contents

# Foreword

## by Sir Eric Smith, F.R.S.

When we look at animals that are unfamiliar and perhaps seen for the first time, certain questions that inevitably come to mind are: what kinds of organisms are they and to what extent do they resemble other more familiar creatures; where and how do they live; how are their several styles of construction and repertoire of organs and parts fitted by their functions to the environments in which they operate, and to what extent are these adapted for the benefit of the whole organism?

Zoology as a science enquires into these questions and seeks to give well-founded and informative answers about them. The growth of zoological knowledge and the progressive enlargement of understanding of the nature, habits and relationships of animals depends, however, not only on finding out as much as possible about individual kinds of organisms, but on knowing more about a greater variety of them. The central theories and generalisations of zoology derive from an appreciation and assessment of the nature and significance of the immense variety of animal form and adaptation and, in turn, provide the principal directives for continuing enquiry and research. The theory of evolution and its pervading influence on biological thought and research, the concept of ecosystems and the interdependence and interaction of organisms in nature, and broadly based comparative studies of aspects of animal function, are founded on this variety and will be progressively developed and refined as the full extent of the variety becomes better known.

Evolutionary change ensures that, so long as life remains on earth, the variation of animals will be unlimited, the panorama slowly but continually changing with the emergence of new forms with adaptational advantages to replace those that are less competitive. The pace of evolutionary change and opportunistic exploitation of new and changing environments varies greatly from one group of animals to another with the predominantly terrestrial insects being the present day pacemakers. In its broader perspectives the panorama of the animal kingdom is seen by zoologists to be made up of between 25–30 distinctive architectural styles (the phyla). All but one of these phyla are comprised of invertebrate animals, with more than half entirely marine and the rest predominantly so. Our knowledge of them is, in most instances, limited both in respect of the number of their species and of their way of life and role in the economy of the sea. Generally speaking, however, the animals that are least well known are those that are small, inaccessible and of puzzling appearance and construction. The, as yet, pioneer explorations of deep-sea submersibles and cameras, and the now wide-ranging surveys and excursions of SCUBA divers within the highly productive coastal waters have opened great areas of the sea to direct viewing. In the past, zoologists have too often had to work with dead material or dredged specimens of limited life and divorced from their natural surrounds. This is no longer so, and the new generation of marine naturalists, amateur and professional, have almost unlimited opportunities for observing animals in their natural habitats and over long periods. The problem of identifying and describing the things seen still remains with creatures that are puzzling and unfamiliar.

*Marine Life: An Illustrated Encyclopedia of Invertebrates in the Sea* has been prepared with patience and skill by David and Jennifer George, both of whom are experienced and expert zoologists of high professional reputation. With its wealth of aesthetically pleasing and scientifically informative colour illustrations, this fine book will be an invaluable field guide.

# Introduction

In *Marine Life* we have provided readers with a combination of text and colour photographs of the living marine animals in their natural environment, which should enable readers to identify and classify the marine invertebrates that they see, both on the shore and underwater, anywhere in the world. In addition, the text includes biological, behavioural and ecological notes of interest.

This book is intended to fill a gap that existed in the literature on marine invertebrates, as well as to provide a book that will be useful to professional and amateur marine biologists alike. Considering their importance in the marine ecosystem, invertebrates have received relatively little attention in the popular literature compared with their vertebrate cousins, the fishes.

We hope that students of biology who require a working knowledge of modern classification of the invertebrates will find particularly useful the provision of classification trees which serve to reflect the evolutionary relationships within the invertebrate groups. Specialists, particularly those who do not dive, will find that the photographs of living animals underwater provide a useful comparison with the dead material that they study in the laboratory. Finally, it is our hope that the book will appeal to the ever increasing number of laymen and SCUBA divers who wish to learn more about animal life beneath the surface of the sea.

Of the total of over one million species of animal that exist in the world today, about 97% are without backbones, and are collectively known as the invertebrates. The remaining 3%, those with backbones, are the well known vertebrate groups: fishes, amphibians, reptiles, birds and mammals. Apart from the fishes, the vertebrates are poorly represented in the sea in terms of number of species. Here the invertebrates hold sway, much as they do on land, occupying every conceivable habitat from high on the shore to the abyssal depths of the ocean.

To an untrained observer, the range of form amongst marine invertebrates must seem almost limitless, yet it is possible, with care, to arrange them in groups based upon similarities of body form and structure. It is one of the prime intentions of this book to acquaint the reader with the range of invertebrates existing in the sea and to enable him to identify many of them and to classify all of them in an orderly manner.

*Often, standing on the shore at low tide, has one longed to walk on and in under the waves, and see it all but for a moment.*

These wistful words of Charles Kingsley in his book *Glaucus; or The wonders of the shore*, published in 1855, reflected the feeling amongst both professional and amateur marine biologists of that period. Happily for the many hundreds since, the techniques of diving and photography were developing at the very time that Kingsley made his memorable statement. It is the rapid advances in recent years in these two important techniques that have contributed to making a book such as this a reality.

As far as can be ascertained, the first biologist to use a diving helmet supplied by air pumped from the water surface was Professor Milne Edwards who, in 1844, ventured several metres below the surface of the Mediterranean sea. Approximately forty years later, Louis Boutan, of the marine laboratory at Banyuls-sur-Mer, took the first successful underwater photographs. However, the diver was still tied to the surface by his air umbilicus and unable to roam freely over the bottom. The invention in the late 1930s of self-contained underwater breathing apparatus (SCUBA) at last enabled man freely to 'walk in under the waves'.

Several advantages accrue from the individual being able to study marine invertebrates *in situ*. Not only can he see the shape and colour of the live animal, but can also observe its behaviour and that of the countless other animals and plants that live in ecological balance with it. Obviously, aquarium studies, although of value, must always be suspect when compared with those carried out in the natural environment.

When it becomes necessary to collect the animals in good condition from below tide level for laboratory studies the surface-based investigator is at a disadvantage. His crude trawls, dredges and grabs damage or destroy many delicate specimens, and are virtually useless for collecting from rocky subtidal areas and from coral reefs. The diver, on the other hand, can search beneath the seaweed canopy and under boulders, probe into crevices and swim into caves. Recently even the exquisite beauty and often complicated lifestyle of oceanic drifting invertebrates has been subjected to scrutiny by diving scientists (Hamner, 1974). Nowhere beneath the sea surface are animals free from the probing eyes of man; the invention of such underwater vehicles as the bathyscaphe, and the proliferation of submersibles in the last decade, has meant that direct observations can be made on marine life below divable depths.

Before photography became a readily available tool, marine biologists relied upon words and drawings to communicate their observations on living organisms to others. Rapid advances in recent years, particularly in the quality of colour film emulsions and in the techniques of underwater photography, have made it possible to obtain first-class photographs of living marine animals in their natural environment, from these positive identifications can be made. This approach cannot be applied equally well to all groups of marine invertebrates, many of which must be removed from the sea and examined closely in the laboratory before they can be identified with any degree of certainty. The importance of obtaining a photographic record of living marine invertebrates cannot be over-emphasised since on preservation soft-bodied organisms in particular may become so distorted as to be almost unrecognisable to anyone but an expert. Even invertebrates with a hard external skeleton tend to assume unnatural attitudes and to lose appendages when placed in preservative. The colour and pigment patterns of living marine invertebrates are sometimes distinctive, and can be used in addition to body shape to facilitate identification. Unfortunately, when animals are preserved their colours frequently change or are lost completely. This fact emphasises still further the desirability of obtaining a photographic record before an animal is preserved for later identification and classification.

The great variety of organisms in the world has necessitated the establishment of a uniform system of naming and classifying them. The system used here is familiar to all professional biologists, but perhaps needs explaining to the layman. Animals constituting a species can loosely be defined as a group of individuals capable of interbreeding, but which cannot breed with individuals from another such group. Each species is given a two-part Latin name, usually printed in italic type: the first (generic) part of the name, which is written with an initial capital letter, is shared by closely related species. The second (specific) part of the name, which is written with a small initial letter, refers only to that particular species. Hence a species of winkle, *Littorina littorea*, is closely related to other species of the same genus, e.g. *Littorina obtusata* and *Littorina neritoides*, but cannot interbreed with them. This binomial system of naming is in universal use today for all living things. Although it may appear unnecessarily complicated to the layman, its one great advantage over the use of common names is that the scientific name stays the same no matter where a species is found in the world. Common names often vary from one geographical region to another. Although the generic and specific names are sufficient to identify any species, strictly the surname of the person who first described the organism should follow the specific name, e.g. *Ostrea edulis* Linnaeus. A species which has been originally described in one genus, and, in the light of further research, is transferred to another, has the name of the original author placed in parentheses, e.g. *Eunicella verrucosa* (Pallas).

Just as several species can be grouped into a genus, so several genera can be grouped into a family unit, and so on. Such a hierarchical system of arranging species into groups of related forms is used in this book. The groups are not merely units of convenience, but serve to reflect the evolutionary relationships and ancestry of the living organisms (phylogeny). The principal divisions employed within the phylogenetic classification of the Animal Kingdom are: Phylum, Class, Order, Family, Genus, Species. In practice, it is often convenient to further subdivide these categories. For example, the winkle *Littorina littorea* (Linnaeus) is classified as follows:

*Subkingdom* Metazoa
*Phylum* Mollusca
*Class* Gastropoda
*Subclass* Prosobranchia
*Order* Mesogastropoda
*Superfamily* Littorinacea
*Family* Littorinidae
*Genus* Littorina
*Species* Littorina littorea (Linnaeus)

Because the experts sometimes disagree over the ancestry of species, the classification above family level in many phyla is not entirely stable. Hence we have made it clear on whose system of classification we have based the phylogenetic trees given in each chapter. It is hoped that professional biologists will find these trees particularly useful for we feel that they are the best available at present (November, 1978). At both higher and lower levels of classification, where a recent name-change has taken place, or where a well known alternative name exists, the name judged to be the less suitable has been given in parentheses after the chosen name. e.g. Scleractinia (= Madreporaria); *Neanthes virens* (Sars) (= *Nereis virens*).

In writing this book, which deals with extant marine invertebrates on a world-wide basis, the size of the task has forced us to set certain limits on our coverage. We have found it possible to deal only with non-parasitic forms, and little more than passing reference has been made to larval stages. Invertebrates that have invaded the shore and sea from land (e.g. insects), or that inhabit inland saline waters, have also been excluded.

Each of the 27 invertebrate phyla with marine representatives is considered in turn. An account of the general morphology and ecology of each phylum is given before it is split into its component classes and orders, using differences between external features of the animals whenever possible and internal features only when absolutely necessary.

The species numbers given for the phyla and their constituent classes refer only to living species and are based on the latest available information. Only those families containing marine species are listed and, when feasible, they are arranged in phylogenetic sequence. Where the phylogenetic sequence is not known, they are listed alphabetically. It has not proved possible to obtain photographs of representative species from each family, but, in the great majority of cases, specimens have been photographed in their natural environment. The captions for the photographs include ecological and behavioural notes as well as the size and geographical distribution of the animal.

For convenience, the world's oceans and seas have been divided into eleven regions: Arctic, Antarctic, North Atlantic, Mediterranean, Caribbean, South Atlantic, North Pacific, Panamanian, South Pacific, Indo-Pacific and Zealandic (Figure 1). In the captions to the plates the animals have been assigned to one or more of these regions, thus indicating the area in which they are most likely to be found. Although the species may occur only in the region(s) indicated, similar species within the same genus may be found elsewhere.

It is not possible to write a text spanning the marine invertebrates without the introduction of a few technical terms. These terms are defined when they are encountered in the text for the first time as well as in a glossary at the end of the book.

It is our hope that amateur and professional marine biologists alike will find this book a useful reference work, as well as deriving considerable aesthetic enjoyment from the wide range of photographs included; these have been taken by some of the world's finest underwater photographers. It is difficult to express the pleasure we have received from examining the many thousands of photographs, sometimes of exquisitely beautiful and bizarre invertebrates, whilst making the selection for this book.

The numbers in the margin against each caption refer to the photographic plates. The first number is that of the plate whilst the second number, following the oblique stroke, relates to the illustration on that plate. (e.g. 18/4 refers to the photograph of *Chrysaora hysoscella* which is illustration number 4 in Plate 18).

**Figure 1** *World map showing the main marine zoogeographic regions*

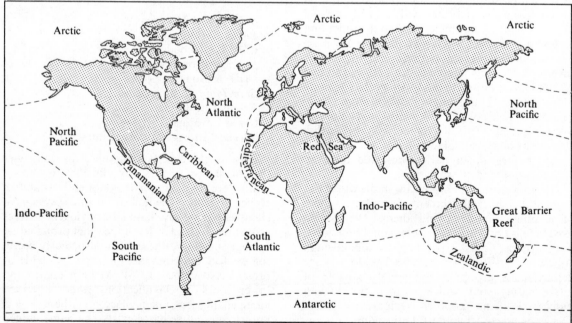

# Kingdom Animalia

The Kingdom Animalia can be grouped into two subkingdoms, the Parazoa and Metazoa, based upon the level of cellular organisation of the animals within them.

# Subkingdom Parazoa

Animals composed of aggregations of cells which retain a high degree of independence from one another and are not consistently differentiated into tissue layers. The subkingdom contains only one phylum, the Porifera.

# Phylum Porifera

[sponges] approx. 10000 species

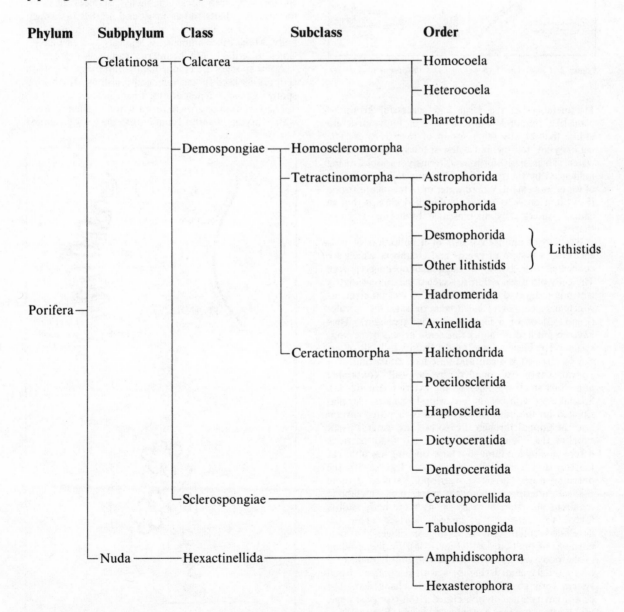

No generally acceptable overall classification of the Porifera exists at the moment. The Calcarea are classified according to Burton (1963) and Vacelet (1970). The arrangement of the Demospongiae is based on the ideas of Levi (1973) and Bergquist (1978), and the Sclerospongiae classified as in Hartman & Goreau (1970, 1975). The arrangement of the Hexactinellida follows Ijima (1927).

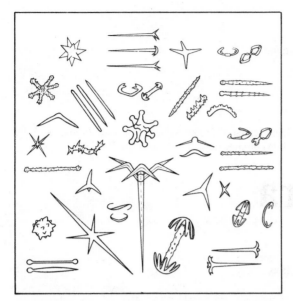

**Figure 2** *A selection of sponge spicules showing a wide range of form.*

The members of this large and successful group of plant-like immobile animals have a body structure unlike that of any other group of invertebrates. The majority are marine, but a few species occur in freshwater. Their body form is extremely variable, being influenced by the type of substratum and the amount of water movement. Where water movements are strong they often grow as round or flattened clumps, but in calmer waters they may assume branching tree-like shapes.

Basically, a sponge consists of a collection of cells enclosing a system of canals and chambers which are connected to the exterior through small openings (pores). The cells often lie within a secreted gelatinous matrix which is supported on a framework of skeletal elements which may consist of calcareous or siliceous spicules (Figure 2) or of a fibrous material (spongin). They have no gut and no noticeable sense organs or nervous system. The basic sponge model from which all others can be derived is small and vase-like, consisting of a central cavity surrounded by a wall containing numerous small openings to the exterior (Figure 3a). Special cells with collars and whip-like hairs (flagella) situated on the inner body wall draw a water current into the animal through the pores. The water current supplies the sponge with oxygen and food particles before passing out through a large opening (osculum) at the top of the vase. Waste material leaves with the outgoing water current. Larger sponges need a more efficient filtering system to meet their metabolic requirements. Thus in many sponges the body wall is folded to increase the internal surface area, the flagellated collar cells occurring in finger-like out-pushings of the body wall (Figure 3b). Further folding of the body wall leads to the formation of small, oval or round flagellated chambers with a complex canal system through which the water is channelled. The central cavity is usually obliterated in this type of sponge, which often develops a number of oscula (Figure 3c).

The majority of sponges contain both male and female sex cells (hermaphrodite), the eggs and sperm being produced at different times. In the few species in which reproduction has been studied in detail, fertilisation of the egg occurs inside the parent body, where it develops into a small flagellated larva. The larva leaves the parent through an osculum and leads a short free-swimming existence before settling to the bottom and developing into a sponge. Sponges have remarkable powers of regeneration, small broken pieces being capable of growing into a complete sponge. Reproduction by asexual budding or by splitting of the body into small parts is common. In some species, asexual reproductive bodies (gemmules) are formed. These are collections of cells and spicules surrounded by a thick wall; when they are released from the parent they develop into a new individual.

Sponges are found in all seas, living mainly in shallow waters although some occur at great depths. The majority attach themselves to any suitable surface such as rock, hard-shelled animals and seaweeds. A few species bore into rocks or shells. Sponges feed on microscopic plants and animals, and on detritus, which they filter from the water current passing through the body. Many other animals use sponges as a surface on which to settle, or use the internal canals as a refuge.

A few sponges are commercially important, e.g. bath sponges are used for cleaning and polishing. The boring sponge, *Cliona*, is a pest of oyster beds in some countries.

The variation of growth forms within a single species with varying habitat conditions makes it almost

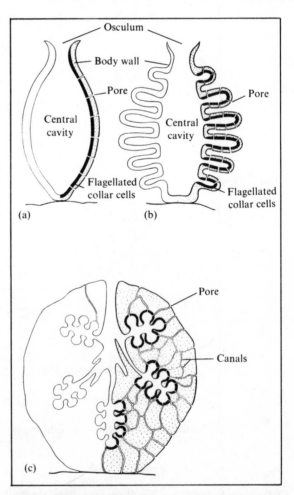

**Figure 3** *Sponge structure.* (a) *Basic sponge model from which all others can be derived.* (b) *Body wall with simple folding to increase the internal surface area.* (c) *Complex folding producing flagellated chambers.*

impossible to identify sponges using only external characters. The type and arrangement of spicules within the body wall is commonly used in identification, but other aspects, such as general shape and consistency, colour, physiology and life-history, are also used.

# Subphylum Gelatinosa

Sponges in which the cells of the body are embedded in a transparent gelatinous matrix in which the spicules and water canals are also situated. The body has a definite layer of cells covering the outer and inner surfaces.

## Class Calcarea

approx. 50 species

An exclusively marine group of sponges, whose members have a skeleton of calcareous spicules always present. The spicules cannot be clearly differentiated into two size-categories. The sponges are usually small, less than 10 cm in height, with vase- or cushion-shaped bodies or with a branching growth form. A few species are attached to the substratum by a stalk. The spicules, which are not fused together (except in some Pharetronida), characteristically have three (sometimes four) rays radiating from a central point. In addition, simple rods may be present. Internal structure ranges from the simplest type with a single large cavity lined with collar cells to complex forms with many branching canals and flagellated chambers.

Calcareous sponges have a world-wide distribution, occurring mainly in the intertidal zone or in shallow waters. They usually live in sheltered situations beneath rocky overhangs, in crevices or amongst algae. Their bodies are most frequently white or a pastel colour.

### Order Homocoela

Simple sponges with the whole of the large central cavity and its outgrowths lined by collar cells.

Genera: *Clathrina, Dendya, Leucosolenia.*

1/1, 1/3   *Clathrina coriacea* (Montagu). In areas of strong water movement this species forms low, lace-like encrustations of anastomosing tubes spreading over large areas of rock (Plate 1/1). In sheltered areas the shape is more upright (up to 5 cm high) and the photograph of a yellow form shows the smooth-walled tubes comprising an individual joining to empty through a single osculum (Plate 1/3). An individual starts life as a simple vase-like form which then branches continuously until the familiar complicated structure is produced. Distribution appears to be cosmopolitan in shaded areas.

1/2   *Leucosolenia botryoides* (Ellis & Solander). This species has an upright branching tubular form up to 2 cm high, rising from an anastomosing mass of rootlets. Attaches to shells, colonial animals and seaweeds. Found in all seas but uncommon in the Indo-Pacific.

1/4   *Leucosolenia* sp. A specimen from the Great Barrier Reef.

### Order Heterocoela

Sponges in which the collar cells do not line the central cavity, but are confined to chambers.

Genera: *Achramorpha, Amphoriscus, Amphroceras, Eilhardia, Hypograntia, Lamontia, Leucettusa, Leuconia, Scypha, Sycettusa, Sycodorus, Sycolepis, Sycyssa, Uteopsis.*

*Leucettusa imperfecta* (Potejaeff). A species with a finely   1/8 stippled surface and a sprawling or upright growth form. Specimens rarely reach a height of more than 5 cm above the rocky substratum to which they are attached. Occurs in the Indo-Pacific, Zealandic and South Atlantic regions.

*Scypha ciliata* (Fabricius). Tubular or barrel-shaped   1/5 individuals up to 10 cm high, sometimes growing in clumps on rocks, shells, colonial animals and seaweeds. The surface has a furry appearance due to protruding spicules, and a crown of long strong spicules occurs around the osculum. It has a cosmopolitan distribution in sheltered intertidal and shallow waters.

*Scypha compressa* (Fabricius) [purse sponge]. This   1/6 laterally flattened species, which may reach a length of 10 cm or more in sheltered areas, is often found in clumps under rocky overhangs and attached to seaweeds, sometimes in the company of *S.ciliata*. It has a cosmopolitan distribution.

*Scypha gelatinosa* (Blaineville). This species, which often   1/9 occurs in branching clumps up to 10 cm high, has a characteristic reticulate pattern on the sides of the body. Found on Indo-Pacific reefs.

### Order Pharetronida

Sponges in which the collar cells are confined to chambers. The skeleton is produced in a number of ways. It may be composed of 3- or 4-rayed spicules cemented together, of 4-rayed spicules fused together, or of a rigid calcareous framework without spicules. In one family (Lepidoleuconidae) the main skeletal support is provided by oval scales (possibly derived from 3-rayed spicules) on the body surface. The osculum is frequently ringed by 4-rayed spicules. Peculiar 3-rayed spicules shaped like tuning-forks are usually present.

Modern representatives of this ancient order have been found only in caves.

Families: Lelapiidae, Minchinellidae, Murrayonidae, Paramurrayonidae, Lepidoleuconidae.

## Class Demospongiae

approx. 9500 species

A widespread group of sponges which, apart from a few freshwater forms, are marine. The skeleton is composed of siliceous spicules, spongin fibres, or a mixture of both, never of calcareous spicules. In a few species the skeleton is lacking altogether. The spicules, which are scattered throughout the body, are variously shaped although fundamentally 4-rayed, never 6-rayed. The spicules can usually be loosely grouped into two categories: large supporting spicules (megascleres) and smaller forms (microscleres). The body is extremely variable in form, ranging from branching, vase- or cushion-shaped forms to encrusting masses which may reach several metres across. The internal canal system has small chambers lined with flagellated collar cells. There is usually no large central cavity; instead one or more wide canals convey the outgoing water to the osculum.

These sponges are common in coastal waters of all

oceans, but also are found to depths of 6000 m or more. They are frequently brightly coloured, especially in shallow water, and many grow in well illuminated places.

2/1, 2/2, 2/3, 2/4, 2/5    Microscleres from several different species of demosponge magnified several thousand times to show the intricacy of their structure.

## Subclass Homoscleromorpha

The skeleton, which is absent in a few species, consists of a large number of small similar-sized spicules, often 3-rayed, (occasionally 2- or 4-rayed) scattered throughout the body of the sponge.

Families: Oscarellidae, Plakinidae.

### Family Oscarellidae

1/7   *Oscarella lobularis* (Schmidt). A soft, lobed, encrusting species without a skeleton, occurring in clumps up to 10 cm across on rocks in the Mediterranean and North-east Atlantic.

## Subclass Tetractinomorpha

Large sponges which may reach a diameter or a height of 50 cm or more, although most species are smaller. The skeleton normally consists of a framework of 4-rayed megascleres, and/or 1-rayed megascleres. In many species, star-shaped microscleres of several different types are also present, although hooked microscleres are never found. Spongin fibres are lacking in most groups in this subclass.

### Order Astrophorida (=Choristida)

Sponges which usually have star-shaped microscleres, as well as an incomplete framework of megascleres, radiating out from the centre of the sponge.

Families: Geodiidae, Calthropellidae, Stellettidae, Pachastrellidae, Theneidae, Thrombidae, Jaspidae.

Problematic family: Epipolasidae.

### Family Geodiidae

2/7   *Pachymatisma johnstoni* (Bowerbank). A grey-brown species which forms thick encrusting masses up to 50 cm across on rocks. The oscula are normally situated on top of the lobes. Found in greatest abundance below the kelp zone in the North-east Atlantic, where it is often found bridging crevices and gaps between boulders.

### Order Spirophorida

Globular sponges with a well developed radiating framework of megascleres and spirally shaped microscleres.

Family: Tetillidae.

### Order Desmophorida [lithistids]

A group of rare sponges living in low-light areas of warm seas. They have 4-rayed megascleres as well as other large spicules with characteristic branching rays which frequently fuse at their tips to form a rigid framework.

Families: Theonellidae, Corallistidae, Pleromidae.

### Other lithistids

Sponges which do not have the typical 4-rayed megascleres, although spicules of varying sizes and shapes occur.

Families: Scleritodermidae, Leiodermatiidae, Desmanthidae, Siphonidiidae, Vetulinidae, Petromicidae.

## Order Hadromerida

Sponges with a radiating skeleton of 1-rayed megascleres which frequently have a knob at one end. The microscleres, if present, are usually star-shaped, or are derived from this basic form.

Families: Tethyidae, Polymastiidae, Suberitidae, Spirastrellidae, Clionidae, Placospongiidae, Timeidae, Stylocordylidae, Chondrosiidae, Latrunculiidae.

### Family Tethyidae

*Tethya aurantium* (Pallas). Forms golf-ball-like growths up to 8 cm in diameter, which are usually yellow or reddish-yellow in colour. The surface is covered with prominent papillae. Found attached to rocks subtidally, sometimes in areas with a high silt content. Appears to have a cosmopolitan distribution.    2/9

*Tethya diploderma* Schmidt. Similar in form and colour to *T. aurantium*. The specimen, which is covered with fine sand, has several oscular openings and is expelling water. It tends to grow in shallow water on reefs and attached to dead shells in sea-grass beds. Found in the Caribbean.    2/6

*Tethya* sp. A specimen from an Indo-Pacific reef.    2/11

### Family Polymastiidae

*Polymastia boletiformis* (Lamarck). Forms flattened, rounded growths up to 6 cm across, with large projecting conical processes, each bearing an osculum. Attached to hard bottoms, often in areas of strong current. Found in the North-east Atlantic.    2/8

### Family Suberitidae

*Suberites domuncula* (Olivi) [sulphur sponge]. The common name of this sponge derives from its pungent odour. It is a rounded, encrusting form with a smooth surface sometimes reaching up to 30 cm across. It is not unusual to find this species, which varies in colour from yellow to reddish-yellow, attached to rock intertidally (Plate 2/10) as well as subtidally (Plate 2/12). It is also found encrusting the shells occupied by hermit crabs. Over a period of time the shell is dissolved by chemical secretions of the sponge, which then provides direct protection for the crab. Plate 3/1 shows the hermit crab, *Pagurus arrosor*, within the sponge. Occurs in the Mediterranean and North-east Atlantic.    2/10, 2/12    3/1

An unknown suberitid forming an encrusting growth several metres wide over shallow rocks in the Caribbean.    3/2

A close-up view of this unknown suberitid from the Mediterranean reveals the spicular sieves protecting the closely packed inhalent openings.    3/6

### Family Spirastrellidae

*Anthosigmella varians* (Duchassaing & Michelotti). An encrusting species of varying thickness and colour, with a rather coarse networked surface. The partially raised oscula are surrounded by paler tissue than elsewhere. Forms growths up to 50 cm across on Caribbean reefs.    3/3

14

3/5   *Spheciospongia vesparium* (Lamarck). When fully mature this cylindrical sponge may reach a diameter of 1 m and a height of over 50 cm. Oscula are situated in a depression at the top of the brown body. Young specimens, however, are very different in appearance, being oval in shape and having a spiky surface. Found on Caribbean reefs.

3/4, 3/7   *Spirastrella cunctatrix* Schmidt. This encrusting sponge may form masses over 30 cm across. It is characterised by the spirally arranged pattern of water-collecting channels leading to the oscula. Although orange is the most usual colour for this species it sometimes has a blue, green or grey hue. Found in shaded areas in the Mediterranean and adjacent Atlantic.

### Family Clionidae

*Cliona caribboea* Carter [boring sponge]. Often forms massive yellow encrusting growths up to 50 cm across. The body surface has many flattened papillae and raised oscula. The sponge achieves its attachment to limestone rocks and shells by chemical dissolution of the limestone and replacement by a branching root system of sponge tissue. Occurs in the Caribbean.

4/1   *Cliona celata* Grant [boring sponge]. Closely resembles *C. caribboea* in form and habit. This species can be a serious pest of commercial shellfish beds in Europe. Reportedly has a cosmopolitan distribution.

4/2, 4/3   *Cliona lampa* de Laubenfels [red boring sponge]. Very similar in habits and form to *C. celata*, although the surface seems always to have a strong red coloration. Recorded on rocks, reefs and shells in the Caribbean.

### Order Axinellida

A group of erect sponges, often branching, which may reach a height of 50 cm or more. They are usually yellowish-orange in colour. The skeletal framework is of two types, either consisting of an axial skeleton of longitudinally arranged 1-rayed megascleres from which a radiating skeleton arises, or, more usually, consisting of an axial network containing spongin fibres which is more dense in the older parts of the sponge. Microscleres may or may not be present.

Families: Axinellidae, Bubaridae, Desmoxyiidae, Hemiasterellidae, Trachycladidae, Rhabderemiidae, Sigmaxinellidae, Raspailiidae, Euryponidae, Agelasidae.

### Family Axinellidae

4/4   *Axinella damicornis* (Esper). The short, flattened, fan-shaped branches, bearing oscula at their tips, reach a height of about 15 cm. Often the branches coalesce. Attached to rock in shaded areas in the Mediterranean.

4/9   *Axinella polypoides* Schmidt. This branching axinellid with attached *Scypha ciliata*, reaches a height of about 20 cm. It is characterised by its furry appearance and star-shaped oscular regions. Attached to rocks in the Mediterranean and North-east Atlantic.

4/6   *Axinella verrucosa* (Esper). The relatively smooth surface of this species compared with *A. polypoides* makes it one of the favourite anchoring points for *Parazoanthus axinellae* (see p. 34). Found on rocks in the Mediterranean and the North-east Atlantic.

4/5   *Phakellia* sp. A cup-shaped specimen about 15 cm high from the Indo-Pacific.

4/7   An unknown axinellid from the Indo-Pacific.

### Family Raspailiidae

*Raspailia hispida* Montagu. A species very similar in form and colour to some axinellid sponges. Can be differentiated with certainty only by microscopic examination of the spicules. Found on rocks in the North-east Atlantic.   4/8

### Family Agelasidae

*Agelas clathrodes* (Schmidt). This sponge, which often reaches a height of 30 cm or more and spreads over large areas, has a typical pock-marked surface (Plate 4/10). Anemone-like zoanthids (see p. 34) can sometimes be seen occupying these holes, rather like tenants in a block of high-rise flats (Plate 4/11). This sponge usually has large oscula directed upwards, the whole colony sometimes giving the appearance of a series of united barrel-shaped pots (Plate 5/6). There is sometimes competition between different sponge species, as can be clearly seen in Plate 5/1, which shows *Agelas* and another sponge overgrowing a coral colony. Found on Caribbean reefs.   4/10, 4/11, 5/1, 5/6

## Subclass Ceractinomorpha

Sponges with a skeleton of spicules and spongin fibres or with spongin fibres only. The megascleres always have one ray (never four) and the microscleres are usually in the form of curved hooks, never star-shaped.

## Order Halichondrida

Encrusting sponges which possess only a small amount of spongin in their skeleton. The megascleres are often arranged haphazardly in the body and the microscleres are usually absent.

Families: Halichondriidae, Hymeniacidonidae.

### Family Halichondriidae

*Halichondria panicea* (Pallas) [breadcrumb sponge]. So called because of the ease with which it crumbles when handled. An encrusting form up to 2 cm thick which sometimes covers large areas of rock, particularly under overhangs and in shaded crevices in the intertidal zone and shallow subtidal region. The surface is covered with numerous conical protruberances bearing the oscula. The colour varies normally between dark green and light yellow. Found on rocks, shells and seaweeds in non-tropical regions.   5/7

### Family Hymeniacidonidae

*Hymeniacidon perleve* (Montagu). Forms irregular encrusting growths, sometimes 30–40 cm across, over hard objects. The body surface may be smooth or furrowed and has oscula scattered over the surface, sometimes on raised papillae. It is commonly orange in colour but may be dark red. Widely distributed in non-tropical intertidal and shallow subtidal zones.   5/2

## Order Poecilosclerida

A large group of sponges with a skeleton of megascleres frequently joined together by spongin fibres. The abundant hooked microscleres are variously shaped and are important in the classification of the families in this order.

Families: Mycalidae, Hamacanthidae, Cladorhizidae,

Biemnidae, Esperiopsidae, Coelosphaeridae, Crellidae, Myxillidae, Tedaniidae, Hymedesmiidae, Anchinoidae, Clathriidae.

## Family Mycalidae

5/9   *Amphilectus fucorum* (Esper). A common encrusting species on rocks and kelp holdfasts in shallow waters of the North-east Atlantic. Could be mistaken for *Halichondria panicea* (see p.15).

5/3   *Hemimycale columella* (Bowerbank). An encrusting form up to 10 cm across, with a prominent reticulated pattern when expanded. Found on rock, shells and calcareous algae in shaded areas of the Mediterranean.

5/5   *Mycale macilenta* (Bowerbank). Forms a thin encrusting growth on rocks which may spread for 20 cm or more. Recorded from the North Atlantic and Zealandic regions.

5/8   *Mycale* sp. An unknown species from the Indo-Pacific. Sometimes referred to as the strawberry sponge because of its vague resemblance in colour, form and size to the fruit of the strawberry plant.

## Family Esperiopsidae

       *Crambe crambe* (Thiele). A common encrusting species of variable colour, forming patches up to 30 cm
5/10  across on subtidal rock surfaces (Plate 5/10). The collecting ducts leading to the oscula are raised above the surface of the surrounding tissue when the sponge
5/4   is fully expanded (Plate 5/4). Common in the Mediterranean and adjacent Atlantic.

## Family Myxillidae

2/4, 6/4  *Myxilla incrustans* (Johnston). An encrusting species with an irregular surface, forming growths several centimetres thick and up to 20 cm across on rocks and other hard surfaces. Colour ranges from yellow to orange. Found both intertidally and subtidally in the North Atlantic.

6/7   *Myxilla* sp. An encrusting form with a reticulate surface. Photographed on the Great Barrier Reef.

## Family Clathriidae

6/1   *Clathraria rubrinodis* Gray. This species forms characteristic antler-like growths on Indo-Pacific reefs.

7/1   *Microciona* sp. A specimen with a smooth surface appearance and small oscula, which encrusts hard objects in the Mediterranean. In the photograph the sponge has overgrown the shell of the Mediterranean jewel box, *Chama gryphoides*.

## Family uncertain

6/2   *Echinoclathria gigantea* (Lendenfeld) [honeycomb sponge]. This species, which reaches a height of 30 cm or more, has a distinctive honeycomb surface and consequently cannot readily be mistaken for any other sponge. It occurs on reefs in the Indo-Pacific.

6/3   This close-up photograph of a poecilosclerid sponge from the Indo-Pacific shows the delicate tracery of the spicules across its surface.

6/5   An unknown poecilosclerid forming a fan-like growth about 8 cm high. Long collecting channels can be seen running over the surface of the sponge. Photographed on rocks off Mombasa.

## Order Haplosclerida

Sponges with a well developed reticulated skeleton of spongin fibres, which often completely envelops the spicules. The megascleres are unusual amongst those in the ceractinomorphs in that they vary little in size within a specimen. The microscleres, when present, are hooked.

Families: Haliclonidae, Renieridae.

### Family Haliclonidae

*Adocia* sp. Two specimens varying greatly in form and colour. Photographed on an Indo-Pacific reef.    6/6, 6/10

*Cribrochalina* sp. An unmistakable vase-shaped sponge occurring on reefs in the Red Sea.    6/9

*Haliclona (Amphimedon) compressa* Duchassaing & Michelotti. A characteristically red-coloured species, with a rubbery consistency, which normally forms upright branching colonies on reefs in the Caribbean.    6/8

*Siphonochalina siphonella* Levi. A tree-like branching tube sponge which can reach a height of 30 cm or more. Found on reefs in the Indo-Pacific.    7/3

*Siphonochalina* sp. Two specimens of tube sponge from the Mediterranean, possibly *S. crassa*, and reaching a height of 10 cm or more.    7/2, 7/5

*Spinosella plicifera* (Lamarck) (= *Callyspongia plicifera*). A tube sponge up to 40 cm in height, shaped like an inverted cone with a large opening at the top. The exterior surface is markedly furrowed and often appears to fluoresce in strong sunlight. Found on Caribbean reefs.    7/7

*Spinosella vaginalis* (Lamarck) (= *Callyspongia vaginalis*) A distinctive species forming clusters of open-ended tubes up to 1 m high. The exterior surface is less corrugated than that of *S. plicifera*. Common on reefs and attached to rocks on open sand in the Caribbean.    7/4

*Xestospongia exigua* (Kirkpatrick) [string sponge]. An elongated, thin-branching form with a branch diameter of 2–3 mm, which sprawls at random across reefs in the Indo-Pacific.    7/6

*Xestospongia* sp. This barrel sponge, which is similar to *Xestospongia muta* (Schmidt) of the Caribbean, is found on Indo-Pacific reefs. The large central cavity provides shelter for night-active species of invertebrates during the day.    7/8

### Family Renieridae

*Petrosia ficiformis* Poiret. An irregularly branching, smooth-surfaced, encrusting form with scattered oscula and with a hard texture. The mauve or blue colouring on the upper surface of this shade-loving form is caused by blue-green algae living in the body wall. Found in the Mediterranean.    7/9, 8/1

*Petrosia testudinaria* (Lamarck). A shade-loving species sometimes encountered in caves or in deeper waters of the Indo-Pacific.    8/2

*Rhizochalina ramsayi* Lendenfeld. A characteristically shaped sponge of variable colour, 10 cm or more across. The oscula are arranged on flat-topped protuberances from the upper surface. Attached to the bottom by a strong rooting system. Found on rocks and reefs in the Indo-Pacific.    8/9

## Order Dictyoceratida

A group of large irregular sponges which have a reticulate

skeleton of laminated spongin fibres. There are no siliceous spicules present, but sand grains and other extraneous matter may be incorporated into the skeleton. The familiar bath sponges belong to this group.

Families: Dysideidae, Spongiidae, Verongiidae.

### Family Dysideidae

8/7   *Dysidea avera* (Schmidt). An irregularly branching form up to 10 cm across, with the 'spiky' appearance typical of the genus. Colour varies from pink to violet. Found on rocks in shaded areas of the Mediterranean.

8/8   *Dysidea etheria* de Laubenfels. A delicate branching form up to 10 cm across with a beautiful violet-blue coloration. Found on Caribbean reefs and grass beds.

8/4   *Dysidea* sp. The large terminal oscula are readily visible in this specimen from the New South Wales coast of Australia.

### Family Spongiidae

8/3   *Hippospongia communis* (Lamarck) [horse sponge]. Forms rounded clumps up to 50 cm across, which are usually grey or dark brown in colour. The oscula are not raised above the level of the surrounding tissue. Although the spongin skeleton is without spicules, it is coarser than that of the bath sponge and consequently is used commercially only for cleaning and polishing. Found on rocky bottoms in the Mediterranean.

8/5   *Ircinia campana* (Lamarck) [vase sponge]. A cup- or bell-shaped species which regularly reaches a height of over 50 cm. The coarse outer surface is often ribbed. Occurs on rock and reefs in the Caribbean and Indo-Pacific.

8/6   *Phyllospongia* sp. This unidentified specimen forms leaf-like growths up to 40 cm high on the Great Barrier Reef.

8/10   A spongiid with two large oscula at the top of the vase-shaped body, seen growing on a reef in the Indo-Pacific.

### Family Verongiidae

8/11   *Aplysina aerophoba* (Schmidt) (= *Verongia aerophoba*). A typically bright yellow sponge, forming clumps of cylindrical vertical pots up to 10 cm high with knobbly walls and flat tops. Occurs on rocks in shallow sheltered areas of the Mediterranean and adjacent Atlantic.

9/1   *Aplysina fistularis* (Pallas) (= *Verongia fistularis*). Similar to *A. aerophoba* from the Mediterranean, but the pots often reach a height of 50 cm or more. Found on rocks and reefs in the Caribbean.

9/2, 9/6   *Aplysina lacunosa* (Lamarck) (= *Verongia lacunosa* and *Luffaria archeri* Higgin). A characteristically-coloured species reaching a height of 30 cm or more, with the roughly ribbed exterior of the pots contrasting with the smooth, light-coloured interior. A Caribbean species.

9/9   *Aplysina longissima* (Carter) (= *Verongia longissima*). A sprawling branching species, with a line of slightly raised oscula along the upper surface of the cylindrical branches. Varies in colour from yellow to violet. The aggressive nature of the sponge–coral interaction can be seen clearly in the photograph – wherever the sponge touches the coral head, the coral is killed. Found on reefs and in grass beds in the Caribbean.

9/7   *Verongula* sp. A bowl-shaped specimen about 10 cm across, attached to a shallow Bahamian reef.

### Order Dendroceratida

Sponges either with a tree-like skeleton of non-laminated spongin fibres or with no skeleton at all. They have no siliceous spicules and no extraneous bodies incorporated into the skeleton.

Families: Aplysillidae, Halisarcidae.

This unknown species of demospongiarian, which has a rubbery papillated body, was found hanging in a pendulous fashion from the walls and roof in the darkest area of a cave in Majorca. Such pendulous growths have been reported for other cave-dwelling sponges, e.g. *Oscarella lobularis*.    9/3, 9/10

## Class Sclerospongiae [coralline sponges]

Sponges with a massive limy skeleton composed of calcium carbonate, siliceous spicules and organic fibres. The surface of the skeleton is covered with small pits. The living tissue fills the pits and forms only a thin layer over the skeleton surface. Exhalent water canals are often raised above the body surface and form a star-shaped pattern which is sometimes impressed on the calcareous skeleton beneath. The growths may be up to 1 m across with dome-shaped, irregular or encrusting forms.

Coralline sponges, which can easily be mistaken for corals, occur in shaded areas and in caves, in the Caribbean, Indo-Pacific, Mediterranean and adjacent Atlantic.

### Order Ceratoporellida

The calcareous skeleton is composed of crystals of aragonite arranged in a radiating column-like microstructure. Siliceous rod-shaped spiny spicules are embedded in this skeleton. The walls of the pits do not have protruding spines.

Six genera have been described: *Astrosclera, Ceratoporella, Goreauiella, Hispidopetra, Merlia*.

*Ceratoporella nicholsoni* (Hickson). The living tissue of the sponge forms only a thin layer over the dome-shaped calcareous skeleton, which is cemented to hard bottoms. Specimens up to 1 m in diameter have been found in caves and on deep reef-faces in the Caribbean.    9/11

### Order Tabulospongida

The calcareous skeleton is composed of calcite crystals arranged in a layered microstructure. The siliceous spicules are smooth rod-shaped structures and/or star-shaped and are not embedded in the calcareous skeleton. The walls of the pits bear spiny outgrowths.

Only one species, belonging to the genus *Acanthochaetetes*, has so far been described.

## Subphylum Nuda

Sponges in which the cells of the body are not situated in a gelatinous matrix, but form an open network supported by the spicules of the skeleton. The surfaces of the body lack definite cell layers, but instead are bounded by a network of strands formed from the cells beneath.

Plate 9                                                                                   Phyla Porifera, Coelenterata

### Class Hexactinellida [glass sponges]
approx. 450 species

Exclusively marine sponges with a skeleton of siliceous spicules, fundamentally 6-rayed. The spicules, whose arrangement and shape are important in classification, are either separate or fused together to form a framework. The body is cylindrical, vase-shaped, or can be massively barrel-shaped. The barrel-shaped forms can be up to 1 m across. Water enters the sponge through pores in the body wall which lead into chambers lined with flagellated collar cells. Often there is a well developed central cavity opening to the exterior through a wide osculum, sometimes covered with a sieve-plate of spicules.

Glass sponges are found in all oceans, being particularly common on the Antarctic sea floor. They are normally found at depths of 500–1000 m, although a few species live at depths as great as 7000 m.

### Order Amphidiscophora

Sponges which always possess an anchoring root tuft of needle-shaped spicules. The skeleton consists of a loose network of spicules. One type of spicule characteristic of the order has its ends developed into umbrella-shaped expansions (amphidisc). The flagellated chambers are irregular in shape and are not distinctly separated from one another.

Families: Pheronematidae, Monorhaphididae, Hyalonematidae.

### Family Hyalonematidae

9/5    *Hyalonema* sp. [glass rope sponge]. Sponges in this genus will grow to a height of 50 cm or more. The sponge body is situated well above the bottom on a stiff stalk of very long spicules, which anchors it firmly in soft sediment. A common feature of all species of glass rope sponge is the presence of colonies of anemone-like zoanthids (see p.34) on the stalk. The genus has a world-wide distribution in deep-water soft sediment.

### Order Hexasterophora

Sponges which are often attached directly to a hard object, although root tufts occur in some forms. The skeleton frequently consists of a rigid framework of spicules. One type of spicule characteristic of the order has six rays meeting at right angles at a common centre (hexaster). The flagellated chambers are thimble-shaped and distinctly separated from one another.

Families: Farreidae, Euretidae, Tretodictyidae, Aulocalycidae, Craticularidae, Aphrocallistidae, Aulocystidae, Diapleuridae, Leucopsacasidae, Euplectellidae, Caulophacidae, Rossellidae.

### Family Euplectellidae

*Euplectella aspergillum* Owen [Venus's flower basket].    9/4, 9/8
Photographs taken in deep water of this species reveal that the white or pale yellow body, which varies from 20 to 40 cm in length, is attached to rock by a tuft of long spicules. The beautiful lattice-like skeleton (Plate 9/4) faithfully retains the shape of the animal after its death (Plate 9/8). The large osculum at the summit of the animal is covered by a sieve plate of spicules. Pairs of shrimps are sometimes found within the central cavity of the sponge, having entered as juveniles to take advantage of the constant supply of water-borne food and then being unable, when adult, to escape past the sieve plates. It probably has a world-wide distribution in deep water.

# Subkingdom Metazoa

Animals composed of cells which are highly dependent on one another and are differentiated into definite tissue layers. The subkingdom contains 26 phyla with free-living marine representatives.

# Phylum Coelenterata

approx. 10000 species

Representatives of this group, which includes the hydroids, jellyfish, sea anemones and corals, are found on the sea bottom from the shore to abyssal depths and also floating in the plankton. Only a few species have colonised freshwater successfully. They range in length from one millimetre to several metres.

Coelenterates, although extremely variable in appearance, all have the same basic body plan, which is symmetrical about a central axis (radially symmetrical). The basically sac-like body has a central stomach cavity. The body wall consists of three layers, inner and outer cell layers being separated by a jelly-like layer known as the mesogloea. A single opening to the central cavity serves as both a mouth and an anus and is surrounded by one or more circlets of food-capturing tentacles. Stinging cells (nematocysts) situated on the tentacles paralyse the prey before it is drawn into the mouth.

Two different structural grades occur within the phylum, a cylindrical 'polyp' form which is usually attached to rocks or other objects, and a 'medusa' form which is usually free-swimming (Figure 4). The polyp, which has its mouth and tentacles directed upwards, often has a skeleton secreted by its external surface or lying within the body wall. The medusa, which is bell-

**Phylum Coelenterata**

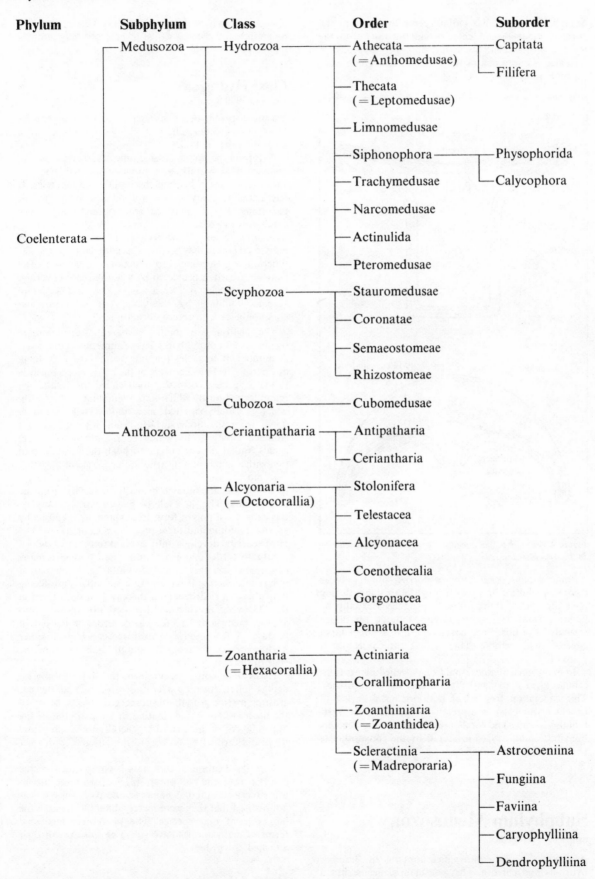

| Phylum | Subphylum | Class | Order | Suborder |
|---|---|---|---|---|
| Coelenterata | Medusozoa | Hydrozoa | Athecata (=Anthomedusae) | Capitata |
| | | | | Filifera |
| | | | Thecata (=Leptomedusae) | |
| | | | Limnomedusae | |
| | | | Siphonophora | Physophorida |
| | | | | Calycophora |
| | | | Trachymedusae | |
| | | | Narcomedusae | |
| | | | Actinulida | |
| | | | Pteromedusae | |
| | | Scyphozoa | Stauromedusae | |
| | | | Coronatae | |
| | | | Semaeostomeae | |
| | | | Rhizostomeae | |
| | | Cubozoa | Cubomedusae | |
| | Anthozoa | Ceriantipatharia | Antipatharia | |
| | | | Ceriantharia | |
| | | Alcyonaria (=Octocorallia) | Stolonifera | |
| | | | Telestacea | |
| | | | Alcyonacea | |
| | | | Coenothecalia | |
| | | | Gorgonacea | |
| | | | Pennatulacea | |
| | | Zoantharia (=Hexacorallia) | Actiniaria | |
| | | | Corallimorpharia | |
| | | | Zoanthiniaria (=Zoanthidea) | |
| | | | Scleractinia (=Madreporaria) | Astrocoeniina |
| | | | | Fungiina |
| | | | | Faviina |
| | | | | Caryophylliina |
| | | | | Dendrophylliina |

Classification of the Hydrozoa follows Petersen (1979 and personal communication) supplemented by Totton (1965) for the Siphonophora. The Scyphozoa are arranged according to Russell (1970) and the Anthozoa as in Moore (1956) supplemented by Friese (1972) for the Actiniaria.

19

or umbrella-shaped, has a mouth centrally placed on the concave underside and tentacles hanging down from the margin of the bell. Since the bulk of the body of medusae is composed of the jelly-like mesogloea, they are commonly known as jellyfish.

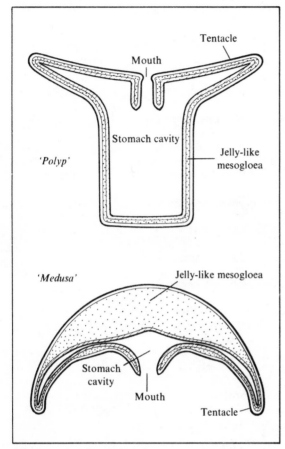

**Figure 4** *The 'polyp' and 'medusa' structural grades occurring in the coelenterates.*

Some species occur only as polyps, some only as medusae, whilst others pass through both stages during their life cycles. Polyps often reproduce asexually by budding from the parent polyp. Medusae reproduce sexually, a fertilised egg developing into a small larva covered with motile hairs (cilia) which is often planktonic.

In many coelenterates, small single-celled plants (unicellular algae – zooxanthellae) live in the body tissues. This association, from which both organisms appear to benefit (symbiosis), is especially prevalent in the reef-building corals. The corals are thought to use the carbohydrates and oxygen produced by the zooxanthellae to aid their own metabolism.

# Subphylum Medusozoa
approx. 3500 species

A group of coelenterates which contains the familiar hydroids and jellyfish. They are mainly marine but a few species occur in fresh and brackish waters.

The medusa, which is regarded as the normal sexual adult, is always basically present in the life cycle.

However, it may be reduced and remain attached to the polyp in many species, whilst in others it may adopt a sedentary way of life.

## Class Hydrozoa
approx. 3000 species

The majority of hydrozoans are marine, but a few species occur in freshwater. There is often both a polyp and a medusa in the life cycle.

A typical polyp is divided into three regions: an attached base, a stalk, and an elongated cylindrical or vase-like structure bearing the mouth and tentacles. It has a simple body cavity not divided by ridges or partitions. Polyps may be solitary, but they occur commonly in fixed colonies connected to one another by tubular extensions of the body cavity. In these colonies individual polyps may become specialised for a particular function, e.g. feeding, reproduction, or defence. The colonies are often protected by a horny outer covering, but some species secrete a solid calcareous skeleton. Asexual budding by the polyps is universal and commonly leads to colony formation.

The medusae are small, transparent, free-swimming forms, seldom more than a few centimetres in diameter. A number of tentacles (varying from 1 to 250) hang down from the bell. The edge of the bell projects inwards in many species to form a shelf called the velum, the muscular action of which aids swimming. The mouth, which is usually situated on a hollow stalk extending down from the undersurface, leads into the central stomach cavity from which four (sometimes more) canals (radial canals) radiate through the mesogloea to join with a canal running around the margin of the bell.

In most medusae the sexes are separate, and the sex cells (gonads) are situated beneath the canals or on the mouth stalk. The life cycle of hydrozoans is complex. Basically, fixed polyps form free-swimming medusae by asexual budding and these produce eggs or sperm. The fertilised egg develops into a planktonic larva which eventually settles and grows into a new polyp. In many cases, however, the medusa stays attached to the parent polyp, although still remaining a sexually reproducing individual. In hydrozoans with only a medusa form in the life cycle, the larva either develops directly into another medusa or, in a few cases, settles on the parent medusa. A few species of medusae can bud off other medusae asexually from the mouth stalk or from the edge of the bell.

Hydrozoan polyps occur from the shore to abyssal depths, often forming a considerable part of the encrusting marine growth on submerged objects. Many of the medusae are found floating in the plankton of the upper layers of the sea. In general, both polyp and medusoid forms feed on organisms captured with their tentacles.

The classification of this class is complicated by the fact that medusae and polyp stages of the same species often received different names in the past, since it was not realised that they were merely different stages in the life cycle of that species. The planktonic medusoid forms of many species have yet to be linked with their attached polyp stage.

## Order Athecata (= Anthomedusae)

Polyps are either solitary or colonial; they usually live attached to the bottom, but in one family, Velellidae, they float freely on the water surface. The stalks of the polyps and the colony branches are often protected by a horny outer layer which does not extend around each polyp. In two families (Milleporidae and Stylasteridae) the polyps are in pits in the surface of a large calcareous skeleton.

The free-swimming medusae are tall and bell-shaped, with the gonads situated on the mouth stalk or on the stomach. At the junction of the mouth stalk with the bell there is a quadrangular stomach which may have pouches. There are usually only a few tentacles around the margin of the bell and the sense organs are simple eye-spots. In some species the medusae remain attached to the parent colony.

## Suborder Capitata

Solitary or colonial hydroids which, apart from some primitive species, have two rings of tentacles – an inner ring around the mouth and an outer ring around the edge of the polyp. The two rings may be equal in size, but often one or the other is reduced. Club-tipped tentacles frequently occur in the mouth ring (occasionally in the outer ring also), but they are thread-like in several species.

The medusae of this suborder do not usually have a stomach with pouches, although they are present in some primitive forms.

Families: Moerisiidae, Sphaerocorynidae, Tricylusidae, Candelabridae, Acaulidae, Euphysidae, Corymorphidae, Paracorynidae, Tubulariidae, Margelopsidae, Halocordylidae, Dicylocorynidae, Corynidae, Velellidae, Cladonemidae, Eleutheriidae, Halocorynidae, Hydrocorynidae, Solanderiidae, Cladocorynidae, Zancleidae, Teissieridae, Milleporidae.

### Family Corymorphidae

10/1, 10/3    *Corymorpha nutans* (Sars). A solitary species with a poorly developed chitinous outer layer. The large polyp (4–5 cm across the tentacle tips) is borne on the end of a long flexible stalk anchored in soft sediment by a bunch of rootlets. The polyp with its cluster of attached medusae faces away from the prevailing current. At regular intervals the polyp is swept over the sediment, and food that attaches to the tentacles is conveyed to the mouth. In sand and mud of deeper waters of the North Atlantic and Mediterranean.

### Family Tubulariidae

10/2    *Tubularia indivisa* Linnaeus. A colonial species with each polyp at the end of a stiff stalk which may be up to 15 cm long. Attached medusae are clustered around the central cones of mature polyps. This species is often found attached to hard bottoms in shallow coastal regions of the North Atlantic, particularly where there is a high percentage of organic matter in suspension.

10/4    *Tubularia mesembryanthemum* Allman. A Mediterranean species with broad polyps (1 cm in diameter) displaying well developed reproductive bodies.

10/5    An unidentified tubulariid from the Caribbean, showing the inner ring of short tentacles, the reproductive medusae, and the outer ring of larger tentacles typical of the family.

### Family Halocordylidae

10/6    *Halocordyle disticha* (Goldfuss). A bushy colony of polyps up to 10 cm high with an outer ring of thread-like tentacles and an inner ring of stout club-shaped tentacles. Found throughout the Atlantic and Mediterranean.

10/7, 10/8    *Pennaria* sp. Bush-like colonies of this Caribbean hydroid are often found growing on gorgonians (Plate 10/8). The polyps have several rings of club-shaped tentacles around the mouth and an outer ring of thread-like tentacles (Plate 10/7).

### Family Velellidae

10/9    *Porpita porpita* (Linnaeus). A freely floating, modified hydroid polyp up to 8 cm in diameter. Numerous club-shaped tentacles, bearing batteries of powerful stinging cells, hang from the rim of the horny central disc-shaped float which supports the animal on the water surface. The reproductive bodies are situated between the tentacles and the centrally-placed mouth, as in attached tubularian polyps. The body tissues contain symbiotic algae. Often in shoals in the Mediterranean and the warmer waters of the North Atlantic.

11/1, 11/3    *Velella velella* (Linnaeus) [by-the-wind-sailor]. A floating polyp up to 10 cm in length with an organisation of body parts on the ventral surface similar to that of *Porpita*. The float is oval and across the top has a translucent sail which protrudes above the water surface and allows the animal to be propelled by wind as well as by currents. Small planktonic organisms are captured by the stinging cells on the tentacles. It is itself eaten by a planktonic snail, *Janthina*, which is immune to its stinging cells (Plate 11/3). Often in shoals in the Mediterranean and the warmer waters of the North Atlantic.

At one time members of this family were considered to be closely related to the siphonophores.

### Family Solanderiidae

11/4    *Solanderia fusca* (Gray). A hydroid forming colonies up to 30 cm across and bearing a superficial resemblance to a sea-fan. The chitinous skeleton is covered by living tissue and small polyps with 12–15 club-shaped tentacles. Found in the Indo-Pacific and Zealandic regions.

### Family Cladocorynidae

11/2    *Cladocoryne* sp. A colonial hydroid with long-stemmed polyps arising at intervals from a creeping stolon. All tentacles are club-shaped. Found in the Caribbean.

### Family Zancleidae

11/5    *Zanclea costata* Gegenbaur. A colonial, branching hydroid which can be seen overgrowing a sponge. Found in the North Atlantic and Mediterranean.

### Family Milleporidae [fire corals]

11/7    *Millepora alcicornis* Linnaeus. A colonial form, whose small polyps are embedded in a branching calcareous skeleton, which may spread over rocks for several metres. The stinging cells of the tentacles are powerful and produce a nettle-like rash on contact with the skin. Found on rocks and reefs in the Caribbean.

11/8    *Millepora complanata* Lamarck. Forms flat, plate-like colonies sometimes extending for several metres on rocks and reefs in the Caribbean.

11/6, 11/9    *Millepora dichotoma* Benett. A dichotomously branching species from the Indo-Pacific. Plate 11/9 shows a mature colony and Plate 11/6 a young colony beginning to establish itself.

12/4    *Millepora platyphylla* Hemprich & Ehrenberg. A massive, plate-like form from the Indo-Pacific.

21

## Suborder Filifera

Colonial hydroids which usually have one ring of thread-like tentacles around the edge of the polyp, although they are scattered over the polyp body in a few species.

The medusa often has a stomach with four pouches, and the edges of the mouth may be drawn out into lips or arms or have tentacles.

Families: Eudendriidae, Calycopsidae, Protiaridae, Pandeidae, Niobiidae, Cytaeidae, Bougainvilliidae, Russelliidae, Rathkeidae, Rhysiidae, Stylasteridae, Hydractiniidae, Ptilocodiidae, Clavidae, Polyorchidae.

### Family Eudendriidae

12/7  *Eudendrium rameum* (Pallas). This species forms bushy colonies up to 10 cm high, with strong stems. It is found attached to rocks, often in caves, in the Mediterranean and North Atlantic.

12/1  *Eudendrium ramosum* (Linnaeus). Colonies are less bushy and taller than those of *E. rameum* (up to 15 cm high). Found in Mediterranean and North Atlantic waters.

### Family Calycopsidae

12/3  *Calycopsis typa* Fewkes. A thin-walled medusa with 16 internal canals, each of which terminates in a long tentacle with a club-shaped tip filled with stinging cells. Occurs in oceanic waters of the North-west Atlantic.

### Family Pandeidae

12/5  *Amphinema rugosum* (Mayer). A bell-shaped medusa, some 5–6 mm high, with a rounded conical apex and two large tentacles with swollen bases at the edge of the bell. There are four characteristic ribbon-like radial canals. The large stomach has gonads attached to its walls. The species is found in coastal waters of the North Atlantic, Caribbean and North Pacific. It migrates from deeper water to the surface at night, returning during the day.

12/2  *Pandea conica* (Quoy & Gaimard). A medusa about 6 mm high, with a conical apex. There are 16–24 fine tentacles with basal swellings, and four ribbon-like radial canals. The large stomach carries a network of gonads on its walls. Widely distributed in the Atlantic, Mediterranean and other seas.

### Family Stylasteridae

12/6, 12/8  *Distichopora* sp. A colonial species up to 5 cm high, with a massive calcareous skeleton branching in one plane. It occurs in caves and under ledges on reefs. Two different colour forms of this Indo-Pacific inhabitant are shown.

12/10  *Stylaster elegans* Verrill [pink lace coral]. A finely branching Indo-Pacific species growing beneath overhangs at right angles to prevailing water currents.

13/3  *Stylaster* sp. An unidentified Indo-Pacific species. The colour of lace corals does not fade when the polyps die, as it is incorporated in the skeleton.

### Family Hydractiniidae

12/9  *Podocoryne carnea* Sars. Forms encrusting colonies, usually not more than 1 cm high, over algae, stones and shells in the Mediterranean and North Atlantic.

### Family Clavidae

13/1  *Clava squamata* (Müller). Colonies of club-shaped polyps situated on stalks 2–3 cm long, arising from a network of attaching roots. Reproductive bodies are in clusters at the junction of polyp and stalk. Found on intertidal brown seaweeds (particularly *Ascophyllum*) in the North-east Atlantic.

## Order Thecata ( = Leptomedusae)

The polyps live in attached colonies. The protective horny outer covering extends beyond the stems, forming a cup around each polyp into which it can withdraw.

The free-swimming medusae are hemispherical to saucer-shaped and bear the gonads on the radial canals. There are usually many tentacles around the edge of the bell, and balancing sense organs occur frequently. Eye-spots are present in only a few species. As in the Athecata, the medusae in some species may be retained permanently on the hydroid colony.

Families: Campanulariidae, Campanulidae, Lafoeidae, Bonneviellidae, Haleciidae, Syntheciidae, Sertulariidae, Plumulariidae, Mitrocomidae, Laodiceidae, Melicertidae, Dipleurosomatidae, Eutimidae, Aequoridae, Phialellidae, Calycellidae, Lovenellidae, Eirenidae, Timoididae, Phialucidae.

### Family Campanulariidae

*Obelia geniculata* (Linnaeus). A colonial form, with erect branching stems up to 5 cm in length. The bell-shaped polyp cups are situated on alternate sides of the stem. Commonly found on brown seaweeds in shallow waters of all seas from polar to tropical regions, and is probably among the most widespread of all marine animals.  13/2, 13/4

### Family Haleciidae

*Halecium halecinum* (Linnaeus). A stiff-stemmed branching colony up to 10 cm high. Found in the North Atlantic and Mediterranean.  13/10

### Family Syntheciidae

*Synthecium evansi* (Ellis & Solander). A colonial form 5–10 cm high, with branches arising from the main stem in opposite pairs. Polyps opposite each other on the branches. Found on hard objects in the Mediterranean.  13/9

### Family Sertulariidae

*Abietinaria abietina* Linnaeus [sea-fir]. Branching colonies up to 25 cm high with polyps arising alternately. Attached to rocks, shells and other hard objects in the North-east Atlantic.  13/5

*Dynamena pumila* (Linnaeus). Colonies up to 5 cm high with upright stems which occasionally branch. The vase-shaped polyps are situated opposite each other. The characteristic large oval reproductive polyps can be clearly seen. A common hydroid on algae and stones in the North-east Atlantic and Mediterranean.  13/6

*Hydrallmania falcata* (Linnaeus) [sickle-coralline]. A colonial form up to 50 cm high with a curved and twisted main stem from which arise feather-like branches. Found on rocks, shells or gravel in the North Atlantic.  13/7

*Sertularella polyzonias* (Linnaeus). Irregularly branching colonies up to 5 cm high with prominent oval reproductive polyps. Usually on rocks, but sometimes attached to other hydroids. North Atlantic and Mediterranean.  13/8

*Sertularia cupressina* (Linnaeus) [white weed, sea-cypress]. Small tufts of branches arise from a main stem which may be up to 50 cm high. Sometimes collected and dyed for decorative purposes. Occurs on stones and gravel in the North-east Atlantic where it is particularly abundant in the southern North Sea.  14/1

## Family Plumulariidae

14/2 *Aglaophenia cupressina* Lamouroux. A stinging colonial hydroid, possessing zooxanthellae, which occurs in fern-like clumps, up to 20 cm high, attached to rocks and corals. Found in the Indo-Pacific.

14/4 *Aglaophenia elongata* Meneghini. Loose feathery clumps, up to 15 cm across, which are attached to rocks. Usually found below 30 m in the Mediterranean.

14/3 *Aglaophenia kirchenpaueri* (Heller). Feathery clumps, up to 10 cm high, which are found frequently in the North-east Atlantic and Mediterranean.

14/5 *Aglaophenia octodonta* (Heller). Clumps up to 5 cm high, but usually smaller in shallow waters. Found in the Mediterranean.

14/6 *Aglaophenia pluma* (Linnaeus). A shallow-water form with branching curved stems. Often found attached to the brown seaweed, *Halidrys siliquosa*, and less frequently to shells and rocks. Occurs in the North and South Atlantic and Mediterranean.

14/7 *Antenella* sp. A small unbranched undergrowth form from the Mediterranean.

14/8 *Gymnangium montagui* (Billard). Closely packed colonies, up to 50 cm across, usually occurring in areas with strong water movements. Found in the North-east Atlantic.

14/10 *Lytocarpus myriophyllum* (Linnaeus). A large colonial hydroid, up to 50 cm high, with thick rigid stems bearing branches at right-angles to the current. It roots in soft sediments in the Mediterranean, where it is rarely found in waters shallower than 30 m.

14/9 *Lytocarpus philippinus* (Kirchenpauer). A colonial hydroid common in the Indo-Pacific, with powerful stinging cells on the fine white branches.

14/11 *Lytocarpus* sp. This specimen, photographed on the Great Barrier Reef, is possibly a variety of *L. philippinus*, an Indo-Pacific species.

15/1 *Nemertesia antennina* (Linnaeus). Bushy colonies, up to 20 cm high, with wiry stems and very fine branches. Commonly found on soft bottoms below 10 m in the North Atlantic and Mediterranean.

15/2 *Nemertesia ramosa* (Lamouroux). A thick-stemmed branching species up to 10 cm high. Occurs on sand and rocks in the North-east Atlantic.

15/3 *Thecocaulus diaphanus* (Heller). In dense colonies up to 5 cm high, with polyps on the stem as well as on the branches. Found in shaded rocky areas in the North-east Atlantic and Mediterranean.

## Family Laodiceidae

15/5 *Orchistoma pileus* (Lesson). This medusa (approximately 3 cm in diameter) and all others in this family have characteristic club-shaped organs between the bases of the tentacles on the bell margin. Found in the North-west Atlantic.

## Family Eutimidae

15/7 *Tima flavilabris* Eschscholtz. The medusa, which has a diameter of up to 10 cm, has convoluted gonads distributed along the radial canals. The mouth has four large frilled lips. Found in the North-west Atlantic.

## Family Aequoreidae

15/6 *Aequorea forskalea* Pèron & Lesueur. A saucer-shaped medusa up to 20 cm in diameter. The numerous radial canals, with their attached gonads, can be seen through the thick jelly of the bell. Found in the North Atlantic, Mediterranean, Caribbean and North Pacific.

*Rhacostoma atlanticum* Agassiz. Similar to *A. forskalea*, 15/9 but the bell is usually about 10 cm in diameter and has more radial canals. Commonly found offshore during the summer and autumn in the North-west Atlantic.

## Order Limnomedusae

A group of marine and freshwater hydrozoans; the polyp stage is known for only five of the twelve genera belonging to this group. Where known, the polyp is often solitary, occasionally colonial, with few or no tentacles.

The medusae have hollow tentacles which, in some of the marine species, bear adhesive pads near their tips that are used for clinging to seaweeds. The gonads are either situated on the walls of the stomach or on the radial canals or on both.

Marine representatives occur mainly in shallow waters.

Family: Olindiidae.

## Family Olindiidae

*Olindias phosphorica* (Delle Chiaje). The bell is about 15/4 5 cm in diameter and bears four radial canals and numerous tentacles. The medusae can be found throughout the Indo-Pacific and Mediterranean clinging to seaweeds and grasses with the aid of adhesive pads on the tentacles.

## Order Siphonophora

Free-swimming, passively floating, or, in rare cases, bottom-dwelling, hydrozoan colonies consisting of modified polyps and medusae. Three types of polyp occur – feeding polyps, which carry a long single tentacle, often branched, laden with stinging cells; stinging polyps with a single unbranched tentacle and no mouth; and reproductive polyps. The medusoid individuals, which differ markedly from the normal hydrozoan medusae, usually remain attached to the colony throughout life, and may function as swimming bells, protective flaps, bladder-like floats or modified reproductive structures.

Siphonophores are generally hermaphrodite, with the fertilised egg developing into a larval form which produces the large colony by asexual budding. They occur in all seas but are more common in warmer waters. They may be carried by winds and currents into shore regions and sheltered bays, where they sometimes accumulate in large numbers. The majority of colonies are small and transparent, but some can reach 30 cm in diameter and have tentacles several metres in length, e.g. *Physalia*, the Portuguese man-of-war.

## Suborder Physophorida

All colonies have a bladder-like float, which can be very large in some families. Apart from this feature the structure of colonies in this suborder is very variable, ranging from forms with swimming-bells and groups of individuals on a stem to forms where individuals are clustered together beneath the large float.

Families: Physaliidae, Rhizophysidae, Apolemiidae, Agalmidae, Pyrostephidae, Physophoridae, Athorybiidae, Rhodaliidae, Forskaliidae.

23

### Family Physaliidae

15/8 *Physalia physalis* (Linnaeus) [Portuguese man-of-war]. This species has a large gas-filled float (up to 30 cm long) that maintains the colony of individuals at the water surface, where it is at the mercy of wind and currents. The stinging tentacles, which may trail below the float for several metres, capture food and transfer it to the feeding polyps. The powerful sting of this cosmopolitan species is a danger to bathers when onshore winds drive it into shallow coastal waters.

### Family Rhizophysidae

16/3 *Bathyphysa sibogae* Lens & van Riemsdijk. A deep-water cosmopolitan species with a small balloon-shaped float, sometimes found entangled amongst cables and ropes of fishing gear.

16/7 *Rhizophysa filiformis* (Forskål). The balloon-like float, which is up to 10 cm long, supports a long contractile stem bearing clusters of feeding and reproductive individuals. Found in the North Atlantic and Mediterranean.

### Family Apolemiidae

16/1 *Apolemia uvaria* (Lesueur). This species has only a very small float, but the contractile stem bearing clusters of individuals may reach a length of 20 m or more. Occurs throughout the Mediterranean.

### Family Agalmidae

16/4 *Agalma okeni* Eschscholtz. This species, which may reach a length of 10–15 cm, is shown here feeding on captured planktonic copepods. The rows of large swimming-bells can be seen. Widely distributed in warm waters.

16/5 *Nectalia* sp. The shortened stem, bearing feeding and stinging polyps, and the large protective flaps are characteristic of this genus.

### Family Physophoridae

17/1 *Physophora hydrostatica* Forskål. This species, which is up to 10 cm long, bears a small sausage-shaped float above the two rows of swimming bells. The long processes below the swimming-bells bear batteries of powerful stinging cells at their tips. There are no protective bracts (cf. *Nectalia*). A warm-water cosmopolitan species.

### Family Athorybiidae

16/2 *Athorybia rosacea* (Forskål). The float, which is relatively large, is rose-coloured. There are transparent protective flaps surrounding the feeding and reproductive polyps. Tentacle-like processes bearing stinging cells at their tips, protrude between the flaps. The fine tentacles of the feeding polyps hang down beneath the body to capture prey. Widely distributed in warm waters.

### Family Rhodaliidae

16/6 *Stephalia corona* Haeckel. A rarely-seen siphonophore living at depths from 100–1500 m in all oceans. The float, about 1 cm in diameter, is surrounded by a circlet of large swimming-bells. It is thought that this species trails its tentacles along the bottom, capturing small organisms from the mud surface.

## Suborder Calycophora

The colony consists of a long stem which bears clusters of individuals repeated along its length. Each group of individuals usually comprises a protective flap, a feeding polyp, and one or more reproductive individuals. At the upper end of the colony there are a number of muscular swimming bells. The stem is protected by a sheath where it emerges from the swimming bells.

Families: Prayidae, Hippopodiidae, Diphyidae, Clausophyidae, Sphaeronectidae, Abylidae.

### Family Prayidae

*Rosacea flacida* Biggs, Pugh & Carré. Members of this genus are characterised by a pair of swimming-bells from which protrudes the stem bearing the feeding and reproductive polyps surrounded by protective flaps. Found in the North Atlantic. 17/2

### Family Diphyidae

*Diphyes dispar* Chamisso & Eysenhardt. There are two similar swimming-bells each about 2 cm long, from between which the body stem emerges. Found in the North Atlantic. 17/3

*Lensia conoidea* (Keferstein & Ehlers). Members of this genus often have one of the bells reduced in size. This specimen, less than 1 cm long, has captured two calanoid copepods. It has been found in the Atlantic and Indian Oceans. 17/8

### Family Abylidae

*Abylopsis tetragona* (Otto). There are two transparent swimming-bells about 5 mm long, each with a cylindrical cavity, seen clearly here. The upper bell bears a chamber (with a pointed tip) from which the body stem originates. One of the most abundant *Abylopsis* species, it has a world-wide distribution. 17/4

## Order Trachymedusae

Medusae in this order range from a few millimetres to several centimetres in diameter. There is no polyp stage. The bell-shaped medusa has an unfrilled rim projecting inwards to form a large muscular velum. The rim is thickly coated with stinging cells and bears solid or hollow tentacles. The mouth is often situated on a stalked extension, with reproductive organs on the radial canals.

The fertilised egg develops into a tentacle-bearing larva which changes directly into a small medusa.

Trachymedusans are found in the open ocean where they have been recorded from the surface to depths as great as 3000 m.

Families: Geryonidae, Ptychogastridae, Petasidae, Halicreatidae, Rhopalonematidae.

### Family Geryonidae

*Liriope tetraphylla* (Chamisso & Eysenhardt). The mouth is situated on a stalked extension arising from the bell, which is about 3 cm in diameter. There are leaf-like gonads on the radial canals. A widely distributed species in warm waters. 17/5

## Order Narcomedusae

The medusae of this group are similar in size to those of the Trachymedusae, but they have firm flattened bells with a scalloped margin. The tentacles are solid and emerge well above the edge of the bell, which bears a dense mass of stinging cells and club-like sense organs.

There is no mouth stalk, the large round mouth opening directly into the stomach. Radial canals are absent and often the canal encircling the edge of the bell is reduced. The reproductive organs are situated on the stomach wall.

In some species, the fertilised egg develops directly into a medusa via a short free-swimming tentacular larval stage, as in the Trachymedusae. However, in many other species the life cycle is more complex, with the egg developing parasitically into the larva within the parent, or alternatively, the swimming larva attaching itself to the parent medusa before transforming into a medusa.

Narcomedusae live in the open ocean where they can be found from the surface waters to great depths.

Families: Cuninidae, Aeginidae, Solmarisidae.

### Family Cuninidae

17/9 *Solmissus incisa* (Fewkes). A soft fragile medusa, with a bell diameter of up to 10 cm, bearing 20–40 marginal tentacles. An oceanic species, photographed here at a depth of 1000 m.

### Family Aeginidae

17/7 *Aegina* sp. (probably *A. citrea* Eschscholtz). The number of tentacles may vary from three to six in this genus, which has a bell diameter of up to 5 cm. Widely distributed in deep waters.

### Family Solmarisidae

17/6 *Solmaris* sp. (probably *S. corona* Keferstein & Ehlers). Members of this genus usually have a bell with more than 20 tentacles and a bell diameter approaching 2 cm. A cosmopolitan species.

## Order Actinulida

A group of very small, free-living, solitary forms, whose length varies from 0.5 to 1.5 mm, and which have features which can be related to both the polyp and medusa form. The upper part of the body bears tentacles and sense organs, and the mouth is situated on the lower part. Adhesive organs may be present on the upper surface. They have retained as adults several larval structures: for instance they are covered with cilia.

Both forms with separate sexes and hermaphrodites occur within the order. The fertilised egg develops into a tentacle-bearing larva which eventually grows into the adult form.

These tiny hydrozoans are found living between sand grains.

Families: Halammohydridae, Otohydridae.

## Order Pteromedusae

Small medusae with a pyramid-shaped bell separated from an inverted pyramid-shaped mouth stalk by a groove. Four swimming lobes (modifications of the velum), which carry balancing sense organs, project from this groove.

Pteromedusans are found in the open ocean.

One genus only: *Tetraplatia*.

## Class Scyphozoa [jellyfish]
approx. 250 species

An exclusively marine group in which the medusa is the dominant form. It is usually free-swimming, but may be attached by a stalk to weeds or rocks. The polyp is either a temporary stage in the life cycle or absent altogether. Jellyfish are usually much larger than hydrozoan medusae, the diameter of the bell ranging from 2 cm to over 1 m. The margin of the bell has no velum and is usually divided into a number of lobes which carry tentacles and sensory structures.

The mouth, which normally is situated on a mobile stalk, leads into a food cavity composed of a central stomach with four compartments extending from it. Canals, often branched, run from the stomach through the mesogloea to the edge of the bell. Water currents, carrying food and oxygen, circulate through these canals.

Jellyfish are transported by ocean currents and maintain themselves in the water column by rhythmic pulsating of the bell.

The sexes are separate in most species, with the gonads developing near the base of the stomach. The fertilised egg develops into a ciliated larva which, in many inshore jellyfish, changes into a temporary polyp stage which attaches to rock. Young medusae bud off from the polyp and grow into adult jellyfish. In open-sea forms the larva develops directly into a small medusa with no transient polyp stage. One deep-sea species broods its young.

Jellyfish are found in all seas from the tropics to polar regions. They feed on almost any animal of appropriate size, with fish forming an important part of their diet. Prey is usually caught by nematocyst-laden tentacles, but a few species feed on planktonic organisms which they trap in mucus on the upper side of the bell. In coastal waters the stings from a few species of jellyfish are a hazard to bathers.

## Order Stauromedusae

A group of polyp-like jellyfish which usually have a diameter of only a few centimetres. They attach by a stalk to seaweeds and rocks. The body is often drawn out into eight arms, each carrying bunches of short tentacles. Adhesive cushion-like structures, the anchors, frequently occur between the arms. Stauromedusae cannot swim and many remain permanently attached, but others can detach and reattach at will, using the anchors and tentacles for grasping. They are found chiefly in the colder waters of the world, where they feed upon small animals.

Families: Eleutherocarpidae, Cleistocarpidae.

### Family Eleutherocarpidae

*Haliclystus auricula* (Rathke). A stalked form up to 5 cm high, whose bell has eight arms, each bearing a cluster of tentacles. It cannot swim but is able to progress with a 'head-over-heels' movement using 'anchors' between the arms as temporary attachment organs. It is found on seaweeds and sea-grasses in the intertidal zone and shallow waters of the North Atlantic and North Pacific.   18/9

## Order Coronatae

Free-swimming medusae, whose diameters range from 2 to 20 cm. The bell is divided into an upper and a lower part by a distinct groove, and has a deeply scalloped margin. Solid tentacles and sensory structures, often bearing single eyes, alternate with one another between the marginal lobes. These medusae, which are often brightly coloured, commonly occur in the deeper colder waters of the ocean. A few species, however, are found near the surface in warmer waters.

Families: Nausithoidae, Atollidae, Atorellidae, Linuchidae, Paraphyllinidae, Periphyllidae.

### Family Nausithoidae

18/6 *Nausithoe punctata* Kölliker. A small flattened medusa up to 2 cm in diameter with eight tentacles. A surface and shallow-water species in warm seas.

18/1 *Nausithoe rubra* Vanhöffen. The scalloped margin characteristic of this order can be clearly seen in these deep-water specimens.

### Family Atollidae

18/2 *Atolla wyvillei* Haeckel. An atoll-shaped bell, up to 15 cm in diameter, with 16–24 tentacles. A long mouth stalk hangs below the bell. A deep-sea luminescent medusa, widely distributed in all oceans.

### Family Periphyllidae

18/3 *Periphylla periphylla* (Pèron & Lesueur). This species has a characteristic conical bell of thick jelly up to 20 cm high with 12 tentacles. It is found in deeper waters of all oceans except for the high Arctic.

## Order Semaeostomeae

These jellyfish are of moderate to large size, usually with diameters from 5 to 50 cm, but the largest known coelenterate, *Cyanea capillata* (the lion's mane jellyfish), over 1 m across and with tentacles up to 30 m long, belongs to this group. They have bowl- or mushroom-shaped bells with scalloped margins. Tentacles may either be distributed evenly around the margin or occur in bunches. The mouth stalk is drawn out into four long frilled lobes and there are usually canals running from the large stomach cavity to the edges of the bell.

They are mainly found in coastal areas, where they may occur in large wind-drifted aggregations. Many capture fish and other animals with the tentacles but there are several which trap small organisms in mucus on the upper surface of the bell. Certain members of this group are well known for their powerful stings.

Families: Pelagiidae, Cyaneidae, Ulmaridae.

### Family Pelagiidae

18/4 *Chrysaora hysoscella* (Linnaeus) [compass jellyfish]. Its common name derives from the distinctive markings often present on the upper side of the bell, which may reach a diameter of 30 cm. Commonly found floating near the surface in coastal waters of the North-east Atlantic and Mediterranean.

18/5, 18/10 *Pelagia noctiluca* (Forskål). This luminescent mushroom-shaped species has a bell about 10 cm in diameter with eight highly contractile marginal tentacles. The upper surfaces of the bell bear warts of stinging cells (Plate 18/10). Often found in surface waters of the North Atlantic, Mediterranean and North Pacific.

### Family Cyaneidae

18/8 *Cyanea capillata* (Linnaeus) [lion's mane jellyfish]. A species with a shallow disc-like bell, occasionally exceeding 1 m in diameter, bearing numerous marginal tentacles which may trail below the bell for several metres. The powerful sting of this northern species is a danger to swimmers.

*Cyanea lamarcki* Pèron & Lesueur. A smaller species than *C. capillata* with the diameter of the purple-blue bell rarely exceeding 30 cm. The sting of this North-east Atlantic species is less virulent than that of *C. capillata*.   18/7

### Family Ulmaridae

*Aurelia aurita* (Linnaeus). A saucer-shaped bell up to 30 cm in diameter with numerous fine marginal tentacles. This species supplements its diet by trapping small planktonic organisms in mucus on the upper side of the bell and transferring them by ciliary action to the mouth. A widely distributed species in coastal waters.   19/1, 48/‹

## Order Rhizostomeae

These medusae, which can attain a diameter of 80 cm, have a firm bell with a dense mesogloea layer. There are no tentacles present. The mouth stalk has four long lobes which divide to form eight thick gelatinous arms. As the medusa grows, these arms fuse together and eventually close off the original mouth opening. The arms usually carry elongated appendages which fulfil the role of tentacles, as they carry nematocysts and are used for food capture. Food is sucked through many small openings, the suctorial mouths, which occur on the arms, and is passed through a complicated canal system to the stomach cavity.

These jellyfish are more commonly found in the shallow warmer waters of the world but a few species occur in temperate seas.

Families: Rhizostomatidae, Stomolophidae, Cassiopeidae, Cepheidae, Mastigiidae, Versurigidae, Thysanostomidae, Lynchnorhizidae, Catostylidae, Lobonematidae.

### Family Rhizostomatidae

*Rhizostoma pulmo* (Macri). A large jellyfish with a dome-shaped bell up to 1 m in diameter. There are no tentacles. It feeds on small plankton which is sucked through small openings in its eight mouth arms. This jellyfish is sometimes accompanied by young horse-mackerel (*Trachurus*).   19/2

### Family Cassiopeidae

*Cassiopeia andromeda* (Forskål). A species with a flattened bell, up to 30 cm in diameter, bearing mouth arms with protruding bladders filled with zooxanthellae. This jellyfish is unusual in that it is found more frequently resting on the bottom with its arms uppermost than swimming in the water (Plate 19/4). This posture enables the zooxanthellae to photosynthesise more effectively than if the jellyfish was bell-uppermost. Common on sandy bottoms in the Indo-Pacific.   19/3, 19/4

### Family Cepheidae

*Cotylorhiza tuberculata* Agassiz. A species with a sombrero-shaped bell up to 30 cm across. The eight bifurcated mouth arms branch repeatedly towards their tips (Plate 19/6). This species is often accompanied by young horse-mackerel (cf. *Rhizostoma pulmo*). Found throughout the Mediterranean.   19/6, 19/9

### Family Mastigiidae

*Phyllorhiza punctata* von Lendenfeld. The characteristically spotted bell reaches a diameter of about 30 cm. The mouth arms branch towards their tips. Occurs in the Indo-Pacific.   19/7

An unidentified rhizostomean from the Red Sea.   19/8

# Class Cubozoa
16 species

Free-swimming medusae which are easily recognised by their cuboid shape with four flattened sides, and by the simple unfrilled margin of the bell which curves inwards. A tentacle or a group of tentacles is situated at each of the four corners of the bell. Although they are characteristic of shallow waters of warm seas they also occur in the open ocean. Cubozoans are strong swimmers and active predators feeding mainly on small fish. The group, which contains the so-called 'sea-wasps' of eastern Australian waters, is noted for its painful stings, which in severe cases cause death of humans in three minutes or less.

## Order Cubomedusae [box-jellyfish]
Characters identical to those of the class.

Families: Carybdeidae, Chirodropidae.

### Family Carybdeidae
19/5 *Carybdea rastoni* Haacke. A small jellyfish up to 5 cm high, with a large tentacle at each corner of the 'box'. They sometimes occur in swarms in the late summer or autumn in the Indo-Pacific. They have powerful stings which may cause persistent weals on the skin of bathers.

### Family Chirodropidae
20/1 *Chironex fleckeri* Southcott [sea-wasp]. The translucent bell of this jellyfish, which may be up to 20 cm high, bears four clumps of barred tentacles which extend to several metres in length. The sting from this species is extremely venomous and sometimes lethal to man. Severe stings can cause death in three minutes, but even small stings cause excruciating pain which may last for ten hours or more. An inhabitant of the Indo-Pacific region, particularly the north-east coast of Australia.

# Subphylum Anthozoa
approx. 6500 species

An exclusively marine group of coelenterates which contains the familiar sea anemones and corals. There is no medusa stage. The polyp, which may possess an internal or external skeleton, has a cylindrical body with the mouth end flattened into a disc surrounded by tentacles. A short tubular gullet extends from the mouth into the stomach cavity. The gullet commonly bears one or more ciliated grooves which direct water currents into the interior of the animal. The stomach cavity is divided into several chambers by partitions which often cross from the body wall to the gullet.

They may have separate sexes or be hermaphrodite. The gonads are situated on the partitions in the body cavity. The planktonic larvae develop directly into young polyps; asexual reproduction by budding is common throughout the group.

Anthozoans, although very common in tropical waters, occur in all seas of the world, particularly in coastal waters. They feed mainly on planktonic organisms but some of the large polyps capture non-planktonic invertebrates and small fish.

# Class Ceriantipatharia
approx. 200 species

Bottom-living solitary or colonial forms whose polyps have simple unbranched tentacles. The stomach cavity is divided by single complete partitions and there are one or two ciliated grooves present in the gullet.

Both forms with separate sexes and hermaphrodites occur within this class.

## Order Antipatharia [black or thorny corals]
Members of this order form branched plant-like colonies which vary in height from a few centimetres to several metres. The slender branches are strengthened by a brown or black skeleton of horny material. The polyps, which are situated in living tissue around this skeleton, are short and cylindrical with 6–24 tentacles, which cannot be retracted. The lower end of the colony may be attached to some firm object by a flattened base or may simply extend into the sediment.

The polyps are either male or female, but colonies may be hermaphrodite.

Antipatharians mainly inhabit deeper waters, particularly in tropical and subtropical areas, although a few species are found at shallow depths in temperate regions. Skeletons of the thickened branches of black corals are often cut and polished and used in the making of jewellery.

Families: Antipathidae, Leiopathidae, Dendrobrachiidae.

### Family Antipathidae
*Antipathes dichotoma* Pallas. A plant-shaped colony over 1 m in height, photographed at a depth of 50 m in the Red Sea. This species is the largest antipatharian in the Red Sea. Probably occurs in deep water throughout the Indo-Pacific. — 20/9

*Antipathes* sp. An unidentified specimen growing in bush-like colonies at a depth of over 50 m in the Red Sea. — 20/2

*Cirrhipathes anguina* Dana [whip coral]. This genus, which occurs in the deeper water of all tropical oceans, contains a number of species that are difficult to determine. *C. anguina* is one from the Great Barrier Reef that can reach a length of 1 m or more (Plate 20/4). The polyps and tentacles are only usually fully expanded at night (Plate 20/3). — 20/3, 20/4

*Cirrhipathes* sp. A whip coral from depths below 50 m in the Red Sea. The bizarre coiling shapes are characteristic of some species in this genus. — 20/7

*Stichopathes luetkeni* Brook. A close-up of the long stem of this Caribbean species showing the expanded polyps. — 20/8

A bushy antipatharian colony from the Indian Ocean. — 20/10

## Order Ceriantharia
Anemone-like forms with an elongated muscular body lacking both a basal disc and a hard skeleton. The body may reach a length of 40 cm in some species. The mouth region bears an outer ring of numerous long slender tentacles and an inner ring of shorter ones.

Most species are hermaphrodite.

They live in sediment-encrusted mucus tubes buried vertically in soft sediments. When disturbed, they withdraw rapidly into their tubes. Ceriantharians are cosmopolitan in distribution and range from shallow to deep waters.

27

Families: Cerianthidae, Botrucnidiferidae, Arachnactidae.

### Family Cerianthidae

20/5, 21/1  *Cerianthus membranaceus* (Spallanzani). This species has an outer ring of 100 tentacles and a similar number of shorter tentacles around the mouth. The body of a mature specimen is usually about 30 cm long, but the tubes found buried in soft muddy ground may be up to 1 m in length. There are both thick-walled (Plate 21/1) and thin-walled tube varieties (Plate 20/5) of this species. The species feeds on plankton and suspended organic debris. Found in the Mediterranean and adjacent Atlantic. A related species, *C. lloydi* Gosse, with fewer tentacles, occurs in the North-east Atlantic.

21/2  *Isarachnanthus* sp. A specimen photographed in coral sand surrounding a Caribbean reef.

20/6  *Pachycerianthus maua* Carlgren. A species with a thin-walled tube occurring in sandy sediments of the Indo-Pacific.

21/4  An unidentified species of cerianthid from the Great Barrier Reef. In most ceriantharians there is great colour variation within a species.

## Class Alcyonaria (=Octocorallia)
approx. 1000 species

The polyps have eight feather-like tentacles whose hollow interiors connect with the stomach cavity, which is divided by eight single complete partitions. The gullet usually possesses one ciliated groove. They are exclusively bottom-living and colonial, forming lobed or branching colonies with the polyps connected to one another by canals. The colonies are usually supported by an internal calcareous or horny skeleton.

In most orders, new polyps are budded off from the connecting canals, but in the sea-pens (Pennatulacea) they arise directly from the sides of older polyps. Some species have separate sexes whilst others are hermaphrodite. Eggs and sperm may be shed directly into the water or fertilisation may occur within the body of the parent.

Alcyonarians are most abundant in warm coastal waters.

### Order Stolonifera

Small attached colonies in which the polyps arise separately from flattened creeping branches (stolons). The skeleton, when present, consists of calcareous spicules occurring separately or fused into tubes.

Stoloniferans are mainly shallow-water forms, occurring in both tropical and temperate waters.

Families: Cornulariidae, Clavulariidae, Tubiporidae.

### Family Cornulariidae

21/7  *Cornularia cornucopiae* Pallas. Colonies consist of creeping branches from which 8-tentacled polyps, about 1 cm high, arise at intervals. Found on rocks, stones and algae in shallow subtidal waters of the Mediterranean.

### Family Clavulariidae

21/5  *Clavularia* sp. An Indo-Pacific species forming small colonies with 2 cm high polyps bearing eight feathery tentacles. The colonies are strengthened by numerous warty calcareous spicules.

### Family Tubiporidae

*Tubipora musica* Linnaeus [organ-pipe coral]. The    21/6, 21/
common name derives from the parallel rows of tubes making up the calcareous skeleton of the colony (Plate 21/8). When the polyps are expanded, the red skeleton may be completely obscured (Plate 21/6). Colonies can reach 30 cm or more in diameter. Found in the Indo-Pacific.

## Order Telestacea

The colonies consist of simple or little-branched stems growing up from a root-like base. The stems, which are modified elongated polyps, bear lateral polyps. The skeleton consists of spicules frequently joined by calcareous and horny secretions.

Telestaceans are found in deeper waters of both tropical and temperate regions.

Families: Telestidae, Pseudocladochonidae.

### Family Telestidae

*Telesto multiflora* Laackmann. The arrangement of    21/10
polyps characteristic of the order can be seen in this specimen from the Indo-Pacific. There are large terminal polyps and smaller lateral ones derived from them.

*Coelogorgia palmosa* (Valenciennes). This genus differs    21/3
from *Telesto* in having a more branched structure resembling in some ways a gorgonian. Found in the Indo-Pacific.

## Order Alcyonacea [soft corals]

The polyps of alcyonaceans protrude from a fleshy mass, sometimes lobed, which is strengthened by calcareous spicules. Deep-water species tend to contain more spicules and are consequently more rigid than shallow-water forms, which are subjected to greater wave action. The polyps can often be completely withdrawn into the body mass.

Soft corals are more abundant in tropical waters, but several species occur in temperate and polar regions.

Families: Alcyoniidae, Astrospiculariidae, Nephtheidae, Siphonogorgiidae, Viguieriotidae, Xeniidae.

### Family Alcyoniidae

*Alcyonium digitatum* (Linnaeus) [dead-man's fingers].    21/9, 21/
Forms lobed colonies up to 20 cm high, on rocks in areas where water movement is vigorous. The fleshy mass of the colony ranges in colour from white through pink to orange, although the retractable polyps are always translucent white. When disturbed, the fleshy lobes contract and can be mistaken for sponges when in this condition. Found in the North Atlantic.

*Alcyonium palmatum* Pallas. Forms upright branching    22/1
colonies up to 50 cm high, normally anchored to stones and shells on muddy and sandy bottoms below 20 m in the Mediterranean and North-east Atlantic.

*Parerythropodium coralloides* (Pallas). This species is    22/2
frequently found overgrowing gorgonians, in this case *Eunicella cavolinii* (Pallas). Occurs in the Mediterranean and adjacent Atlantic.

*Sarcophyton trocheliophorum* (Marenzeller). Large soft    22/3, 22/8
colonies up to 1 m across with a characteristic convoluted    22/9
appearance. The surface is very smooth when the polyps are contracted (Plate 22/3). The photographs show successive stages in the opening of the polyps, which are

unusual in that they are frequently expanded during the day. Found on Indo-Pacific reefs.

22/4 *Sinularia* sp. A specimen with widely spaced white polyps emerging from a surface rather similar to that of *Sarcophyton*. Found on Indo-Pacific reefs.

22/5 An alcyoniid from reefs in the Maldive Islands, Indo-Pacific.

### Family Nephtheidae

6, 22/10, 23/1 *Dendronephthya klunzingeri* (Studer). Colonies of this soft coral may reach a height of over 1 m when fully expanded at night (Plate 22/6). The polyps appear as clusters with protruding calcareous spicules when the colony is contracted (Plate 22/10), but are more evenly distributed when the colony is expanded (Plate 23/1). Supporting calcareous spicules are clearly visible in the fleshy walls of a fully expanded colony. They are found on deep parts of reefs in the Indo-Pacific.

2/7, 23/6 *Dendronephthya rubeola* Henderson. This species anchors itself in sand rather than on to rock. It does so by means of a bunch of rootlets, as seen in the specimen pulled from the sediment (Plate 23/6). Colonies are found in Indo-Pacific sediments.

### Family Siphonogorgiidae

23/2 *Cactogorgia simpsoni* Thomson & Dean. Members of this family may resemble gorgonians, although their internal structure is somewhat different. This example occurs on reefs in the Indo-Pacific.

23/9 *Scleronephthya* sp. Members of this genus are often confused with *Dendronephthya*. These specimens were photographed at a depth of 70 m in the Red Sea.

### Family Xeniidae

23/4 *Anthelia glauca* Savigny. The non-retractile polyps have large tentacles with feathery side-branches. The tentacles open and close rhythmically. Found in clumps, 10 cm or more in height, on Indo-Pacific reefs.

23/5 *Xenia elongata* Dana. Species of *Xenia*, unlike those of *Anthelia*, have a stout trunk from which clumps of polyps arise, but like *Anthelia* the tentacles open and close every few seconds. Found on Indo-Pacific reefs.

## Order Coenothecalia [blue coral]

The Indo-Pacific blue coral, *Heliopora*, is the sole member of this order. It has a massive lobed skeleton perforated by cylindrical canals. The blue tint of the skeleton, masked in life by the brown polyps, is due to the presence of iron salts.

Family: Helioporidae.

### Family Helioporidae

23/7 *Heliopora coerulea* de Blainville [blue coral]. The characteristic blue colour of the skeleton of this species is often obscured by living tissue. Colonies can reach a diameter of 1 m on Indo-Pacific reefs.

## Order Gorgonacea [horny corals]

This order contains the sea-fans and sea-whips, which usually have a plant-like growth form. The main stem is firmly attached to a hard surface by a plate or a tuft of creeping branches. The stem contains a central strengthening rod, which may be calcareous but which commonly consists of a horny material (gorgonin). The short polyps occur all over the branches of the colony, being absent only on the main stem.

Gorgonians are cosmopolitan, although they are most abundant in warmer waters. The colonies are often brightly coloured and may reach a height of 3 m. Sponges, hydroids, bryozoans and brittle stars commonly adhere to the colonies.

Families: Briareidae, Anthothelidae, Subergorgiidae, Paragorgiidae, Coralliidae, Melithaeidae, Parisididae, Keroeididae, Acanthogorgiidae, Paramuriceidae, Plexauridae, Gorgoniidae, Ellisellidae, Ifalukellidae, Chrysogorgiidae, Primnoidae, Ainigmaptilidae, Isididae.

### Family Briareidae

23/8 *Briareum asbestinum* (Pallas). The erect, finger-like processes of spongy texture and appearance, which reach 20 cm or more in height, could be mistaken for a purple-grey sponge when the polyps are retracted. It is found growing commonly as an encrusting form on Caribbean reefs.

### Family Subergorgiidae

23/3 *Subergorgia hicksoni* Stiasny. This species of sea-fan may reach a height of over 1 m. It can be found attached to deep, wave-exposed reef fronts or rocks on sand in the Indo-Pacific.

24/2 *Subergorgia* sp. When the polyps are expanded, the fan becomes an efficient sieve, removing plankton and organic particles from the water. Found on deeper parts of Indo-Pacific reefs.

### Family Coralliidae

24/3 *Corallium rubrum* (Linnaeus) [precious coral]. The calcareous skeleton of this gorgonacean, which ranges in colour from pink to bright red, is collected extensively for use in the jewellery trade. The colonies, with their retractile white feathery polyps, branch in any plane. The species is found attached to rocks in poorly lit areas in the Mediterranean and off the coast of Japan, but overcollecting in the past has made it scarce in accessible areas.

### Family Melithaeidae

23/10, 24/1 *Melithaea squamata* (Nutting). The fans, which may reach 50 cm in height, are found protruding from vertical coral and rock faces in the Indo-Pacific.

24/5 *Mopsella ellisi* Haddon. A sea-fan with a tree-like growth form from the Indo-Pacific and Zealandic regions.

### Family Paramuriceidae

24/6 *Echinomuricea klavereni* Carpine & Grasshoff. A relatively little-branched colony about 20 cm high. The species often tolerates silty conditions. Found in the Mediterranean and adjacent Atlantic.

24/4, 24/8, 24/9 *Paramuricea clavata* (Risso). Often occurs in dense clumps in rocky areas with strong currents (Plate 24/8). In more protected deeper areas, the colonies grow larger and may reach a height of 1 m (Plate 24/9). As in other gorgonians, an excess of silt settling on the colony stimulates production of cleansing mucus (Plate 24/4). Found in the Mediterranean.

24/10 *Paramuricea macrospina* (von Koch). The polyp cups of this species are lined with large spicules. Found below 30 m in the Mediterranean.

25/1 *Paramuricea* sp. A specimen from the Great Barrier Reef.

29

## Family Plexauridae

24/7   General view of a Caribbean reef, showing several finger-like plexaurid species, many of which contain symbiotic zooxanthellae.

25/3   *Eunicea clavigera* Bayer. A dull-brown form with a few stout finger-like branches, which may reach a height of 30 cm or more on Caribbean reefs.

25/2, 25/4, 25/7   *Eunicella cavolinii* (Koch). This Mediterranean species, the fan of which reaches a height of 50 cm, is often found growing in clusters in shaded areas with a good current flow. The fans usually branch in a plane at right angles to the direction of water flow thus presenting the maximum area for trapping suspended food particles (Plate 25/4). Bush-like growths occur in caves where currents are prone to change in direction (Plate 25/7).

25/8, 25/9   *Eunicella singularis* (Esper) (=*E. stricta*). The whip-like branches tend to grow parallel to each other and up to 75 cm in length. This phenomenon can be seen well in the photograph, which shows re-growth after a fan has toppled over (Plate 25/9). The greenish-brown coloration of the polyps indicates the presence of zooxanthellae (Plate 25/8). Found in the Mediterranean.

25/2, 25/10   *Eunicella verrucosa* (Pallas). Colonies of this North-east Atlantic and Mediterranean species may reach a height of 50 cm or more, and rarely occur in water shallower than 15–20 m. The normal coloration of this species is a pink-orange, but occasionally completely white colonies are found (Plate 25/10).

25/5   *Muricea muricata* (Pallas). Colonies, which occur on Caribbean reefs, have stout finger-like branches reaching a height of 30 cm. The polyp cups formed from large spicules give the surface of the arms a knobbly appearance.

25/6   *Plexaurella dichotoma* (Esper). A dichotomously branching species with large dull-brown polyps which are usually expanded during the day. Found on Caribbean reefs.

26/1   *Rumphella* sp. An unknown species from Indo-Pacific reefs.

## Family Gorgoniidae

26/2   *Gorgonia flabellum* Linnaeus. This sea-fan, with closely anastomosing box-section branches, may reach a height of 1 m. Sometimes called the Bahamian or Venus fan, it is found on shallow reefs in the Caribbean.

26/3, 26/7   *Gorgonia ventalina* Linnaeus. Very similar to *G. flabellum* except that the branches are flattened in the plane of the fan. Tends to occur deeper than *G. flabellum* on Caribbean reefs.

26/9   *Lophogorgia sarmentosa* (Esper). A species forming spindly bushy growths which can withstand heavy sedimentation. It attaches to stones and shells on muddy bottoms in the Mediterranean.

26/8   *Pseudopterogorgia acerosa* (Pallas). The colony has large feathery branches reaching a height of 1 m or more. It is dry to the touch. Found on shallow reefs in the Caribbean.

26/6   *Pseudopterogorgia americana* (Gmelin). Very similar to *P. acerosa*, but the branchlets tend to be longer and more drooping. It is slimy to the touch. Found on shallow reefs in the Caribbean.

## Family Ellisellidae

26/4   *Ellisella erythrea* Kükenthal. A whip-like colony up to 1 m long with a hooked end. From below 50 m in the Indo-Pacific.

*Juncella fragilis* Ridley. A stout, whip-like form with short polyps, occurring on deeper parts of reefs in the Indo-Pacific.   26/10

An unidentified ellisellid from the Great Barrier Reef.   26/5

## Family Primnoidae

*Primnoella australasiae* Gray. The stems of members of this family are stiff and heavily calcified. The polyp bases are composed of overlapping spicules and are arranged in whorls around the stem. Found on Indo-Pacific reefs.   26/12

## Family Isididae

*Mopsea australis* Thomson & MacKinnon. A small gorgonian with parallel upright branches. It can be quite abundant on rocks below 20 m in the Zealandic region.   26/11

A close up photograph of a gorgonian with its polyps fully extended.   27/1

# Order Pennatulacea [sea-pens]

Fleshy colonies consisting of a modified central polyp with short polyps arising from its sides. The lower end of the main polyp forms a stalk which is thrust into soft sediments. A skeleton of calcareous spicules is present.

Sea-pens, which may reach a height of 1 m, are found on muddy and sandy bottoms, most frequently in shallow waters of sheltered bays and harbours. Several species can retract into the mud when disturbed, and many colonies are noted for their luminescence.

Families: Veretillidae, Echinoptilidae, Renillidae, Kophobelmnidae, Anthoptilidae, Funiculinidae, Protoptilidae, Stachyptilidae, Scleroptilidae, Chunellidae, Umbellulidae, Virgulariidae, Pennatulidae, Pteroeididae.

## Family Veretillidae

*Cavernularia obesa* Valenciennes. Similar features to *V. cynomorium* but the polyps can be retracted (Plate 27/8). Found in sand in the Indo-Pacific and Zealandic regions.   27/3, 27/

*Cavernularia* sp. A cluster of veretillids from sand on the Great Barrier Reef.   27/2

*Veretillum cynomorium* (Pallas). Colonies may be up to 30 cm high when fully extended but very much smaller when contracted. The long, non-retractable polyps are irregularly arranged on the stem. White, yellow or orange colonies occur. Found anchored in sand or mud below 20 m in the Mediterranean and adjacent Atlantic.   27/4

## Family Funiculinidae

*Funiculina quadrangularis* (Pallas). These colonies, which may reach a height of over 1 m, are supported by an internal skeleton with a rectangular cross-section. The polyps are arranged in longitudinal rows. The species is found, sometimes in high densities, anchored in muddy bottoms below 25 m in the North-east Atlantic and Mediterranean.   27/6

## Family Virgulariidae

*Scytaliopsis ghardagensis* Gravenhorst. Colonies up to 30 cm high. The polyps, and side branches on which they   27/5 are arranged, are characteristically translucent. Found in soft sediments in deeper waters of the Indo-Pacific.

27/9    *Virgularia mirabilis* Müller. Colonies up to 50 cm high with numerous side branches bearing polyps. Found anchored in muddy bottoms, sometimes in quite high densities, in the North-east Atlantic and Mediterranean.

### Family Pennatulidae [sea-pens]

7/7, 28/1    *Pennatula phosphorea* Linnaeus. The pinkish-red colonies may reach 40 cm or more in height. The white polyps are arranged in a single row on regular side-branches from the main stem (Plate 27/7). As in many other pennatulaceans, the colonies orientate themselves with the polyps on the leeward side of the body with respect to prevailing currents (Plate 28/1, current from the left), in order that they may take advantage of the zone of microturbulence that exists downstream to bring suspended food particles within range of the tentacles. Found anchored in muddy bottoms, usually below 20 m, in the North-east Atlantic and Mediterranean.

28/2    *Ptilosarcus gurneyi* (Gray). Colonies may reach a height of 50 cm when fully expanded. The polyps are arranged in single rows on leaf-like branches from the main stem. In this photograph the foot can clearly be seen anchoring the stem in the sediment. Occurs in the North-east Pacific.

### Family Pteroeididae

28/4    *Pteroeides spinosum* (Ellis). The thick fleshy stems may reach a height of 50 cm when fully expanded. The numerous rows of polyps are borne on leaf-like branches which are supported by pairs of spicules. Found anchored in soft sediments below 20 m in the Mediterranean and adjacent Atlantic.

28/8    *Pteroeides* sp. A 25 cm high specimen from the Indo-Pacific. In common with most sea-pens, it exudes a luminescent slime of unknown function when disturbed. This specimen has a small *Virgularia* species in its shadow.

28/3    *Sarcoptilus* sp. (= *Sarcophyllum* sp.). A partly expanded specimen anchored in sand in the Indo-Pacific.

28/5    A pteroeidid, possibly *Sarcoptilus*, with resident crabs.

# Class Zoantharia (= Hexacorallia)
approx. 5300 species

The polyps have six (or multiples of six), simple, usually unbranched, tentacles. The stomach cavity is divided by partitions which are situated in pairs. There are typically two ciliated grooves present in the gullet. The skeleton, when present, is an external calcareous mass lying outside the polyp body.

Some species have separate sexes, whilst others are hermaphrodite.

This class, which contains both solitary and colonial bottom-living forms, is represented in all areas, but its members are particularly common in warmer coastal waters.

## Order Actiniaria [sea-anemones]

Solitary polyps without a skeleton; their base is either modified for burrowing or is attached to hard objects by means of a sucker-like disc. The tentacles may be arranged in distinct rings around the mouth or may be crowded over the whole oral surface. Most anemones can move by creeping on the basal disc.

Anemones may be hermaphrodite or have separate sexes. In hermaphrodites the male sex cells usually develop first. In most species fertilisation takes place in the seawater and a planktonic larva is produced. In a few forms the young develop within the body of the parent. Some species reproduce asexually by the parent anemone splitting into two individuals. In others, new anemones develop by budding from the basal disc.

Anemones are cosmopolitan, but are larger in the tropics where forms with a diameter approaching 1 m may sometimes be found. Some species have been located at considerable depths but they are more common in shallow water and in pools and crevices on rocky shores. They are often brightly coloured and a species may have several different colour forms.

Most anemones capture living prey, including fish, with the stinging cells of the tentacles, although a few trap organic particles in mucus streams propelled towards the mouth by cilia on the tentacles.

Symbiotic algae (zooxanthellae) occur beneath the skin of some species.

Families: Gonactiniidae, Boloceroididae, Edwardsiidae, Halcampidae, Halcampoididae, Ilyanthidae, Andresiidae, Actiniidae, Aliciidae, Phyllactidae, Bunodidae, Stoichactidae, Minyadidae, Aurelianidae, Phymanthidae, Actinodendridae, Thalassianthidae, Discosomidae, Actinostolidae, Isophelliidae, Paractidae, Metridiidae, Diadumenidae, Aiptasiidae, Sagartiidae, Hormathiidae.

### Family Boloceroididae

*Boloceroides mcmurrichi* Kwietniewski. An anemone, 1–3 cm in diameter, which contains zooxanthellae. Unusual among anemones in that it is able to swim by beating the tentacles downwards in unison. It commonly reproduces asexually by fragmenting the body; as a result it is often found in clumps on shallow reefs. From the Indo-Pacific.    28/6

*Bunodeopsis globulifera* Verrill. A small anemone, with a body less than 1 cm in diameter, covered by bladders. Found attached to rocks and other hard objects in the Caribbean.    28/7

### Family Ilyanthidae

*Ilyanthus mitchelli* Gosse. A burrowing anemone with a body which is completely buried in the sediment. The slender retractile tentacles extend to form a circle up to 10 cm in diameter. Found in the North-east Atlantic.    28/9

### Family Andresiidae

*Andresia parthenopea* Andres. A burrowing anemone which lies buried in soft sediments with its tentacles exposed. It is possible that this species catches waterborne particles in mucus strings on the tentacles. Occurs in the Mediterranean.    29/7

### Family Actiniidae

*Actinia equina* (Linnaeus) var. *mesembryanthemum* (Ellis & Solander) [beadlet anemone]. A form with several rings of short tentacles around the mouth, reaching 5 cm in diameter (Plate 29/1). The colour of the body and tentacles is usually deep crimson, but may vary considerably, being striped in some individuals (Plate 29/2) and spotted in others. Some are green, still others    29/1, 29/2, 29/3 brown. The tentacles can be fully retracted into the body when this, primarily intertidal, variety is exposed at low water. The species has characteristically prominent bright blue sacs heavily armed with stinging cells at the bases of the outside tentacles. These enable the anemone to repel other anemones encroaching on its territory (Plate 29/3). Very common on rocks in the intertidal zone in the North and South Atlantic and the Mediterranean.

29/4 *Actinia equina* (Linnaeus) var. *fragacea* (Tugwell) [strawberry anemone]. A variety which is usually bigger than var. *mesembryanthemum*, and which invariably has a column spotted with green. It tends to occur in its greatest abundance rather lower on the shore than the beadlet anemone. Found in the North-east Atlantic and Mediterranean.

29/5, 80/2 *Anemonia sargassiensis* Hargitt. A species which forms part of the specialised community associated with *Sargassum* seaweed floating in the surface waters of the North-west Atlantic.

29/8, 29/9 *Anemonia sulcata* (Pennant) [snakelocks anemone]. A species with many sinuous tentacles, up to 15 cm in length, which cannot be fully retracted. It has a preference for strong light because of the zooxanthellae in the tentacles, and thus is often found clustered just below the water line in rock pools (Plate 29/8).

It is also frequently found attached to kelp fronds, pointing its tentacles towards the light. It often reproduces asexually by dividing into two. A specimen can be seen starting to divide in Plate 29/9. Found in the North-east Atlantic and Mediterranean.

29/6 *Anthopleura thallia* (Gosse). The column is up to 5 cm long, with rows of warts that are larger near the tentacles. A species which is predominantly intertidal in the North-east Atlantic.

29/10 *Anthopleura xanthogrammica* Brandt. The diameter of the emerald green crown of tentacles may reach 15 cm. The green coloration results from the symbiotic algae lodged within its tissues. A common intertidal species in the North-east Pacific.

29/11 *Bunodactis verrucosa* (Pennant) [wartlet or gem anemone]. The distinctly-patterned crown of tentacles is up to 5 cm in diameter. The column bears longitudinal rows of conspicuous warts. This species is one of those which broods its young in the stomach cavity, eventually expelling them through the mouth. Several juveniles that have recently been expelled can be seen here attached to rocks in the vicinity of the adults. Found in rock-pools and shallow waters of the North-east Atlantic and Mediterranean.

30/1 *Bunodosoma cavernata* (Bosc). The column has many closely-set rows of warts. This species, which expands mainly at night, is found in shallow Caribbean waters.

30/4 *Condylactis aurantiaca* (Delle Chiaje). A large anemone whose column height may reach 30–40 cm. The short tentacles, which are seen here partly contracted, have characteristically violet tips. It is usually found in shallow Mediterranean waters, attached to rocks and shells buried beneath the sand, with only its circlet of tentacles exposed.

30/2 *Condylactis gigantea* (Weinland). A large anemone up to 30 cm across the tentacles. The tentacles are very often tipped with purple but the colour of the rest of the body varies according to habitat. Common on reefs in the Caribbean.

*Physobrachia douglasi* Kent. This large actiniid species is found in the Indo-Pacific, often accompanied by clown fish (*Amphiprion* spp.), which dart amongst the tentacles, immune from the stinging cells of the anemone. It is thought that a coating of mucus over the fish prevents stimulation of the discharge of the stinging cells. When threatened by predators, the fish retreats amongst the tentacles.   30/6

*Tealia felina* (Linnaeus) var. *coriacea* (Cuvier) [dahlia anemone]. A large anemone which may reach a diameter of 15 cm across the tentacle crown. There are usually over 100 stout tentacles situated around the beautifully patterned oral disc (Plate 30/3). The tentacles can be fully retracted. The body is covered with warts to which shell fragments and sand particles adhere (Plate 30/5). Found on the shore in rock crevices adjacent to the sand and subtidally adhering to the open rock. Widely distributed in the North Atlantic and North-east Pacific.   30/3, 30/5

*Tealia felina* (Linnaeus) var. *lofotensis* (Danielssen). A variety more usually found attached to rocks at greater depths than the previous variety. The column has no warts and is free from adhering shell fragments.   30/7

An actiniid from the Antarctic.   30/9

## Family Aliciidae

*Alicia mirabilis* Johnson [berried anemone]. A night-active venomous anemone with long tapering tentacles and prominent warts on the column. The warts contain batteries of stinging cells which are probably used to defend the exposed column from predators at night. During the day it withdraws into a tube in the sand. Found in the Indo-Pacific and Caribbean.   31/1

*Alicia* sp. (possibly *A. mirabilis*). The column of this specimen, from 45 m depth in the Mediterranean, displays the prominent berry-like bunches of warts, typical of *A. mirabilis*.   31/2

*Triactis producta* Klunzinger. This blue species, up to 2 cm in diameter, is night active. Consequently, a clump of them, as seen in the photograph, may be mistaken for polyps of corals, many species of which are similarly closed during the day. Found in the Indo-Pacific.   30/8

## Family Phyllactidae

*Lebrunia danae* (Duchassaing & Michelotti). The two different types of tentacle characteristic of this family can be seen in this well camouflaged contracted specimen. The sting of this anemone is virulent. Found on rocks and amongst algae in the Caribbean.   31/7

## Family Stoichactidae

*Gyrostoma helianthus* Hemprich & Ehrenberg. Like all anemones in this family, it possesses a short broad column and a large oval disc, up to 50 cm in diameter, with numerous tentacles. Found in crevices on reefs in the Indo-Pacific.   31/3

*Gyrostoma quadricolor* Leuckart, in Ruppel. Found in crevices on reefs in the Indo-Pacific.   31/9

*Radianthus koseirensis* Klunzinger. This species is found on sand-covered rock and amongst reef corals in the Indo-Pacific.   31/8

*Radianthus ritteri* (Kwietniewski). The beautiful lilac column of this specimen contrasts with the colour of the accompanying clown fish. It tends to occur, on shallow   31/6 reef fronts and in lagoon areas where there is a moderately strong current, in the Indo-Pacific.

32/1 *Radianthus simplex* (Haddon & Shackleton). The knobbly appearance of the tentacles makes this species readily identifiable. It usually occurs in sandy areas adjacent to rocks and coral in the Indo-Pacific.

32/3 *Stoichactis helianthus* (Ellis). A species with a virulent sting, which is commonly found inhabiting shallow reefs in the Caribbean.

1/4, 31/5 *Stoichactis kenti* Haddon & Shackleton. Probably one of the largest anemones in this family, regularly reaching 40–50 cm across the tentacle crown. When the tentacles are retracted the top of the anemone is seen as a convoluted disc (Plate 31/5). An Indo-Pacific species.

### Family Phymanthidae

32/4 *Phymanthus crucifer* (Lesueur). A well camouflaged anemone, which lives attached to rocks, shells and other hard objects in shallow sandy areas of the Caribbean.

32/6 *Ragactis pulchra* Andres. A species with a flattened oral disc up to 5 cm in diameter. There is an inner ring of a few tentacles and an outer ring of many tentacles. Attached to hard objects, usually beneath sand and mud, at depths greater than 25 m in the Mediterranean.

### Family Isophelliidae

2/5, 32/7 *Telmatactis americana* Verrill. A species of isophelliid shown with tentacles expanded (Plate 32/7) and retracted (Plate 32/5). It is found on shallow reefs in the Caribbean.

### Family Metridiidae

/9, 32/11 *Metridium senile* (Linnaeus) [plumose anemone]. When fully grown, this species may expand to more than 30 cm in height. It is crowned by a mass of fine tentacles (Plate 32/11). The colour varies from pure white to a deep orange-brown. The food is primarily zooplankton and water-borne organic particles, but larger animals are caught as well.

Young specimens in which the tentacles are not finely divided are sometimes confused with other anemones. However, *M. senile*, always has a distinct collar (Plate 32/9). Found in coastal waters of the North Atlantic, North Pacific and Mediterranean.

### Family Aiptasiidae

32/8 *Aiptasia mutabilis* (Gravenhorst). This species has over 150 tentacles, patterned with irregular brown spots. The column, which may be up to 20 cm high, is attached to shells or rocks. Common in shallow Mediterranean waters.

32/2 *Bartholomea annulata* (Lesueur) [ringed anemone]. This species has long fine tentacles when fully expanded. When partly contracted, the characteristic white rings around the tentacles become more obvious. Found in shallow waters on rocks and other hard objects in the Caribbean.

32/10 *Heteractis lucida* Duchassaing & Michelotti. This shallow-water species from the Caribbean has the tentacles covered with small sacs containing stinging cells.

### Family Sagartiidae

32/12 *Actinothoe sphyrodeta* (Gosse). The column of this small species is rarely more than 2 or 3 cm high, and often has a striped appearance. The tentacles are usually white, but the oral disc varies in colour from white to bright orange. Found intertidally and subtidally attached to rocks, kelp holdfasts, etc. in clean waters of the North-east Atlantic.

*Anthothoe* sp. A delicate anemone with a virulent sting, photographed on the Great Barrier Reef.  33/10

*Cereus* sp. A specimen from coral sand in the Indo-Pacific.  33/1

*Sagartia elegans* (Dalyell) var. *nivea* (Gosse). The column of this species reaches a height of over 6 cm and bears up to 200 tentacles. The species can be readily distinguished from *A. sphyrodeta* by the wart-like suckers on the column. The tentacles, oral disc and sometimes even the column, are white in this variety. Occurs on rocks and in crevices in the North-east Atlantic.  33/2

*Sagartia elegans* (Dalyell) var. *venusta* (Gosse). Similar to the above variety, but with the oral disc orange and the tentacles white. Found on rocks and in crevices in the North-east Atlantic.  33/4

*Sagartiogeton undata* (Müller). A small anemone with the column length rarely exceeding 5 cm, and bearing about 100 long, translucent, drooping tentacles. Found attached to stones and shells on gravel and sand in the Mediterranean and North-east Atlantic.  33/3

### Family Hormathiidae

*Adamsia palliata* (Bohadsch). This species is often found in association with the hermit crab, *Pagurus prideauxi*. The base of the anemone completely envelopes the shell occupied by the crab. As the crab increases in size so the anemone secretes a horny substance from its base and artificially extends the limits of the shell, thus obviating the need for the hermit crab to change shells. The tentacles are well situated, on the underside of the shell just behind the crab's mouth-parts. Associated with hermit crabs on muddy gravel in the Mediterranean and North-east Atlantic.  33/8

*Calliactis parasitica* (Couch). This anemone has several hundred short tentacles and a column up to 10 cm long. It is often found attached to mollusc shells occupied by pagurid hermit crabs. In these situations the crab is partly protected from its predators by the stinging cells of the anemone. The anemone is able to feed on the scraps of food left by the crab and in addition gains a mobile base. From the North-east Atlantic and Mediterranean.  33/8

Hormathiids attached to a dead gorgonian in the Indo-Pacific.  33/5

## Order Corallimorpharia

Solitary or colonial forms with polyps resembling those of the true corals, but having no hard external skeleton. The tentacles which are arranged in rings around the mouth frequently have clubbed tips.

They have a cosmopolitan distribution.

Families: Corallimorphidae, Actinodiscidae.

### Family Corallimorphidae

*Corynactis australis* Haddon & Duerden. The tips of the club-shaped tentacles are characteristic. It occurs in closely packed colonies on subtidal rocks in the Zealandic region.  33/9

33/6, 33/7  *Corynactis viridis* Allman [jewel anemone]. A small translucent polyp whose body is not more than 1 cm high. The club-tipped tentacles are arranged in three rings (Plate 33/6). It sometimes occurs in very large numbers on rock faces in the shallow subtidal zone. There is much colour variation in the species, but asexual reproduction by division of the parent polyp produces clumps in which all are of one colour (Plate 33/7). Found in the North-east Atlantic.

### Family Actinodiscidae [false corals]

34/1  *Actinodiscus nummiformis* (Leuckart). The flat, plate-like surface of a polyp may reach 10 cm in diameter in large specimens. The tentacles of this specimen are retracted. Occurs on reefs in the Indo-Pacific.

34/2  *Rhodactis sanctithomae* (Duchassaing & Michelotti). This species can be readily distinguished from *Ricordia florida* since the short tentacles on the flattened disc are branched. Often found in clumps on rocks and reefs in the Caribbean.

34/4  *Ricordia florida* (Duchassaing & Michelotti). A cluster of polyps, each about 3 cm across, photographed here with their short tentacles expanded. Zooxanthellae occur commonly in the tissues. Found on rocks and reefs in the Caribbean.

### Order Zoanthiniaria (=Zoanthidea)

A group of mostly colonial forms which resemble small anemones. There is no skeleton or basal disc and the polyps possess one or two rings of smooth slender tentacles. The polyps of colonial forms are united by creeping stolons which encrust rocks and other surfaces.

Zoanthiniarians are most common in warm shallow waters, where they frequently occur on rocks and on other animals such as sponges, corals and gorgonians.

Families: Epizoanthidae, Zoanthidae.

### Family Epizoanthidae

34/7  *Epizoanthus paxii* Abel. The polyps of the colony arise from a common tissue mass to a height of about 1 cm. Found in the Mediterranean, below the waterline on rocks in caves and in crevices where there is sufficient water movement to ensure a steady food supply.

34/5  *Epizoanthus* sp. A specimen from the Caribbean.

34/3, 34/6  *Parazoanthus axinellae* (Schmidt). This species, with polyps about 1 cm high, often invades the tissue of the sponge *Axinella*. It is frequently encountered on rock faces and on cave roofs. In adverse conditions the colony gradually increases in length and pieces break off, establishing themselves elsewhere. Found in the Mediterranean and North-east Atlantic.

34/8  *Parazoanthus* sp. An as yet undescribed species, similar to *P. axinellae* but having bigger polyps with radiating brown stripes on the oral disc. This species also differs in never being found on axinellid sponges. Found on rocks in the Mediterranean.

### Family Zoanthidae

34/9  *Palythoa mammillosa* (Ellis & Solander). This species forms small domed colonies about 10 cm across on rocks and dead coral around low tide level in the Caribbean.

34/10, 34/11  *Palythoa tuberculosa* Esper. This Indo-Pacific species, shown with the polyps open (Plate 34/10) and closed (Plate 34/11), is very similar in form and habitat to *P. mammillosa*.

*Palythoa* sp. A specimen of *Palythoa* photographed on a shallow Caribbean reef.  35/1

*Zoanthus sociatus* (Lesueur). Forms mat-like colonies 10 cm or more across. Usually found in shallow sandy areas in the Caribbean.  35/2

A zoanthid from the Great Barrier Reef.  35/3

### Order Scleractinia (=Madreporaria) [true or stony corals]

Colonial and solitary forms that possess a hard calcareous skeleton which lies outside the polyp body. The anemone-like polyps may have many tentacles and can withdraw into cups with vertical, radially arranged, ridges (sclerosepta) within the skeleton (Figure 5). The skeleton, which is being continually formed by the bases of the polyps, often grows into a structure which is massive compared with the thin layer of living tissues connecting the polyps above it. Skeletons range in shape from flat or round masses to branching tree-like forms. Zooxanthellae are common in the tissues of nearly all shallow water corals and aid in their metabolism.

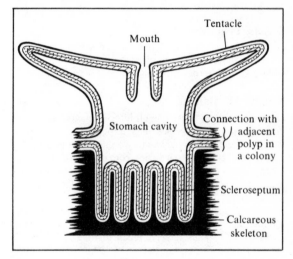

**Figure 5** *Coral polyp within its skeletal cup.*

Asexual reproduction by budding or by splitting of the original polyp is common, although sexual reproduction also occurs.

Corals are cosmopolitan but are more abundant in the tropics, where they may form reefs. Reef-building corals require warm shallow waters and are thus found mainly in the Indo-Pacific and the Caribbean. In most corals, the polyps are retracted during the day and expand to feed on plankton at night.

View across the exposed reef table at Heron Island, Great Barrier Reef.  35/4

### Suborder Astrocoeniina (=Astrocoeniida)

Mostly colonial forms with the small polyps usually bearing less than 12 tentacles in a single ring. Small polyp cups (1–3 mm in diameter) with few skeletal elements in the vertical ridges.

Families: Thamnasteriidae, Astrocoeniidae, Pocilloporidae, Acroporidae.

## Family Astrocoeniidae

35/7　*Stephanocoenia michelini* (Milne Edwards & Haime). Occurs as rounded boulder-like growths, usually less than 30 cm in diameter, with closely packed brown polyps. Found in the Caribbean.

## Family Pocilloporidae

35/5　*Pocillopora damicornis* (Pallas). Forms dense knobbly clusters up to 30 cm in diameter and pink or brown in colour. Found on reef flats in the Indo-Pacific.

35/9　*Seriatopora hystrix* Dana [needle coral]. Normally forms fragile, dense, bushy growths up to 50 cm across. Colour varies from pink-brown to yellow-green. The dried skeleton is unusual among true corals in being pale pink, almost all others being white. Found in the Indo-Pacific.

35/6　*Stylophora pistillata* (Esper). A common Indo-Pacific coral, occurring as stoutly branched clumps up to 30 cm in diameter. This species is known to contain zooxanthellae.

## Family Acroporidae

/8, 35/10, 36/3　*Acropora cervicornis* (Lamarck) [staghorn coral]. Occurs as loosely branched colonies which may grow to 2 m or more in height or sprawl in a haphazard fashion over the reef surface (Plate 35/8). Polyps protrude from tubular cups over all the colony surface (Plate 35/10). An abundant species in the Caribbean.

36/1　*Acropora humilis* (Dana). The polyp cups are arranged in tightly packed rows around the branches. Found in the Indo-Pacific.

36/2, 36/5　*Acropora palmata* (Lamarck) [elkhorn coral]. Massive tree-like growths with flattened branches expanding horizontally. Colonies of this abundant species may attain a height of 3 m or more. Found in the Caribbean.

36/7　*Acropora pulchra* (Brook). This species is very difficult to distinguish from others in the genus. Its plate-like growths over 1 m in diameter are common in semi-protected reef areas in the Indo-Pacific.

36/4　*Acropora variabilis* (Klunzinger). Considerable variation in colour and growth form occurs in this Indo-Pacific species.

36/6　*Acropora* sp. A specimen with delicate branching photographed on the Great Barrier Reef.

36/10　*Montipora foliosa* (Pallas) [leaf coral]. Forms flat, leafy colonies which may reach 2 m or more across. Often grows in dense masses over reef slopes in the Indo-Pacific.

5/8, 36/11　*Montipora* sp. There are many species and the growth form is very varied. Some forms grow in flat plates (Plate 36/8), whilst others encrust the reef surface (Plate 36/11). Colour varies from yellow-green to pink-brown. Common on Indo-Pacific reefs.

## Suborder Fungiina (=Fungiida)

Solitary and colonial forms, with the polyps usually having more than two rings of tentacles. The polyp cups are larger than 2 mm in diameter and the ridges have numerous skeletal elements with many perforations between them. Skeletal rods join adjacent ridges.

Families: Agariciidae, Siderastreidae, Fungiidae, Poritidae.

## Family Agariciidae

*Agaricia lamarcki* (Milne Edwards & Haime). Forms leaf-like growths up to 30 cm across, with rows of polyp cups between concentric ridges on the upper surface only. Brown is the usual coloration of this species, which is found on Caribbean reefs.　36/9, 36/12

*Leptoseris* sp. A specimen from the Indo-Pacific forming a leaf-like encrusting growth, 20 cm across.　37/1

*Pachyseris rugosa* (Lamarck). A species forming brown-tinted irregular clumps 20–30 cm across, with prominent ridges. Found on reef slopes in the Indo-Pacific.　37/3

*Pachyseris* sp. A species forming plate-like growths up to 30 cm across with concentric ridges. Found on reef slopes in the Indo-Pacific.　37/2

*Pavona varians* Verrill. This species, as the name implies, is variable in growth form, size, and colour. Found on reef flats in the Indo-Pacific.　37/7

## Family Fungiidae

*Cycloseris cyclolites* (Lamarck). A small coral comprising a single free-living polyp up to 5 cm in diameter but usually smaller. The two photographs show one closed (Plate 37/5) and one open (Plate 37/6). Found on sand around reefs in the Indo-Pacific.　37/5, 37/6

*Fungia actiniformis* (Quoy & Gaimard). This free-living species, which is up to 20 cm across, is unusual in not retracting its tentacles during the day. Found on reef flats and in lagoons in the Indo-Pacific.　37/10

*Fungia echinata* (Pallas). This Indo-Pacific species has prominent tooth-like edges to the radiating lines of septa.　37/11

*Fungia fungites* (Linnaeus) [mushroom coral]. A common species reaching 20 cm in diameter, which is found on reefs and lagoons in the Indo-Pacific. As adults, all *Fungia* species are free-living but when young they are often seen in clumps, each polyp attached to the bottom by a stalk (Plate 37/4).　37/4, 37/8

*Fungia scutaria* (Lamarck). A stalked juvenile stage, less than 2 cm in diameter, photographed on an Indo-Pacific reef.　37/9

*Herpolitha limax* (Esper) [slipper coral]. An elongated form, up to 30 cm or more in length, comprising several polyps whose mouths open into a common central groove in the skeleton. Occurs on reefs and adjacent sand in the Indo-Pacific.　38/3

*Parahalomitra robusta* Quelch [basket coral]. A normally dome-shaped colonial fungiid which grows to over 30 cm in diameter. The colony is basket-shaped when viewed with the underside uppermost. Found on reefs and adjacent sand in the Indo-Pacific.　38/4

## Family Poritidae

*Goniopora lobata* Milne Edwards & Haime. Forms lobed colonies which can reach 1 m in diameter. *Goniopora* is unusual amongst corals in that it has very long polyps which are often expanded during the daytime, the elongated polyps can thus be confused with those of a zoanthid if the skeleton is obscured. A detailed photograph of the colony surface shows some expanded polyps (Plate 38/7). Found on Indo-Pacific reefs.　38/6, 38/7

*Goniopora* sp. Colonies of this unknown species are often less lobed than those of *G. lobata* (Plate 38/8) although the polyps are similar (Plate 38/5). Found on Indo-Pacific reefs.　38/5, 38/8

35

38/1    *Porites astreoides* Lamarck. This species usually forms yellow-green encrusting mounds which may reach a diameter of over 50 cm on Caribbean reefs.

38/2    *Porites lutea* Milne Edwards & Haime. Colonies of this species often grow in shallow lagoon conditions. At spring tides the tops of the colonies are sometimes killed by prolonged exposure to air and growth continues in a lateral direction only: hence the common micro-atoll formations which may be several metres in diameter. The smooth appearance of these corals is due to the small diameter and shallowness of the polyps cups. Found on reefs and in lagoons in the Indo-Pacific.

38/9    *Porites porites* (Pallas) [clubbed finger coral]. This species forms low, short, branching clumps in shallow water. The polyps are often expanded during the day giving the colony a furry appearance. Found on hard bottoms in the Caribbean.

39/1    *Porites* sp. A specimen with branching finger-shaped colonies from Indo-Pacific reefs.

## Suborder Faviina (= Faviida)

Solitary and colonial forms with the polyps usually having more than two rings of tentacles. The numerous skeletal elements of the ridges form an almost continuous sheet and the edges of the ridges are toothed. Adjacent ridges are rarely joined by rods.

Familes: Faviidae, Rhizangiidae, Oculinidae, Meandrinidae, Merulinidae, Mussidae, Pectiniidae.

### Family Faviidae

39/2    *Cladocora cespitosa* (Linnaeus). Forms low mounds which may extend for 1 m or more beneath overhanging rocks. In sheltered regions the polyp cups may extend for several centimetres above the surrounding tissue and the colonies are more bushy in appearance. From shallow to deep water in the Mediterranean.

39/3, 39/8    *Colpophyllia natans* (Houttuyn). Forms lightweight convex masses. The shallow valleys (elongated polyp cups) are winding and interconnected (Plate 39/3). The polyps, as in most other corals, expand at night (Plate 39/8). Found on Caribbean reefs.

39/5    *Cyphastrea japonica* Yabe & Sugiyama. The surface of this encrusting form is often very knobbly due to the budding of young polyps from the bases of older ones. Found on Indo-Pacific reefs.

39/7    *Diploria clivosa* (Ellis & Solander). Forms irregular low encrusting growths on Caribbean reefs. The valleys are shallow and winding and rarely interconnect.

39/4, 39/6    *Diploria labyrinthiformis* (Linnaeus) [brain coral]. Mature colonies may form massive boulder-like growths up to 3 m in diameter on Caribbean reefs. The interconnecting valleys are winding and deep, resembling the convolutions of the surface of the human brain.

39/9    *Diploria strigosa* (Dana). Forms dome-shaped masses which are smaller than in *D. labyrinthiformis* and do not have a longitudinal groove running along the top of the walls. This specimen has its polyps expanded. Found on Caribbean reefs.

39/10    *Echinopora horrida* Dana. The branching colonies, which may reach 20 cm or more in height, have large protruding polyp cups. Found on Indo-Pacific reef slopes.

40/1    *Echinopora lamellosa* (Esper). Colonies are variable in structure but often form vertical plate-like growths, up to 20 cm high, which may join to form funnels. Found on Indo-Pacific reef slopes.

*Favia* sp. The species in this common genus are difficult to distinguish. Their rounded colonies are of variable colour (usually brown or green) and occasionally grow to more than 1 m in diameter. The skeletal polyp walls do not merge together. Photographed in the Indo-Pacific.    40/2, 40/

*Favites abdita* Ellis & Solander [honeycomb coral]. The extraordinary brown and white coloration of the polyps in this specimen shows off their polygonal outline. Found on Indo-Pacific reefs.    40/4

*Favites* sp. This specimen from Indo-Pacific reefs shows the fusion of the walls of adjacent polyps which distinguishes *Favites* from *Favia*. The fluorescent green polyps of this specimen can be clearly seen.    40/5

*Hydnophora exesa* (Pallas). This species forms colonies not usually more than 5–10 cm across on Indo-Pacific reefs. Cone-shaped projections over the surface characterise species in the genus *Hydnophora*.    40/6

*Leptoria* sp. An Indo-Pacific brain coral which has formed a massive boulder-like growth 1 m across.    40/7

*Manicina mayori* Wells. Colonies are round to oval and grow up to 15 cm in diameter. The surface is deeply convoluted. Found on Caribbean reefs.    40/8

*Montastrea annularis* (Ellis & Solander). Sometimes forms massive growths up to 2 m across, although the polyp cups themselves are less than 3 mm in diameter. The plate-like growth form seen in the photograph is one that is typical of many corals when growing in deeper water. A principal reef-builder in the Caribbean.    40/9

*Montastrea cavernosa* (Linnaeus). This species forms massive boulders up to 2 m across. The polyp cups are about 8 mm in diameter, considerably larger than in *M. annularis*. The photographs show three successive stages in the opening of the polyps on a Caribbean reef.    41/1, 41/ 41/10

*Oulophyllia crispa* (Lamarck). Usually occurs as domed colonies up to 30 cm across with deep, broad valleys between the walls. The colour of the colonies varies considerably, as seen in the three examples. Found on Indo-Pacific reef slopes.    41/4, 41/ 41/9

*Platygyra* sp. A common brain coral which, like *Leptoria*, forms massive growths up to 2 m across on reefs in the Indo-Pacific.    41/3

*Plesiastraea urvillei* Milne Edwards & Haime. An emerald-green encrusting species which is indigenous to the temperate waters of the Zealandic region.    41/5

*Trachyphyllia geoffroyi* Audouin. A colonial form about 10 cm in diameter, consisting of a small number of polyps. Found on reef rubble and sand in the Indo-Pacific.    41/7

### Family Rhizangiidae (= Astrangiidae)

*Astrangia* sp. A specimen with large solitary polyps that sometimes have a thin encrustation joining the bases of the cups together. Found attached to the bases of reef corals and to rocks in the Indo-Pacific.    41/8

*Phyllangia mouchezii* (Lacaze-Duthiers). Solitary cups up to 10 mm in diameter growing in clusters on rock in the Mediterranean and South-east Atlantic.    42/1

### Family Oculinidae

*Acrohelia horrescens* (Dana). Colonies, which may reach a diameter of 20 cm, have a fragile skeleton forming a    42/2 low growth which is grey or brown in colour. Occurs on Indo-Pacific reefs.

42/3 *Galaxea clavus* Dana. The polyp cups are widely separated, as in all *Galaxea* species. Forms low colonies on reefs in the Indo-Pacific.

2/5, 42/7 *Galaxea fascicularis* (Linnaeus). The polyp cups of members of this genus have radially arranged vertical ridges which characteristically protrude above the surrounding skeleton (Plate 42/5). The low, rounded colonies may reach 30 cm in diameter on Indo-Pacific reefs.

42/9 *Oculina patagonica* de Angelis. This rock-dwelling species is commonly found on the Atlantic coast of South America and has recently been discovered in the Mediterranean. Possibly it was transported from South America on the hulls of ships.

### Family Meandrinidae (= Trochosmiliidae)

42/4 *Dendrogyra cylindrus* Ehrenberg [pillar coral]. Forms heavy, cylindrical pillars, sometimes over 1 m in height. The pillars have a furry appearance when the polyps are expanded during the daytime as well as at night. Found on Caribbean reefs.

42/10 *Dichocoenia stokesi* Milne Edwards & Haime. Forms heavy boulders, sometimes over 30 cm in diameter. The polyp cups are round to elongate in this Caribbean species.

2/6, 42/8 *Meandrina brasiliensis* (Milne Edwards & Haime). Forms flat, boulder-like colonies up to 30 cm across (Plate 42/6). The polyp walls separate large, well defined, skeletal ridges protruding from the deep valleys (Plate 42/8). Found on Caribbean reefs.

3/1, 43/2 *Meandrina meandrites* (Linnaeus). The boulder-like colonies are usually larger than those of *M. brasiliensis*. Zig-zag, steeply sided walls separate adjacent valleys. The photographs show a small colony with the polyps retracted (Plate 43/1) and expanded (Plate 43/2). Common on Caribbean reefs.

### Family Merulinidae

43/4 *Merulina ampliata* (Ellis & Solander). This species forms irregular, plate-shaped colonies up to 50 cm across. The plate surfaces have radiating vertical ridges with occasional raised nodules. Occurs on Indo-Pacific reefs.

### Family Mussidae

3/3, 43/5 *Isophyllia sinuosa* (Ellis & Solander). Dome-shaped colonies, with a short stalk, up to 20 cm across. Often bright green in colour but sometimes slate blue. The photographs show a colony with polyps retracted (Plate 43/5) and expanded (Plate 43/3). Found on Caribbean reefs.

43/9 *Lithophyllia vitiensis* (Brüggemann). Solitary polyps that may reach 15 cm or more in diameter. Many are green but red and brown forms also exist. They fluoresce strongly in ultra-violet light. Found on Indo-Pacific reef slopes.

43/6 *Lobophyllia corymbosa* (Forskål). The colonies are composed of large, fleshy, circular polyps each measuring up to 3 cm in diameter. Found on Indo-Pacific reefs.

3/7, 43/8 *Lobophyllia hemprichii* (Ehrenberg). The large polyps of these colonies are more irregular in outline than those of *L. corymbosa* (Plate 43/7). The polyp cups are supported on long stalks (Plate 43/8). Common on Indo-Pacific reefs.

43/10 *Mycetophyllia daniana* Milne Edwards & Haime. Forms

encrusting growths over dead coral and rock. The walls dividing the valleys are low and disappear in mature colonies. Found on Caribbean reefs.

*Mussa angulosa* (Pallas). Forms boulder-like growths of large, irregularly shaped polyps, 10 cm or more across. This species fluoresces strongly in ultraviolet light. The photograph shows a young colony. Found on Caribbean reefs. **44/1**

*Protolobophyllia* sp. A coral with large irregularly shaped cups occurring on Indo-Pacific reefs. **44/2**

*Scolymia cubensis* (Milne Edwards & Haime). The solitary cups of members of this genus can be mistaken for the young stages of *Mussa angulosa*. However, they are always solitary, never in clusters. Found on deeper Caribbean reefs. **44/3**

*Scolymia lacera* (Pallas). The photograph, taken at night, shows the large polyp (up to 15 cm across) with its stubby tentacles expanded (Plate 44/9). It is closed normally during the daytime (Plate 44/10). Usually less abundant than *S. cubensis* on Caribbean reefs. **44/9, 44/10**

*Symphyllia* sp. A beautifully coloured colony from the Great Barrier Reef with the polyps partly expanded, revealing the mouths in the valleys. This genus differs from the closely related *Lobophyllia* in that the walls of adjacent polyp cups are fused together to form a continuous sheet of coral. **44/8**

### Family Pectiniidae

*Echinophyllia* sp. Grows as irregular sheets, with the polyps facing upwards, on reef slopes in the Indo-Pacific. **44/4, 44/11**

*Mycedium elephantotum* (Pallas). Colonies form plate-like growths (Plate 44/5) with the polyps facing outwards towards the growing edge of the colony (Plate 44/6). This characteristic enables the genus to be readily distinguished from *Echinophyllia*. Found on Indo-Pacific reef slopes. **44/5, 44/6**

*Pectinia lactuca* (Pallas) [carnation coral]. Occurs in colonies up to 50 cm across, with a characteristic convoluted appearance, on reef slopes and in lagoons in the Indo-Pacific. Brown is the most usual colour form. **44/7**

*Pectinia* sp. This deeply convoluted coral grows on Indo-Pacific reefs. **45/1**

## Suborder Caryophylliina (= Caryophylliida)

Solitary and colonial corals with usually more than two rings of tentacles. The numerous skeletal elements of the ridges form a continuous sheet and the edges of the ridges are untoothed. There are no rods connecting adjacent ridges.

Families: Caryophylliidae, Flabellidae.

### Family Caryophyllidae

*Caryophyllia smithi* Stokes [Devonshire cup coral]. A solitary coral, with oval polyp cups up to 2 cm high and 3 cm long. The cups have pronounced radiating ridges which are masked when the polyp is fully expanded. Thus it can be readily mistaken for an anemone (Plate 45/2). Occurs in sometimes quite dense aggregations on silt-covered rock surfaces (Plate 45/3) in the North-east Atlantic. **45/2, 45/3**

*Euphyllia* sp. When the short greyish-green tentacles of this species retract, the polyp skeleton with its prominent radiating ridges is revealed. Cups may reach a diameter of 5 cm or more. Occurs on Indo-Pacific reefs. **45/5, 45/9**

45/8, 45/10    *Eusmilia fastigiata* (Pallas) [flower coral]. Forms domed, branching colonies up to 50 cm across, with round to oval polyp cups (Plate 45/10). Polyps expand at night to reveal the stubby tentacles typical of the family. Found on Caribbean reefs.

45/4    *Physogyra liechtensteina* Milne Edwards & Haime [bubble coral]. The short, bulbous tentacles of this colonial species are characteristic of the bubble corals as a whole. Found on Indo-Pacific reefs.

45/6    *Pterogyra sinuosa* (Dana) [bubble coral]. The bubble-like expansions of the tentacles are thought to provide protection for the colony during the day. The bubbles contract somewhat at night, and the tentacle tips, which capture plankton, become more obvious. Found on reefs and in lagoons in the Indo-Pacific.

45/7    A bubble coral with sausage-shaped expansions of the tentacles forming a protective umbrella over the colony. Occurs on Indo-Pacific reefs.

### Suborder Dendrophylliina
(= Dendrophylliida)

Solitary and colonial corals with usually more than two rings of tentacles on the polyps. The numerous skeletal elements of the ridges form an almost continuous sheet and rods connect adjacent ridges.

Family: Dendrophylliidae.

### Family Dendrophylliidae

46/1    *Astroides calycularis* (Pallas). This species forms rounded, cushion-like masses which could be mistaken for *Cladocora*. However, the polyp branches are densely packed and fused together right up to their cups. Occurs on rocks and stones, often in shaded areas, in warmer parts of the Mediterranean.

46/3    *Balanophyllia bairdiana* Milne Edwards & Haime. Usually solitary corals, which occasionally bud off new individuals. They are able to tolerate a great deal of silt deposition. Found on rocks in the Zealandic region.

46/2, 46/5    *Balanophyllia gemmifera* Klunzinger. Found on Indo-Pacific reefs in closely packed clumps of polyps as a result of budding from larger individuals (Plate 46/5).

Clumps may be up to 50 cm across. Polyps are normally expanded at night (Plate 46/2).

*Balanophyllia italica* Michelin. An oval solitary coral. This Mediterranean species is unusual in that it possesses zooxanthellae.    46/6

*Balanophyllia regia* Gosse. Solitary individuals up to 2 cm high. Found aggregated in rock gulleys and clefts in the intertidal zone and shallow water in the North-east Atlantic.    46/7

*Dendrophyllia gracilis* Milne Edwards & Haime. Clumps of individuals are formed by budding. Polyps can be seen fully retracted into the polyp cups (Plate 46/8) and expanded (Plate 46/4). Found on Indo-Pacific reefs.    46/8, 46/

*Dendrophyllia* sp. A specimen from the Great Barrier Reef, with fully expanded polyps.    46/11

*Heteropsammia michelini* Milne Edwards & Haime [button coral]. A solitary polyp over 2 cm in diameter, which is found on coral sand near reefs in the Indo-Pacific. The coral often has a sipunculan worm boring in its base. The feeding activities of the sipunculan can move the coral polyp slowly over the bottom.    46/9

*Leptopsammia pruvoti* Lacaze-Duthiers. The brilliant yellow colour of this species is distinctive. When fully expanded, the coral polyps resemble anemones. Found below tide level on rocks in caves and under overhangs in the Mediterranean and North-east Atlantic.    46/10

*Tubastrea aurea* Milne Edwards & Haime. Forms domed clumps up to 10 cm across. The polyps protrude from a common encrusting base. Usually found in low-light conditions in caves and beneath rocky overhangs on Indo-Pacific reefs.    47/6

*Tubastrea micrantha* (Ehrenberg) (= *Dendrophyllia micrantha*). Forms tree-like branching colonies up to 50 cm in height (Plate 47/7). The colour of colonies varies between dark brown and green. The polyp tentacles when expanded at night provide a mass of contrasting colour (Plate 47/3). Occurs on deeper parts of reefs in the Indo-Pacific.    47/3, 47/

*Turbinaria heronensis* Wells. Forms plate-like colonies with long polyp cups protruding from the main mass. Occurs on reefs and in lagoons in the Indo-Pacific.    47/1

*Turbinaria* sp. Two examples of a coral resembling *T. peltata* which forms massive leaf-like growths, several metres across, on Indo-Pacific reefs.    47/2, 47/

# Phylum Ctenophora
[comb-jellies] approx. 80 species

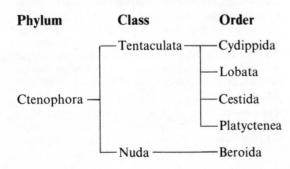

| Phylum | Class | Order |
|---|---|---|
| | Tentaculata | Cydippida |
| | | Lobata |
| Ctenophora | | Cestida |
| | | Platyctenea |
| | Nuda | Beroida |

An exclusively marine group of delicate animals which are mainly found swimming in the open water, although a few genera have become modified for a bottom-living existence. They are primarily radially symmetrical with a secondary bilateral symmetry. Their transparent, gelatinous, variously-shaped bodies have their longest dimension ranging from a few millimetres to over one metre in the cestids. The body possesses eight characteristic bands made up of many short transverse rows of fused cilia called combs. The beating of the comb rows propel the animals through the water. It was originally thought that their gut possessed only one opening, the mouth, but it is now known that, at least in some species, food remains pass out through one or more anal pores. A balancing sense organ is situated near the anal pore(s) on the opposite side of the body to the mouth. The two tentacles usually present contain special 'lasso' cells which secrete a sticky material that adheres to the prey. Stinging cells are not present (cf. Coelenterata).

Ctenophores are hermaphrodite. Eggs and sperm are normally shed into the water where fertilisation occurs, although a few species brood their eggs. A free-swimming larva is usually present.

Ctenophores occur in the surface plankton of all seas, where they are sometimes found in large swarms. Some planktonic species, however, live at great depths and non-planktonic forms creep over the bottom. They are active predators feeding on small animals including fish and, in some cases, other ctenophores. They are noted for their luminescence which originates beneath the comb rows. Many ctenophores bear a superficial resemblance to medusae and the group at one time was included within the phylum Coelenterata.

## Class Tentaculata

Ctenophores with tentacles.

## Order Cydippida

The body is round, oval, or slug-like with two long tentacles which can be retracted into sheaths.

Families: Mertensiidae, Callianiridae, Pleurobrachiidae.

### Family Pleurobrachiidae

47/9 *Hormiphora* sp. The body of this unidentified species (possibly *H. plumosa*) is about 2 cm long. The two long feeding tentacles can be clearly seen. A warm-water ctenophore.

47/10 *Pleurobrachia bachei* Agassiz [sea-gooseberry]. The sub-spherical body is about 2 cm long, but the plume-like tentacles may trail for 15 cm or more behind the body. The filaments of the tentacles increase many-fold the chances of the tentacles contacting prey organisms in the plankton. This species is very abundant in the North-east Pacific at certain times of the year.

## Order Lobata

The body is laterally compressed, with two rounded lobes on each side of the mouth and four ciliated processes projecting above the mouth. The two main tentacles lack sheaths and are usually branched and fairly short. Small secondary tentacles are present in special grooves which run from the main tentacles to the bases of the ciliated processes.

Families: Leseuriidae, Bolinidae, Deiopeidae, Eurhamphaeidae, Eucharidae, Mnemiidae, Calymmidae, Ocyroidae.

### Family Bolinidae

7/8, 47/11 *Bolinopsis infundibulum* (Müller). A large delicate species, which may reach 15 cm in length, with oral lobes about half the body length. Sometimes very abundant in the North Atlantic and adjacent Mediterranean.

### Family Eurhamphaeidae

47/5 *Eurhamphaea vexilligera* Gegenbaur. This torpedo-

shaped species may reach a length of 5 cm or more. It is characterised by the projections at the front of the body on to which the comb rows extend. A tropical Atlantic species which extends northwards in the Gulf Stream.

### Family Eucharidae

*Leucothea multicornis* Eschscholtz (=*Eucharis multicornis*). This species, which is usually about 10 cm long, has very large oral lobes which may extend like the wings of a butterfly. The large conical papillae over the body surface are another notable feature. Found in the Mediterranean and tropical Atlantic. — 48/6

### Family Mnemiidae

*Mnemiopsis maccradyi* (Mayer). A large, pear-shaped ctenophore up to 10 cm in length. The large oral lobes give the animal a distinctly bilobed appearance. Found, often in large numbers, in the warm waters of the Atlantic and the Caribbean. — 48/1

### Family Ocyroidae

*Ocyropsis crystallina* (Rang). The laterally flattened body is about 5 cm long. The two large oral lobes flanking the body drive the animal through the water by their flapping movements. An inhabitant of the tropical Atlantic, it is sometimes carried north in the Gulf Stream. — 48/2

## Order Cestida

The body is expanded laterally into a ribbon-like form, with the main tentacles, although retaining their sheaths, very much reduced. Numerous short tentacles occur along the entire lower edge of the body. Four of the comb rows are reduced and swimming is achieved instead by muscular undulations of the body. Beating of the cilia in the remaining comb rows, which are situated along the upper surface, serves only to assist orientation of the animal in the water.

Family: Cestidae.

### Family Cestidae

*Cestum veneris* Lesueur [Venus's girdle]. The laterally extended ribbon-like body may reach a width of 1.5 m. It swims by sinusoidal movements of the body (Plate 48/3) and, although an oceanic form, is sometimes found in coastal waters. In Plate 48/4 it can be seen amongst the jellyfish, *Aurelia aurita*. A Mediterranean, tropical Atlantic and Indo-Pacific species. — 48/3, 48/4

## Order Platyctenea

Flattened, oval ctenophores whose comb rows are reduced or absent altogether. A pair of lateral tentacles, which can be retracted into sheaths, is present.

Representatives, which are often green or brown in colour, have been found creeping on corals and alcyonarians in warmer waters and on sea-pens off Greenland.

Families: Ctenoplanidae, Coeloplanidae.

### Family Coeloplanidae

*Coeloplana* sp. A flattened slug-like animal without comb rows, which is sometimes found creeping over soft corals. The upper surface of animals in this genus have a few characteristic rows of small papillae. This specimen from the Red Sea can be seen fishing for prey with a long filament-bearing tentacle. — 48/5

39

Plate 48                                                                                    Phyla Ctenophora, Platyhelminthes

## Class Nuda

Ctenophores without tentacles.

## Order Beroida

The conical body has a wide mouth opening into a large gullet which occupies most of the interior of the animal. Members of this order are active swimmers.

Family: Beroidae.

## Family Beroidae

*Beroe cucumis* Fabricius. The flat, helmet-shaped body, whose branched inner canals are visible, may reach 15 cm in length. A wide mouth enables it to engulf prey almost as big as itself. Found in the North Atlantic and Mediterranean.     48/7

*Beroe ovata* Eschscholtz. The flattened body, up to 15 cm in length, has its branched inner canals fused into a network. Occurs in the Mediterranean, North Atlantic and Caribbean.     48/8

# Phylum Platyhelminthes

approx. 25 000 species

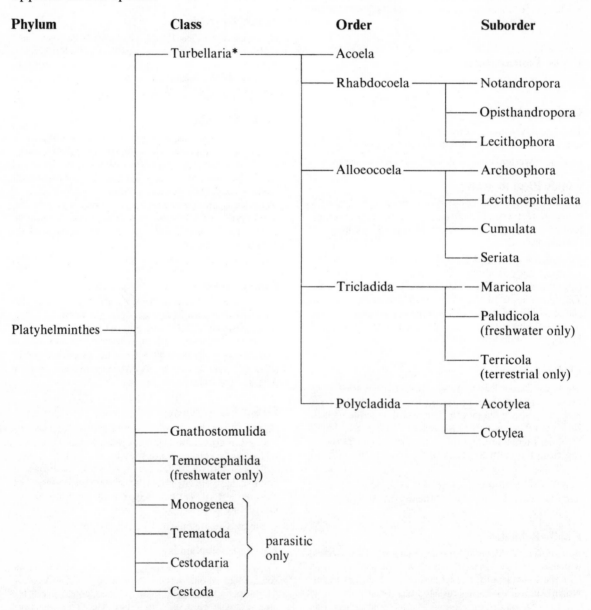

*The traditional arrangement of the Turbellaria has been followed in this classification. Several more recent classifications have been made, but as yet none have been generally accepted by taxonomists.

Members of this phylum, which contains the free-living flatworms and the parasitic flukes and tapeworms, are bilaterally symmetrical, soft-bodied animals. The body, which is usually flattened dorsoventrally, has a recognisable head with sense organs, although these are often reduced in the parasitic species. There is no body cavity and the gut has one opening serving as both mouth and anus. There are no special respiratory structures, respiration occurring through the general body surface.

Platyhelminthes, with few exceptions, are hermaphrodite and the reproductive system is often very complex. Exchange of sperm occurs between individuals, and the fertilised eggs which are released to the exterior may develop into a larval form or directly into a juvenile worm. Complicated life histories, with a succession of larval forms, frequently occur in the parasitic species.

Only two classes, the Turbellaria and the Gnathostomulida, contain free-living marine species.

# Class Turbellaria [flatworms]
approx. 4000 species

The flatworms are mainly marine, but a few species occur in freshwater and in damp situations on land. Several live in close association with other animals, whilst a few species are entirely parasitic. The majority are small, with a body length of not much more than 1–2 cm, but some land-dwellers reach lengths of 50 cm. The flattened body, which is usually elongated or leaf-shaped, is covered with cilia. The cilia are often longer on the ventral surface and, together with the muscles, play an important role in the characteristic gliding movement of flatworms. Numerous single or bundles of rods (rhabdites), of uncertain function, are present in the skin. The head usually possesses one or several pairs of eyes, and tentacles and other sensory structures are frequently present. The mouth opens on the underside and a muscular tube, the pharynx, is frequently protruded through the mouth to grasp food.

Sperm are exchanged between hermaphrodite individuals with the aid of a complex copulatory apparatus consisting of a penis, muscular sac, and various glands. A brief mating ritual, consisting of head and body contacts, occurs prior to copulation in most species. The fertilised egg usually develops directly into a small worm, but in some, a planktonic larva is produced. Some species can reproduce asexually by splitting the body into several parts.

Turbellarians occur in all seas, where they are mainly bottom-dwellers living in the intertidal zone and in shallow waters. A few species swim freely in the water. The majority are thought to be carnivorous, feeding on a variety of small invertebrates, including other turbellarians. Most have sombre colouring, although some display striking colour patterns. A few species are green, due to the presence of symbiotic algae in the body tissues.

# Order Acoela

An entirely marine group of small flatworms which are rarely more than a few millimetres in length. The elongated or oval body may possess tail lobes. There is no distinct brain or eyes, although balancing organs are present in the head region. There is no true gut, the mouth or pharynx opening into a network of digestive cells.

Exchange of sperm occurs between individuals, and the fertilised eggs are enclosed within capsules and attached to rocks and other objects.

Acoelans are more common in temperate and arctic seas, where they live under rocks, amongst algae, or on the bottom mud in both shallow and deep waters. A few species swim freely in the open water, whilst others live in the gut or body cavity of sea cucumbers and sea urchins. They are usually whitish in colour but some species are coloured green by symbiotic algae.

Families: Anaperidae, Convolutidae, Hallangiidae, Otocelididae, Nemertodermatidae, Haploposthiidae, Myostomullidae.

# Order Rhabdocoela

Small flatworms, a few millimetres in length, which live in marine, freshwater and damp terrestrial environments. The elongated body usually lacks tentacles and projections, but the head region may bear a snout. They usually have a distinct brain, eyes, and occasionally possess balancing organs. A gut is normally present.

Fertilised eggs are laid in capsules which either lie free on the bottom or are attached to objects.

Rhabdocoels commonly occur on sandy and muddy shores, although a few are free-swimming. Several species live on other invertebrates or within their body cavity or gut.

# Suborder Notandropora (=Catenulida)

A mainly freshwater group of small rhabdocoels with a simple pharynx.

Chains of individuals are frequently formed by asexual budding. Sexual reproduction is rare.

In sexually mature individuals, the male reproductive system develops first and has a dorsal opening to the exterior and a penis without bristles. The female system is simple with no special opening, and the fertilised egg is released by rupture of the body wall. The few marine species mainly occur in coastal caves.

Family: Retronectidae.

# Suborder Opisthandropora
(=Macrostomida)

Marine and freshwater rhabdocoels with a simple pharynx. The animals have complete male and female reproductive systems with separate openings to the exterior; the penis usually carries a long bristle.

The few free-swimming species belong to this group.

Families: Macrostomidae, Microstomidae.

# Suborder Lecithophora (=Neorhabdocoela)

Marine, freshwater and terrestrial forms with a barrel- or rosette-shaped muscular pharynx. The well developed male and female reproductive systems may have a common opening to the exterior. Asexual reproduction does not occur within this group.

Some of these rhabdocoels inhabit sandy bottoms in coastal waters, whilst others live on or inside the body of other invertebrates.

Families: Provorticidae, Graffillidae, Umagillidae, Pterastericolidae, Typhloplanidae, Kalyptorhynchidae, Schizorhynchidae, Trigonostomidae, Polycystididae, Koinosystididae.

Plate 48																								Phylum Platyhelminthes

## Order Alloeocoela

A predominantly marine group of small to medium-sized worms with a more or less cylindrical body, which may reach lengths of 1 cm. The body surface is sometimes characteristically pigmented. Eyes are often lacking, but sensory pits and grooves commonly occur in the head region. A pharynx is present and the gut, which is visible through the body wall, is occasionally branched.

Alloeocoels are mainly found living in mud and sand and amongst algae in the intertidal zone, although a few occur in deeper waters.

## Suborder Archoophora

Small worms about 1 mm in length, with a gland at the anterior end, no female reproductive ducts and a simple male copulatory apparatus opening posteriorly.

The only known species, *Proporoplana jenseni*, is found in muddy bottoms of the North Atlantic.

## Suborder Lecithoepitheliata

Marine and freshwater flatworms with a simple pharynx. The female reproductive system has either only a single duct or no duct at all. The penis is usually armed with a sharp bristle.

Family: Prorhynchidae.

## Suborder Cumulata (=Holocoela)

A mainly marine group of flatworms in which the pharynx is bulbous or tubular. In several species the common reproductive opening enters the pharynx cavity. The penis usually lacks bristles.

Although they are most commonly found amongst seaweed and debris, a few species occur on the gills of other invertebrates.

Families: Pseudostomidae, Cylindrostomidae, Plagiostomidae, Hypotrichinidae, Multipeniatidae, Scleraulophoridae, Protomonotresidae.

## Suborder Seriata

Marine and freshwater alloeocoels with a tubular pharynx and a branched gut. Sensory pits, a statocyst and bristles are frequently present. Adhesive structures commonly occur at the posterior end of the body.

Families: Monocelididae, Otoplanidae, Nematoplanidae.

## Order Tricladida [planarians]

Marine, freshwater and terrestrial flatworms whose length ranges from a millimetre or two to over 50 cm in the large land-living forms. The body, which is often elongated and flattened, may have adhesive organs. Two or more eyes and a distinct brain are generally present. The mouth is situated in the middle of the ventral surface and leads into a tubular pharynx from which a three-branched gut extends throughout the body.

The male and female reproductive systems open to the exterior through a common pore situated posterior to the mouth. The muscular penis is sometimes armed with sharp bristles.

Triclads are grouped into three suborders depending on the environment in which they live.

## Suborder Maricola

Marine triclads up to 2 cm in length, and white, brown or grey in colour. There is one pair of eyes, and adhesive structures occur near the edge of the body on the ventral surface.

Sexual reproduction appears to be seasonal, and the egg capsules are usually attached to rocks or other objects. They occur mainly under rocks or amongst gravel on the shore in temperate regions. A few species attach to other animals such as the king crab and the skate.

Families: Bdellouridae, Uteriporidae, Procerodidae, Micropharyngidae.

## Order Polycladida

Marine turbellarians whose length ranges from a few millimetres to several centimetres. The body, which is usually oval in outline and greatly flattened, is covered with cilia and has many rhabdites in the skin. Tentacles are frequently present and numerous eyes occur in the head region. A tubular pharynx leads into a gut which has many branches extending throughout the body.

The male and female reproductive systems may have separate openings to the exterior, or open through a common pore. The penis frequently carries a long bristle.

Fertilised eggs are laid in capsules and either develop into a planktonic larva or directly into a small worm.

Many polyclads live in the intertidal zone, although several species are free-swimming planktonic forms. They feed on small invertebrates by everting their pharynx over the prey and partially digesting it with the pharynx still out. Warm-water species are often brightly coloured.

## Suborder Acotylea

Polyclads without an adhesive organ behind the female reproductive opening. Tentacles, when present, are situated in the middle of the head region. Eyes may be present in a band across the anterior end as well as in clusters near the tentacles and brain region.

Both free-swimming forms and bottom-dwellers occur in this group, which includes the 'oyster leech' *Stylochus*, a well known pest in oyster beds.

Families: Discocelididae, Polyposthiidae, Plehniidae, Stylochidae, Latocestidae, Cryptocelididae, Diplopharyngeatidae, Leptoplanidae, Hoploplanidae, Theamatidae, Callioplanidae (=Diplosoleniidae), Planoceridae, Enantiidae, Stylochocestidae, Cestoplanidae, Emprosthopharyngidae, Apidioplanidae.

### Family Stylochidae

*Kaburakia excelsa* Bock. A large oval species reaching 5 cm or more in length. It is found gliding over stones, shells and general undergrowth in the intertidal and shallow subtidal zones of the North-east Pacific.			49/8

### Family Leptoplanidae

*Stylochoplana pallida* (Quatrefages). A species without tentacles, found in the intertidal and subtidal rocky areas of the North Atlantic.			49/9

## Suborder Cotylea

Polyclads with an adhesive organ situated behind the female reproductive opening. The tentacles, when present, arise at the edges of the head region. Eyes may be present on the tentacles, in the region of the tentacles, or in a band across the front end of the body.

The majority occur in warmer waters.

Families: Boniniidae, Stylochoididae, Prosthiostomidae, Anonymidae, Pericelididae, Opisthogeniidae, Pseudocerotidae, Euryleptidae, Laidlawiidae, Chromoplanidae, Diposthidae.

### Family Pseudocerotidae

The genus *Pseudoceros* is particularly common on Indo-Pacific reefs and contains many brightly coloured species usually up to 5 cm in length. Little is known about their feeding habits, although they are thought to be carnivorous, feeding mainly at night on small invertebrates.

49/10　*Pseudoceros bedfordi* Laidlaw. An Indo-Pacific species which is frequently found gliding over rocks and sand, or occasionally swimming.

49/1　*Pseudoceros buskii* (Collingwood). Occurs on reefs in the Indo-Pacific.

49/2　*Pseudoceros corallophilus* Hyman. Occurs on reefs in the Indo-Pacific.

49/3, 49/7　*Pseudoceros dimidiatus* Graff, in Saville-Kent. A brightly-coloured Indo-Pacific species in which a pair of tentacles can be seen clearly at the anterior end of the animal.

50/1　*Pseudoceros hancockanus* (Collingwood). The specimen is seen gliding over a sponge possibly in search of small crustaceans inhabiting the sponge's water canals. Found in the Indo-Pacific.

49/4　*Pseudoceros pardalis* Verrill. A species with an ocellated appearance similar to that of the previous species, but it is found on reefs in the Caribbean.

49/5　*Pseudoceros splendidus* Stummer-Traunfels. An Indo-Pacific species bearing a superficial resemblance to *P. hancockanus*, but with the margin edged in purple.

50/2, 50/3　*Pseudoceros zebra* (Leuckart). The species name is derived from the startling yellow and black transversely striped pattern seen in many specimens (Plate 50/2). Sometimes other colour varieties are found (Plate 50/3). Found on Indo-Pacific reefs.

49/6　*Pseudoceros* n.sp. An as yet undescribed species occurring on Indo-Pacific reefs.

50/4　*Thysanozoon brocchii* (Risso). An easily recognisable species with papillae on its back and with a body less flattened than most species in this genus. It has been found amongst undergrowth on the shore and in shallow waters in the Mediterranean, South-east Atlantic and Zealandic region.

50/5　*Thysanozoon flavomaculatum* (Graff, in Saville-Kent). An easily recognisable Indo-Pacific species with brilliant yellow spots.

50/6　*Thysanozoon* n.sp. An as yet undescribed species, with conical papillae on the upper surface, which is found on Indo-Pacific reefs.

### Family Euryleptidae

50/7　*Prostheceraeus giesbrechtii* Lang. This 3 cm long specimen from the Mediterranean can be seen feeding on sea-squirts (ascidians).

50/8　*Prostheceraeus vittatus* (Montagu). An easily recognisable species occurring on shores, in mussel beds, and amongst undergrowth in shallow waters of the North-east Atlantic and Mediterranean.

### Class Gnathostomulida
approx. 100 species

An entirely marine group of worm-shaped animals, rarely exceeding 1 mm in length, which are considered by some authorities to constitute a separate phylum. The body surface is covered with cilia but these are fewer in number and larger than those in the Turbellaria. A mouth cavity bearing a hardened base plate leads into a pharynx with jaws which precedes a straight gut. Food such as bacteria, fungi and small algae is shovelled up from the bottom by the base plate aided by the grasping jaws.

Gnathostomulids are hermaphrodite and live between sand grains on the shore and in shallow seas all around the world.

Genera include: *Gnathostomula, Pterognathia, Austrognathia*.

# Phylum Nemertea

[ribbon worms] approx. 800 species

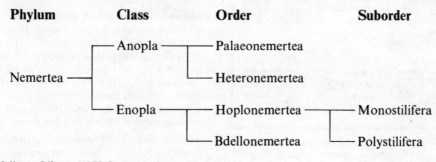

| Phylum | Class | Order | Suborder |
|---|---|---|---|
| Nemertea | Anopla | Palaeonemertea | |
| | | Heteronemertea | |
| | Enopla | Hoplonemertea | Monostilifera |
| | | Bdellonemertea | Polystilifera |

Classification follows Gibson (1972 & personal communication).

Nemerteans are mostly thin, unsegmented worms whose length varies from a few millimetres to several metres. One inhabitant of the North-east Atlantic, *Lineus longissimus*, is reputed to reach a length of 30 m when fully extended. The majority are marine but a few live successfully in freshwater and on land.

The contractile body, which may be either cylindrical or flattened, is covered with cilia. The musculature of the body wall is usually well-developed, consisting of two or three layers. In most species the head is not obvious, but in some nemerteans there is either a distinct head lobe or the head region is much more slender than the trunk. Eyes and a pair of sensory grooves are frequently present at the anterior end. The most characteristic feature of the group is a long tubular proboscis which can be thrust out of the body, either through a special pore lying in front of the ventral mouth or through a common mouth-proboscis opening at the anterior tip of the worm. The proboscis, which may bear piercing barbs (stylets), is used for capturing prey and for defence. The anus is either terminal or dorsally situated near the posterior end of the body which may be drawn out into a slender tail (cirrus) or a flattened fin in swimming species. There are no special respiratory structures in nemerteans.

The sexes are separate in most marine species. Eggs and sperm are often discharged into the seawater, but in some species two or more worms surround themselves with a mucous sheath into which the eggs and sperm are released. Copulation of a primitive type is supposed to occur in the few swimming species. Eggs either develop directly into a young worm or into larval forms which may be free-swimming. Asexual reproduction may occur by splitting of the body into several parts.

Nemerteans are found in all seas, where the majority live under rocks, amongst algae and corals, or in sediments. A few floating or swimming species occur in the open ocean. They are mostly carnivores or scavengers feeding on a variety of invertebrates and small fish. Several species are brightly coloured and have characteristic markings. Colour changes frequently occur in the breeding season.

## Class Anopla

Nemerteans in which the mouth lies posterior to the brain and the proboscis is a simple tube with no stylets. The mouth and proboscis have separate openings.

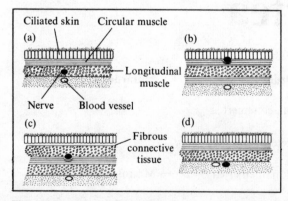

**Figure 6** *Arrangement of body wall musculature in* (a) *and* (b) *Palaeonemertea,* (c) *Heteronemertea, and* (d) *Hoplonemertea.*

44

## Order Palaeonemertea

The muscles of the body wall consist of two (outer circular, inner longitudinal) or three (outer circular, middle longitudinal, inner circular) layers, and the connective tissue beneath the skin forms a thin gelatinous layer (Figure 6*a, b*). The brain and nerve cords lie either in the inner longitudinal muscles or outside the body wall muscles altogether.

Representatives occur mainly in the intertidal zone.

Families: Carinomidae, Cephalothricidae, Hubrechtidae, Tubulanidae.

### Family Tubulanidae

*Tubulanus annulatus* (Montagu). A distinctly marked     51/6
nemertean, usually about 10–15 cm long but which can reach 50 cm. It is found under stones and in rock crevices in the North-east Atlantic and Mediterranean, and is unusual in that it can sometimes be seen scavenging on the lower shore whilst the tide is out.

## Order Heteronemertea

The muscles of the body wall consist of three distinct layers (outer longitudinal, middle circular, inner longitudinal) and the connective tissue beneath the skin is fibrous (Figure 6*c*). The brain and nerve cords are situated in the middle circular muscle layer.

Many common intertidal nemerteans belong to this order.

Families: Baseodiscidae, Lineidae, Poliopsiidae*, Pussylineidae*, Valencinidae. (*Family names as Gibson (1972).)

### Family Baseodiscidae

*Baseodiscus quinquelineatus* (Quoy & Gaimard). A     50/10
smooth-bodied animal, up to 50 cm or more in length, with five characteristic black lines. The head of this form can be retracted completely into the body. Found on coral sand around reefs in the Indo-Pacific.

### Family Lineidae

*Lineus bilineatus* (Renier). This species, which reaches a     50/11
length of about 30 cm, is characterised by a double white line along its back. It is found under stones or shells and amongst undergrowth on the shore and in shallow waters of the North-east Atlantic and Mediterranean.

*Lineus geniculatus* (Delle Chiaje). A species with     50/9
characteristic white rings around the body, which may reach a length of over 50 cm when fully elongated. It is a predatory and scavenging form which is found under stones and amongst undergrowth in the Mediterranean and South-east Atlantic. Also reported from the Californian coast.

*Lineus ruber* (Müller). A smaller nemertean, specimens     51/1, 51/
of which usually do not exceed 15 cm in length. A ventral view of the anterior end of the animal (Plate 51/8) clearly shows that the simple tubular proboscis is everted from an opening in front of the mouth, a characteristic feature of the anoplan class. It occurs under stones on muddy sand and gravel where it hunts for worms on which to feed. Found in the intertidal and subtidal zones of the North Atlantic and Mediterranean.

*Micrura fasciolata* Ehrenberg. A small species which     51/4
rarely reaches more than 2 cm in length. The body has a characteristic pattern of transverse lines, a white snout

on which the eyes are visible, and a small white tail. Found amongst undergrowth below tide level in the Mediterranean.

## Class Enopla

The mouth lies anterior to the brain, and the proboscis, which is divided into distinct regions, may or may not bear stylets.

### Order Hoplonemertea

The proboscis is armed with one or more stylets. The muscles of the body wall consist of two layers (outer circular, inner longitudinal) and the brain and nerve cords lie internal to these muscle layers (Figure 6*d*).

### Suborder Monostilifera

Proboscis with one central stylet. This group, which includes many shallow-water marine forms, also contains the few freshwater and terrestrial species.
Families: Amphiporidae, Cratenemertidae, Carcinonemertidae, Emplectonematidae, Ototyphlonemertidae, Prosorhochmidae, Tetrastemmatidae.

### Suborder Polystilifera

Proboscis with numerous small stylets situated on a central pad.

The deep-water floating and swimming nemerteans belong to this group, as well as more common shallow-water species.
Families: Siboganemertidae, Uniporidae, Paradrepanophoridae, Drepanogigantidae, Drepanophorellidae, Drepanophoringiidae, Drepanophoridae, Brinkmanniidae, Coellidae, Drepanobandidae, Armaueridae, Balaenanemertidae, Burgeriellidae, Chuniellidae, Dinonemertidae, Nectonemertidae, Pelagonemertidae, Phallonemertidae, Planktonemertidae, Protopelagonemertidae.

### Family Drepanophoridae

*Drepanophorus spectabilis* (Quatrefages). This animal has a characteristically flattened body about 10 cm long, with a pointed head and tail. Five longitudinal white lines are also a noticeable feature. It preys on small worms and shelled animals in muddy sand under stones and in crevices in shallow waters. Found in the Mediterranean and North-east Atlantic.          51/5

### Family Nectonemertidae

*Nectonemertes* sp. Members of this genus are deep-water, free-swimming species, with the tail region modified into a fin. Mature males have a pair of head tentacles for grasping the female during copulation. Of the three specimens in the photograph the two outer ones are males. The centrally placed specimen has its proboscis everted. These specimens were recovered from very deep water in the Atlantic.          51/2

### Family Phallonemertidae

*Phallonemertes* sp. (=*Bathynectes* sp.). A deep-water swimming form with the red coloration typical of many invertebrates from this habitat. The proboscis, which is inverted in this specimen, emerges from a common opening with the mouth, a feature which is found only in the 'Class Enopla. An Atlantic species.          51/3

### Order Bdellonemertea

The proboscis, which is not armed with stylets, opens into the fore-gut. A posterior ventral sucker is present. The brain and nerve cords lie inside the outer circular and inner longitudinal layers of the body wall.

Representatives live attached to the wall of the mantle cavity of bivalve molluscs, feeding on small organisms which are swept in with the incoming water currents.
Family: Malacobdellidae.

# Phylum Rotifera

approx. 1500 species

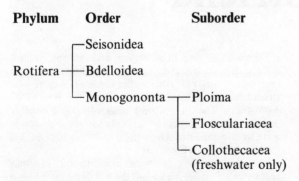

| Phylum | Order | Suborder |
|--------|-------|----------|
| Rotifera | Seisonidea |  |
|  | Bdelloidea |  |
|  | Monogononta | Ploima |
|  |  | Flosculariacea |
|  |  | Collothecacea (freshwater only) |

A group of minute, unsegmented animals whose body length does not exceed 2 mm. They live mainly in freshwaters, but there are a few free-swimming, attached or creeping marine species. The body, which is usually elongated, is composed of a short head region, a trunk

and a terminal foot. A characteristic ciliated structure, the wheel organ, is present on the head. The organ is so called because the rhythmic beating of its cilia causes it to resemble a rotating wheel. In less specialised rotifers it consists of a single ring of cilia around the head and a patch of cilia around the mouth, but in the more advanced forms the cilia are arranged in several rings, tufts, or situated on lobes. The wheel organ propels the animal through the water and also, in filter-feeding forms conveys water-borne food particles towards the mouth. The mouth opens into a muscular pharynx containing a number of hardened areas which are used to macerate the food. In carnivorous and scavenging species the pharynx is everted through the mouth to grasp food. The trunk varies greatly in form and is often protected by thickened plates bearing spines or bristles. The foot frequently has movable projections (toes) which are used for creeping, but in attached rotifers the foot is modified into a long stalk. In the free-swimming forms the foot is usually reduced or absent altogether.

The sexes are separate and often the males are much smaller than the females. In one group, males are unknown and the egg develops without being fertilised. In free-swimming forms the young resemble the adults, but in attached forms the eggs hatch as free-swimming juveniles with a short foot which elongates into the stalk when the young rotifer sinks to the bottom to take up an adult existence.

Marine rotifers occur in all seas, where attached and creeping forms are confined mainly to the intertidal zone.

## Order Seisonidea

Marine rotifers that are commonly found living on the gills of small crustaceans. The elongated body has an extended neck and an attachment disc at the end of the foot. The wheel organ is reduced, having only a few tufts of bristles. They move by alternately attaching the mouth and attachment disc. Seisonids feed mainly on organic debris, although they suck out the contents of the host crustacean's eggs when they are available.

Family: Seisonidae.

## Order Bdelloidea

A predominantly freshwater group with elongated jointed bodies and a wheel organ normally with two rings of cilia. Males have never been found. The only truly marine species lives attached by a foot disc in pits in the skin of sea-cucumbers and lacks the toes typical of other bdelloids.

Family: Philodinidae.

## Order Monogononta

Marine and freshwater rotifers with a variable body form. The foot, when present, often has two toes, but they are absent in some species. Males frequently occur but are much smaller than the females.

Swimming, creeping and attached forms are found in this group.

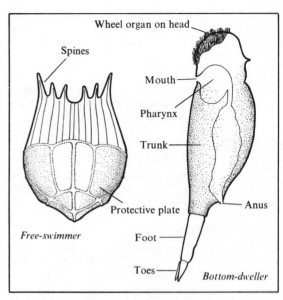

**Figure 7** *Diagrammatic representation of the difference in form between free-swimming and bottom-dwelling rotifers.*

### Suborder Ploima

Most of the known marine rotifers occur within this suborder, which includes bottom-dwellers with two toes on the foot and free-swimming forms with no foot at all (Figure 7).

Families with marine species include: Brachionidae, Lecanidae, Notommatidae, Lindiidae, Trichocercidae, Dicranophoridae, Synchaetidae.

### Suborder Flosculariacea

Bottom-dwellers in this group are without toes on the foot. In free-swimming forms the foot, when present, often ends in a ciliated cup.

Family: Testudinellidae.

# Phylum Gastrotricha

approx. 150 species

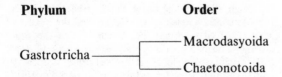

| Phylum | Order |
|---|---|
| Gastrotricha | Macrodasyoida |
| | Chaetonotoida |

A group of minute animals less than 2 mm in length, which live in the sea and in freshwaters. The body is elongated with a convex dorsal surface and a flattened ventral surface. Bristles and scales frequently occur on the body. In some species the head end is separated from the trunk region by a neck. Simple eyes and tentacles are sometimes present. The posterior end of the body may be rounded, forked, or have a long tail. Cilia

are present on the head region and on the ventral surface, the ventral cilia causing a gliding movement reminiscent of that of a flatworm. Characteristic structures, the adhesive tubes, occur in nearly all gastrotrichs. These cylindrical tubes, of varying number and position, project from the body surface and secrete a sticky substance which is used to attach the animal to the substratum (Figure 8).

Some gastrotrichs are hermaphrodite, but in other species only females occur and the egg develops without being fertilised. There is no larval stage.

Gastrotrichs are mainly bottom-dwellers, although a few free-swimming forms exist. They feed on bacteria and on microscopic plants and animals which they suck up with a muscular pharynx.

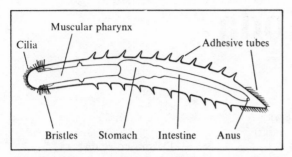

**Figure 8** *Marine gastrotrich.*

### Order Macrodasyoida

Exclusively marine gastrotrichs with adhesive tubes situated anteriorly, posteriorly, and on the sides of the body. They can move by gliding on the cilia or in a leech-like fashion by alternately attaching and detaching the anterior and posterior adhesive tubes.

They inhabit mainly sand in coastal waters, although a few occur amongst seaweeds and barnacles.

Families: Dactylopodolidae, Lepidodasyidae, Macrodasyidae, Thaumastodermatidae, Turbanellidae.

### Order Chaetonotoida

A primarily freshwater group which contains a few marine species. There is usually a well defined neck region and the adhesive tubes are situated on the posterior end of the body. The marine forms commonly occur in sand in coastal waters.

Families: Chaetonotidae, Neodasyidae, Xenotrichulidae.

# Phylum Kinorhyncha

(=Echinoderida) approx. 100 species

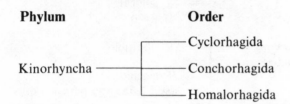

A group of minute marine animals which are less than 1 mm in length. The body, which bears spines, is superficially segmented (Figure 9). The head region (first segment) carries several rings of spines which are used for pushing the animal through the bottom sediment. The mouth is situated on a cone and the entire head, and sometimes the neck region (second segment) also, can be withdrawn into the rest of the body. Special plates, situated on either the neck or the third segment, close the resultant opening. The ventral surface is flattened and usually has one pair of adhesive tubes, similar to those of gastrotrichs. Cilia are completely lacking.

The sexes are separate, and internal fertilisation apparently takes place. The egg develops into a larva which undergoes several moults before the adult stage is reached.

The majority of kinorhynchs live in muddy bottoms (a few amongst algae) of shallow coastal waters, although some have been found at great depths in the ocean. They feed mainly on organic debris which is sucked up by a muscular pharynx as they burrow slowly through the mud. Mature individuals are normally yellow or brown in colour.

### Order Cyclorhagida

Kinorhynchs in which only the head can be withdrawn into the rest of the body. Large plates on the neck region close the opening. Eye-spots are frequently present on the head.

Family: Echinoderidae.

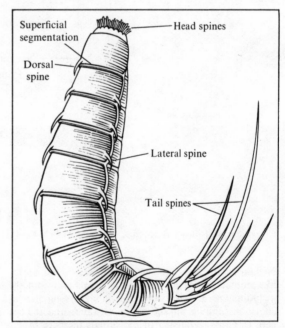

**Figure 9** *Marine kinorhynch.*

### Order Conchorhagida

Kinorhynchs in which both the head and neck regions can be withdrawn into the body. The third segment bears one pair of curved plates which protect the retracted segments.

Family: Semnoderidae.

### Order Homalorhagida

Kinorhynchs which can withdraw both the head and neck into the body, the opening being protected by one dorsal and three ventral plates on the third segment. Members possess fewer body spines than those in the other orders.

Family: Pycnophyidae.

47

# Phylum Nematoda

[roundworms] approx. 15000 species

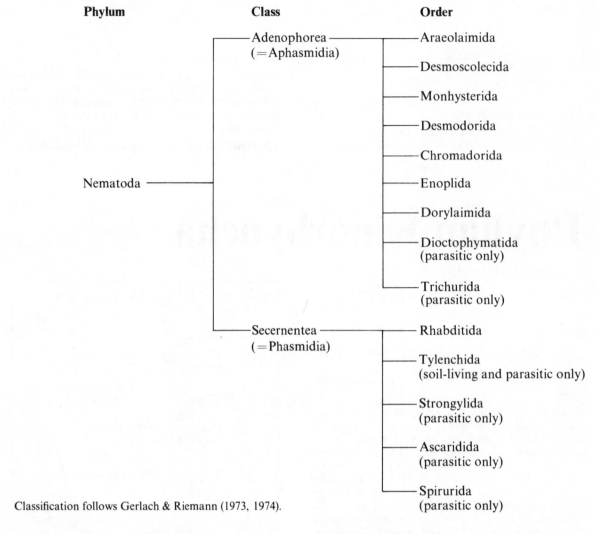

| Phylum | Class | Order |
|---|---|---|
| Nematoda | Adenophorea (=Aphasmidia) | Araeolaimida |
| | | Desmoscolecida |
| | | Monhysterida |
| | | Desmodorida |
| | | Chromadorida |
| | | Enoplida |
| | | Dorylaimida |
| | | Dioctophymatida (parasitic only) |
| | | Trichurida (parasitic only) |
| | Secernentea (=Phasmidia) | Rhabditida |
| | | Tylenchida (soil-living and parasitic only) |
| | | Strongylida (parasitic only) |
| | | Ascaridida (parasitic only) |
| | | Spirurida (parasitic only) |

Classification follows Gerlach & Riemann (1973, 1974).

Nematodes are a versatile group of unsegmented worm-like animals. Some are parasitic in plants and animals, including man. Most are free-living, occurring in enormous numbers in the soil, freshwaters and the sea bed. In terms of number of individuals they are by far the most abundant group of metazoan animals. Most free-living forms are from 1–3 mm in length, although a few marine species reach a length of several centimetres. The body of nematodes is remarkably constant even though they have become adapted to living in many different habitats. The colourless translucent body may be cylindrical, frequently tapering at both ends, or long and thread-like with a uniform thickness throughout. Only a few species, mainly parasites, depart from these two basic body forms. The body surface may be smooth or covered with bristles and various types of ornamentation. In many marine species the surface is transversely striated at regular intervals giving the animal a segmented appearance (Plate 52/1). Roundworms have a characteristic writhing, sinusoidal, motion which enables them to glide between sediment grains and through organic debris, the bristles,

surface ornamentation and striations possibly aiding traction.

The head is not distinct from the rest of the body. The terminal mouth is surrounded by lips and sensory papillae or bristles. A pair of lateral sense organs, the amphids, are present at the anterior end of all nematodes. These structures, which are particularly well developed in marine forms, are variously shaped inpushings in the body wall, with openings to the exterior (Plate 51/7). They function probably as detectors of chemical changes in the environment. In some nematodes, a pair of sense organs, the phasmids, structurally and probably functionally similar to the amphids, are located posteriorly near the anus. Such sense organs are usually absent in marine species, which have special glands in the tail region whose secretion aids attachment of the animal to the substratum.

The sexes are separate in nematodes, males and females usually being about the same size in marine forms. The male, which has a curved tail, possesses sickle- or needle-shaped structures on either side of the anus which aid in copulation (Plate 51/9). After internal fertilisation, the eggs are normally shed and develop into young worms which moult four times before the adult stage is reached. In a few species the young develop within the body of the female before being shed (viviparous). Complex life histories, involving both parasitic and free-living stages, which are common in other nematodes do not occur in marine species.

Nematodes are present in all seas from the shore to great depths. Many feed on organic debris, occurring in vast numbers in organically-rich sediments (100 million/m²) where they play a key role in breaking down and recycling organic matter. Some eat living organisms, including diatoms, small animals, and other nematodes.

## Class Adenophorea (= Aphasmidia)
approx. 5000 species

Mainly free-living nematodes which frequently have bristles on the head region. There are no phasmids, although tail glands are often present. The amphids, which are located posteriorly in the head region, may be circular, coiled, or pouch-shaped with a slit-like opening to the exterior.

### Order Araeolaimida

Marine, freshwater and soil nematodes with transverse striations on the body surface, which may bear a few bristles. Coiled amphids usually occur.

Families: Rhabdolaimidae, Haliplectidae, Leptolaimidae, Axonolaimidae.

### Order Desmoscolecida

Small, often stout, nematodes living mainly in the sea. The body surface frequently bears protruberances and bristles, and has pronounced striations, giving the animal a segmented appearance. The amphids are usually circular, rarely coiled.

Families: Meyliidae, Desmoscolecidae.

### Order Monhysterida

Mainly marine nematodes with a smooth or transversely striated body surface. The amphids are usually circular although coiled forms sometimes occur. The small mouth may occasionally bear small teeth.

Families: Siphonolaimidae, Linhomoeidae, Monhysteridae, Scaptrellidae, Sphaerolaimidae.

### Family Monhysteridae

One of the most commonly represented families in the sea.

51/9  *Cylindrotheristus normandicus* (de Man). A detail of a male, showing the sickle-shaped structures situated near the anus which are protruded to aid in copulation. The transverse striations on the body surface can also be seen. From intertidal sand in the North-east Atlantic.

### Order Desmodorida

Predominantly marine forms with a smooth or transversely striated surface. In some species the body is shaped like a figure 3 with walking bristles on the ventral surface. Coiled amphids usually occur.

Families: Aponchiidae, Desmodoridae, Xennellidae, Ceramonematidae, Monoposthiidae, Richtersiidae, Draconematidae, Epsilonematidae.

### Family Desmodoridae

*Spirinia parasitifera* (Bastian). This detail of the head end  51/7 clearly shows the pair of laterally placed, spirally coiled sense organs (amphids). It is not unusual to find this species with groups of minute ciliated protozoa clinging to the body surface near the anus, where they possibly benefit from the waste material voided by the animal. This species is very common in intertidal soft sediments of the North-east Atlantic.

## Order Chromadorida

A mainly marine group, whose members frequently have a transversely striated body surface containing numerous depressions and/or protruberances. The prominent amphids often have many coils.

Families: Comesomatidae, Chromadoridae, Cyatholaimidae, Choniolaimidae, Selachinematidae.

### Family Chromadoridae

*Chromadora macrolaima* de Man. A detailed view of this  52/1 species reveals the transverse striations as well as lines of protruberances on the body surface. Found in intertidal fine sand in the North-east Atlantic.

## Order Enoplida

Predominantly marine nematodes with a smooth or transversely striated body surface and conspicuous sensory bristles in the head region. Many have prominent teeth around the mouth opening. The simple pouch-like amphids open to the exterior through slits.

Families: Ironidae, Tripyloididae, Trefusiidae, Oxystominidae, Lauratonematidae, Leptosomatidae, Triodontolaimidae, Anticomidae, Phanodermatidae, Enoplidae, Rhabdodemaniidae, Anoplostomatidae, Oncholaimidae, Enchelidiidae, Rhaptothyreidae.

### Family Ironidae

*Dolicholaimus marioni* de Man. A large species,  52/3 sometimes over 5 mm long, which can be found amongst coralline algae in rock pools on the shore in the North-east Atlantic.

### Family Leptosomatidae

*Thoracostoma coronatum* (Eberth). A large species  52/4 which has been found amongst the branches of the calcareous algae *Corallina* in the intertidal zone of the North-east Atlantic.

### Family Enoplidae

An unknown enoplid, with beautiful iridescence, from  52/2 intertidal sand in the North-east Atlantic.

### Family Oncholaimidae

*Viscosia* sp. This close-up of the head region of a  51/10 predatory species reveals the teeth within the large mouth. Found in intertidal sediment in the North-east Atlantic.

## Order Dorylaimida

The free-living nematodes in this order are commonly

found in the soil or in freshwater, but one species is marine. The body surface is usually smooth, without bristles. The anterior part of the gut (buccal cavity) has either a hollow spear or teeth which can be protruded to pierce plants or animal prey.

Only marine species: *Thorneella teres.*

### Class Secernentea (= Phasmidia)
approx. 10 000 species

The free-living members of this class, which contains many plant and animal parasites, live mainly in the soil,

rarely in the aquatic environment. There is only one known marine species. The head usually has no bristles and the amphids are located on the lips surrounding the mouth. Phasmids are a characteristic feature of the group.

### Order Rhabditida
Mainly soil-living nematodes which have teeth in the buccal cavity. The amphids are often reduced to two small pockets.

Only marine species: *Rhabditis marinus.*

# Phylum Nematomorpha

[hair-worms] approx. 250 species.

| Phylum | Order |
|---|---|
| Nematomorpha | Nectonematoidea |
| | Gordioidea (freshwater and terrestrial only) |

A mainly freshwater and terrestrial group of long thin worms which may reach 1 m in length. Only one genus, *Nectonema*, is found in the sea. The unsegmented body is generally of uniform diameter (1–3 mm) throughout its length and has a terminal mouth and anus.

The sexes are separate, and copulation normally occurs. The egg develops into a larva which seeks out and parasitises another animal, usually an arthropod (p. 62), in order to complete its development. The free-living adults do not seem to feed. The group has a world-wide distribution.

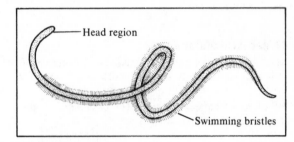

**Figure 10** *Marine nematomorph*, Nectonema.

### Order Nectonematoidea
Marine worms about 20 cm in length which have numerous swimming bristles on the body (Figure 10). The adults are found swimming in the surface waters of coastal regions, whilst the larva, which closely resembles the adult, spends its life inside a crab.

Family: Nectonematidae.

# Phylum Entoprocta

approx. 60 species

A group of small, attached, solitary or colonial animals, rarely exceeding 5 mm in length, which bear a superficial resemblance to hydroid polyps. They are mainly marine, although one genus is found in freshwater.

The round or oval body has a crown of tentacles and a stalk which attaches to the substratum. The gut is U-shaped, and both the mouth and anus open to the exterior inside the ring of tentacles. In solitary species, the stalk attaches by an adhesive disc or by cement, whilst in colonial forms numerous stalks arise from horizontal creeping branches or by branching from a common stem (Figure 11).

Both species with separate sexes and hermaphrodite

forms occur. Internal fertilisation usually takes place, and the fertilised eggs are brooded within the body until a free-swimming larva is released. After a short period of swimming, or creeping along the bottom, the larva attaches to a suitable substratum and changes into the adult. Both solitary and colonial forms can reproduce asexually by budding off new individuals from either the body wall or from the stalk.

Entoprocts are widely distributed in coastal waters, living attached to rocks, shells, and animals such as sponges, tubeworms and crabs. They feed by filtering organic debris and small organisms from the water with cilia on the tentacles. Some of the solitary species

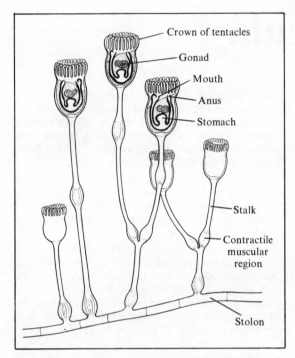

**Figure 11** *Part of an entoproct colony showing the main structural features.*

change their position by looping movements of the body using both the tentacles and stalk disc alternatively for attachment.

Families: Loxosomatidae, Pedicellinidae.

# Phylum Priapulida

4 species

Worm-like marine invertebrates, whose length varies from a few millimetres to over 8 cm. The stout cylindrical body is divided into two parts, a short barrel-shaped anterior proboscis and a longer posterior trunk region. The proboscis has a terminal mouth edged with spines and is covered with rows of spiny projections. The proboscis can be withdrawn into the trunk, which is superficially divided into segments and covered with warts and small spines. Tail processes of unknown function, consisting of small bladders attached to a hollow stalk, occur near the terminal anus in two species (Figure 12).

The sexes are separate, and fertilisation of the eggs takes place in the water. A small larva, similar to the adult but with protective plates covering the trunk, emerges from the fertilised egg. It appears to live in the bottom sediment for about two years before becoming an adult.

Priapulids live buried in the sediments of colder seas from the intertidal zone to depths in excess of 500 m. They are predators, feeding particularly on slow-moving worms which they seize with the mouth spines.

Three genera only: *Priapulus, Halicryptus, Tubiluchus.*

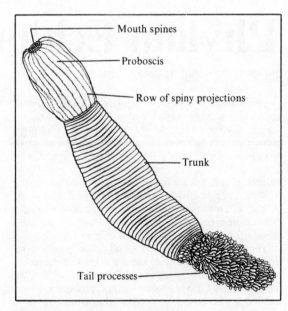

**Figure 12** *External features of* Priapulus.

# Phylum Sipuncula

[peanut-worms] approx. 325 species

An exclusively marine group of unsegmented worm-like animals whose length varies from a few millimetres to over 50 cm. The bilaterally symmetrical body is divided into two regions, a plump trunk region and a thinner, highly extensible, anterior region (the introvert) which can be withdrawn completely into the trunk. Spines and warts are frequently present on the body surface. The mouth, which is situated terminally on the introvert, is often surrounded by tentacles. The anus is positioned prominently on the anterior dorsal side of the trunk. The cylindrical or flask-shaped trunk normally ends bluntly or in a point, but some deep-sea species have a slender tail.

Apart from one hermaphrodite species, the sexes are separate and fertilisation of eggs and sperm occurs in the water. The egg develops into a ciliated larva which swims in the plankton for some time before sinking to the bottom and changing into a young worm.

Sipunculans are in all seas from the intertidal zone to considerable depths. Some are found in such protected situations as crevices, empty shells, mussel beds or seaweed holdfasts. Others burrow in sand, mud or calcareous rock. Sediment-dwellers feed by extracting organic material from large quantities of sand and mud ingested during burrowing, whilst others appear to collect food particles from the water or from the surface of the surrounding substratum using their ciliated tentacles.

Families: Sipunculidae, Golfingiidae, Aspidosiphonidae, Phascolosomatidae.

## Family Golfingiidae

52/6   *Golfingia* (*Golfingia*) *elongata* (Keferstein). This highly contractile species may reach a length of 10 cm or more. It is found in open muddy sand both intertidally and subtidally, and occasionally in muddy accumulations in crevices. It feeds by extracting organic matter from ingested bottom deposits. Common in the Arctic, North Atlantic and Mediterranean.

*Golfingia* (*Phascoloides*) *minuta* (Keferstein). A small   52/5 species not often more than 2–3 cm long. Found most commonly in rock crevices in the intertidal zone and in tufts of coralline algae. Occurs everywhere other than in the Indo-Pacific.

*Phascolion strombi* (Montagu). This small sipunculan   52/9 lives in the mud- and sand-filled shells of dead molluscs (occasionally also in the calcareous tubes of worms) lying on the surface of offshore sediment deposits. The upper of the two *Dentalium* shells in the photograph has the introvert of the sipunculan protruding from it. A cosmopolitan species.

## Family Aspidosiphonidae

*Paraspidosiphon* sp. A specimen whose introvert is as   52/7 long as the trunk. It is found in burrows in calcareous rock or coral in the Caribbean.

## Family Phascolosomatidae

*Phascolosoma* (*Phascolosoma*) *nigrescens* Keferstein. The   52/8, 52 papillated trunk is up to 3 cm long (Plate 52/12). The introvert is longer than the trunk, and has characteristic red transverse bands (Plate 52/8). Occurs under stones or in crevices on reefs. Widely distributed, but not recorded in the North Atlantic, Arctic or Antarctic.

# Phylum Echiura

[spoon-worms] approx. 130 species

A group of marine worm-like animals with unsegmented, muscular, sausage-shaped bodies which can reach a length of over 50 cm. At the anterior end above the mouth there is an extensible proboscis which cannot be withdrawn into the body. The proboscis, which may be broad and flat, or long and slender with a spoon-shaped or forked end, has a ciliated food groove on its ventral surface. The body is either smooth or covered with warts. A pair of curved bristles (setae) are situated ventrally behind the mouth and there are often one or two rings of setae around the terminal anus (Figure 13).

The sexes are separate and usually superficially alike. In one family (Bonelliidae) there is a marked difference in size and shape between the male and female; here the male is usually much smaller with no proboscis, and is carried on or inside the body of the female. In these species internal fertilisation may occur, but in most the egg is fertilised in the water, and a free-swimming ciliated larva develops.

Echiurans are cosmopolitan bottom-dwellers, ranging from the intertidal zone to depths of 10 000 m. They are found burrowing in sand and mud, under rocks, and in rock and coral crevices. Many feed by scooping up organic material with the proboscis, although some species trap food particles in a slime tube secreted along the wall of the burrow.

Families: Bonelliidae, Echiuridae, Urechidae, Ikedaidae.

## Family Bonelliidae

*Bonellia viridis* Rolando. The female normally lies with   52/10, 5 her trunk (up to 15 cm long) hidden in rock fissures below tide level. The highly contractile proboscis with its forked tip forages for food particles which are

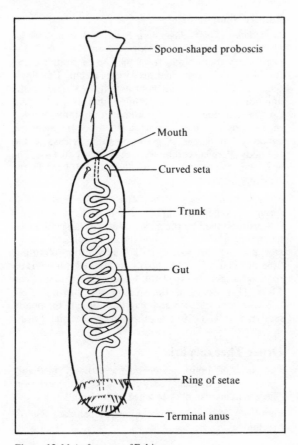

- Spoon-shaped proboscis
- Mouth
- Curved seta
- Trunk
- Gut
- Ring of setae
- Terminal anus

**Figure 13** *Main features of* Echiurus.

transferred to the mouth along a ciliated food groove (Plate 52/10). The proboscis may extend for as much as 1 m in search of food. Plate 52/11 shows a specimen that has been removed from its habitat in order to display the complete animal.

# Phylum Pogonophora

approx. 100 species

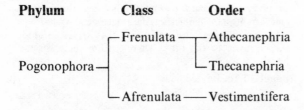

| Phylum | Class | Order |
|---|---|---|
| Pogonophora | Frenulata | Athecanephria |
| | | Thecanephria |
| | Afrenulata | Vestimentifera |

Classification based on Webb (1969) and Ivanov (1963).

The pogonophores are an entirely marine group of little-known, worm-like invertebrates living within tightly fitting cylindrical chitinous tubes. Two of the most striking features of these animals are the extreme attenuation of the body and the complete absence of a gut. The body, which varies in length from 5 cm to 80 cm depending on the species, is divided into a number of recognisable regions. The most anterior region bears a crown of tentacles varying in number from one to several hundred according to the species. With few exceptions there is a raised ridge of tissue called the bridle running obliquely around the body behind the tentacular region. The following body division forms the majority of the animal's length and contains the gonads within it. The body surface in this region frequently bears papillae, either in rows or scattered randomly, and in the majority of species the region is divided into two sections by one or two rings of bristles (setae). The girdles of short toothed setae serve to anchor the worm in its tube. The short posterior region of the animal is segmented and usually also bears setae (Figure 14).

The sexes are separate, and in all species so far studied the eggs are fertilised within the parent tube by sperm released by males into the water. The eggs, which are brooded in the anterior end of the female's tube, pass through a larval stage there, and are only discharged when they are capable of forming their own tubes.

The majority of species live with their tubes buried vertically in silt or mud, or wedged within mud-filled crevices between stones. However, some species dwell in stiff tubes standing proud of the sediment or in tubes winding amongst bottom debris. They feed on dissolved organic matter, and possibly also on fine particulate matter, that is absorbed through the body surface.

Plate 53                                                                                      Phyla Pogonophora, Annelida

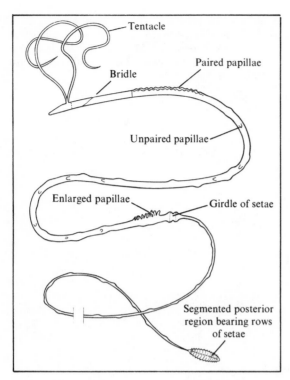

**Figure 14** *External features of a generalised pogonophore.*

### Family Siboglinidae

*Siboglinum ekmani* Jägersten. Plate 53/1 is a close-up photograph of the anterior region of the body protruding about 3 mm from the top of its tube. The total length of the animal may reach 10 cm. The single tentacle, which coils on contraction, is visible at the anterior end. The red coloration is due to the oxygen-carrying haemoglobin in the blood. A photomicrograph of the body of a male specimen (Plate 53/2) reveals the cylindrical packages of sperm awaiting release within the body. Papillae can be seen on the body surface. The species is found in mud, usually below 100 m, in the North Atlantic.      53/1, 53/

*Siboglinum fiordicum* Webb. The anterior end of this species is somewhat similar to that of *S. ekmani*, but it is distinguished by the presence of a hump behind the tentacle (Plate 53/4). Two oblique girdles of setae can be seen holding the worm in its tightly fitting chitinous tube in Plate 53/3. A highly magnified photograph reveals toothed heads of the setae in the girdles (Plate 53/5). This species is the shallowest-occurring of all known pogonophores and has been found in muddy sand in as little as 20 m depth in the Norwegian fjords.      53/3, 53/  53/5

### Order Thecanephria

The anterior body cavity has a pair of medianly situated ciliated ducts connecting it with the exterior. The sperm are usually packaged in flat bundles.

Families: Polybrachiidae, Lamellisabellidae, Spirobrachiidae.

### Class Afrenulata
2 species

This class was erected to contain two recently discovered species which differ from other pogonophores in not possessing a bridle.

### Order Vestimentifera

Members of this order possess, instead of a bridle, two lateral folds of tissue that meet in the dorsal midline of the body and extend for some distance near the anterior end. Amongst the tentacles at the anterior end of the body there is a plug of hardened tissue that serves to close the entrance to the tube. There are no setae on the body.

Family: Lamellibrachiidae.

They have a world-wide distribution at depths below 20 m. The majority occur on the continental slopes although they also live at abyssal depths.

### Class Frenulata
approx 100 species

The great majority of pogonophorans belong to this class, which is characterised by the presence of a bridle.

### Order Athecanephria

The anterior body cavity has a pair of laterally situated ciliated ducts, possibly with an excretory function, connecting it with the exterior. The sperm are often packaged into cylindrical bundles before release from the body.

Families: Oligobrachiidae, Siboglinidae.

# Phylum Annelida

approx. 14 000 species

A large group of worms, the majority of which are marine, but which includes a substantial proportion of terrestrial and freshwater forms. A few species are parasitic. The familiar earthworms and leeches are in this phylum, whose members range in length from a few millimetres to over 1 m.

Annelids are characterised by the possession of an elongated soft body divided into a series of similar segments. The muscular body wall surrounds a fluid-filled body cavity through which passes a more or less straight gut with an anterior mouth and a posterior anus. Although annelids possess no rigid skeleton, they are able to use the fluid-filled body cavity as a hydrostatic skeleton upon which the muscular body wall acts to

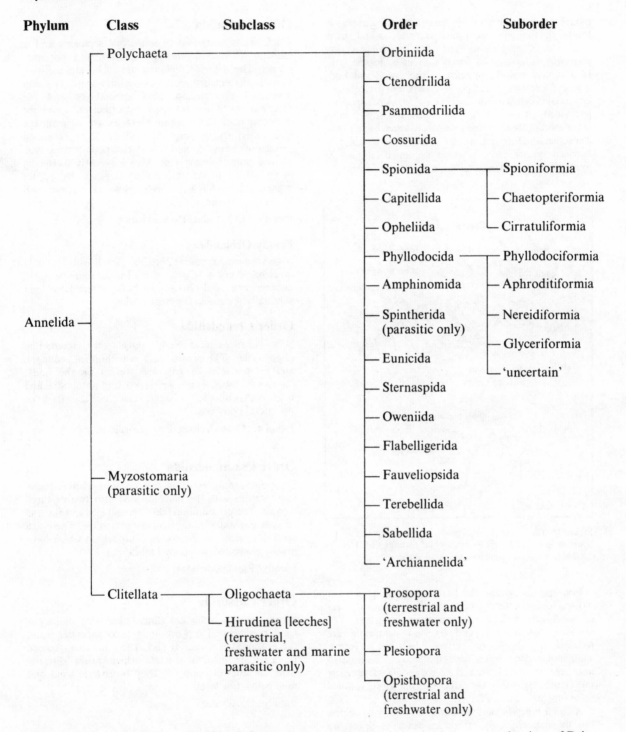

| Phylum | Class | Subclass | Order | Suborder |
|--------|-------|----------|-------|----------|
| | Polychaeta | | Orbiniida | |
| | | | Ctenodrilida | |
| | | | Psammodrilida | |
| | | | Cossurida | |
| | | | Spionida | Spioniformia |
| | | | Capitellida | Chaetopteriformia |
| | | | Opheliida | Cirratuliformia |
| | | | Phyllodocida | Phyllodociformia |
| | | | Amphinomida | Aphroditiformia |
| Annelida | | | Spintherida (parasitic only) | Nereidiformia |
| | | | Eunicida | Glyceriformia |
| | | | Sternaspida | 'uncertain' |
| | | | Oweniida | |
| | | | Flabelligerida | |
| | | | Fauveliopsida | |
| | Myzostomaria (parasitic only) | | Terebellida | |
| | | | Sabellida | |
| | | | 'Archiannelida' | |
| | Clitellata | Oligochaeta | Prosopora (terrestrial and freshwater only) | |
| | | Hirudinea [leeches] (terrestrial, freshwater and marine parasitic only) | Plesiopora | |
| | | | Opisthopora (terrestrial and freshwater only) | |

Classification of the Polychaeta follows that adopted by Fauchald (1977) which takes into account the views of Dales (1962), Storch (1968) and Clark (1969).

elongate or shorten the body. Movement in most species (not in the leeches) is aided by segmentally arranged bristles (chaetae or setae) which protrude from the body wall. The segmentation of the body is not merely superficial but also affects the worm internally so that the musculature, blood vessels, nerves, excretory organs and gonads are often repeated in many segments.

Both hermaphrodite species and species with separate sexes occur within the annelids.

## Class Polychaeta
approx. 8000 species

An almost exclusively marine group of annelids with very few freshwater or terrestrial species. The anterior end of the body has two zones, collectively called the head, which are unlike those that follow. The anteriormost zone (prostomium) sometimes bears eyes, antennae, and palps, whilst the second zone surrounding the

**Plate 53**

**Phylum Annelida**

mouth (peristomium) may have further sensory or food-collecting equipment such as tentacles and palps. The anterior part of the gut often has an eversible pharynx (proboscis) which may sometimes possess jaws. In a typical polychaete each of the cylindrical body segments behind the head possesses a pair of fleshy lateral appendages called parapodia (Figure 15a). A parapodium is divided into upper (notopodium) and lower lobes (neuropodium) each bearing numerous setae (Figure 15b). The setae, which are important features for differentiating between species, may be either unjointed (simple) or jointed (compound) and have a wide range of form.

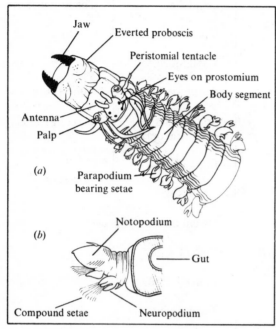

**Figure 15** *The active polychaete worm,* Neanthes virens. *(a) Anterior end showing the main external features. (b) A cross-section to show the structure of a parapodium.*

Few species possess the basic structural plan of a 'typical' polychaete, the body showing various degrees of modification depending on their life style. For instance, the head appendages and parapodia are reduced or absent in actively burrowing forms. Tubiculous forms often have the body differentiated into regions, with reduced parapodia frequently represented by transverse ridges bearing short, toothed setae (uncini).

Asexual reproduction is known in some polychaetes, but the majority only reproduce sexually. Sexes are usually separate but a few species are hermaphrodite. The liberation of eggs and sperm is sometimes associated with the habit of swarming. A few species brood their eggs and developing larvae. Larvae, which may be planktonic or bottom-living, metamorphose into small worms.

Some polychaetes are planktonic throughout the whole of their life cycle, but the majority are bottom-dwellers ranging from the intertidal zone to abyssal depths. Many live in tubes or burrows in soft sediments. Others inhabit tubes encrusting rocks, live in crevices, under stones or amongst seaweed. Some bore into rocks and shells. In soft sediments in particular they are probably the commonest macroscopic invertebrate.

## Order Orbiniida

The body is composed of numerous segments and is indistinctly divided into an anterior and posterior region. The anterior region tends to be more dorso-ventrally flattened than the posterior region. The well developed prostomium lacks appendages and the eversible proboscis is either tube-like or a ventral muscular pad. There are no tentacles arising from the body anteriorly, but finger-like gills occur on the dorsal surface of many segments. The parapodial lobes bear various types of simple setae. They are mostly burrowing forms which ingest soft bottom-deposits to extract organic matter. A few species live amongst undergrowth on rocky bottoms.

Families: Orbiniidae, Paraonidae.

### Family Orbiniidae

*Naineris quadricuspida* (Fabricius). Small worms, rarely exceeding 5 cm in length, which live amongst seaweed undergrowth and in sand, both intertidally and subtidally. Found in the North Atlantic.

53/6

## Order Ctenodrilida

Very small sand-dwelling polychaetes resembling oligochaetes. The prostomium is without appendages, and no tentacles or gills are present on the body. Parapodial lobes are not developed, and notopodial and neuropodial bundles of simple setae arise directly from the lateral body wall.

Families: Ctenodrilidae, Parergodrilidae.

## Order Psammodrilida

Only one family exists in this order, which contains small worms with the body divided into two or three regions. The prostomium lacks appendages and there is at least one anterior segment without setae. There are several segments in the middle body region which have greatly elongated notopodial lobes.

Family: Psammodrilidae.

## Order Cossurida

Worms in this order are thread-like, with numerous similar segments. The prostomium lacks appendages and the proboscis is a ventral pad. They are characterised by a single median tentacle which arises dorsally from an anterior segment. They burrow in sand and mud below tide level.

Family: Cossuridae.

## Order Spionida

Burrowing or tube-dwelling worms with a distinct prostomium which lacks appendages. A pair or two groups of feeding tentacles are present on the peristomium or on an anterior part of the body. The proboscis is either a ventral muscular pad or is tubular. Jaws are never present. The parapodia can be either well developed or reduced, and bear simple setae.

### Suborder Spioniformia

The feeding tentacles are present on the peristomium or at the junction between the prostomium and peristomium. The parapodia are often well developed and always lack uncini. Members of this suborder are very common in bottom-deposits throughout the world, and also sometimes bore into shells and soft rock.

Families: Apistobranchidae, Spionidae, Magelonidae, Trochochaetidae, Poecilochaetidae, Heterospionidae.

### Family Spionidae

53/7 *Polydora* sp. Members of this genus can be readily distinguished from other spionids by the presence of a line of large setae on the fifth segment. These help the animals to bore into shells and calcareous rocks which are used as a base for their filter-feeding activities. Their boring activities may seriously debilitate shellfish, since the molluscs are continuously having to repair the damage caused by the worms. They are a common pest on oyster beds. They have a cosmopolitan distribution.

53/8 The tentacles of a large spionid protruding from the sediment at a depth of 40 m off southern Spain. The tentacles of bottom-dwelling spionids collect food particles from the sediment surface and from the over-lying water. The food is trapped in mucus and passed along a ciliated groove to the mouth.

### Suborder Chaetopteriformia

Tube-dwelling worms whose body is divided into three distinct regions and which possess various structural modifications for feeding. A pair of tentacles originate on the peristomium but they are not used for food gathering in members of this group. The pharynx is a simple tube that cannot be everted. The parapodia of the posterior region possess uncini. Their tubes are either buried in soft sediment or lie attached to stones or rocks on the surface.

Family: Chaetopteridae.

### Family Chaetopteridae

53/9 *Chaetopterus variopedatus* (Renier) [parchment tube-worm]. This species, which may reach a length of 25 cm or more, lives in a parchment-like tube open at both ends. The specimens in the photograph have been removed from their tubes to show how the body varies greatly along its length. Paddle-like parapodia in the middle of the body draw water through a food-collecting mucus net secreted at the front end of the worm. The food-laden net is transferred to the mouth at regular intervals. This species has a cosmopolitan distribution.

### Suborder Cirratuliformia

Worms with either one pair of grooved tentacles arising from the peristomium or with several pairs appearing to arise from segments further back. The peristomium is fused with at least two segments. The proboscis is a muscular ventral pad.

Families: Cirratulidae, Acrocirridae.

### Family Cirratulidae

54/4 *Cirriformia tentaculata* (Montagu). This worm, which is up to 20 cm long, lives buried in mud or coarser deposits between stones and rocks. Only its feeding tentacles and long retractile gills gorged with blood can normally be seen above the surface of the deposit. Found in the North-east Atlantic.

### Order Capitellida

Burrowing worms with a cylindrical body that may be divided into several regions. The prostomium lacks appendages. There is an eversible thin-walled proboscis. The parapodia are poorly developed, although always divided into notopodial and neuropodial sections. Simple setae, which include finely tapering forms as well as uncini, are present on all anterior segments except for the first one or two. Members of this order are common in bottom deposits of all seas.

Families: Capitellidae, Arenicolidae, Maldanidae.

### Family Arenicolidae

*Arenicola marina* (Linnaeus) [lugworm]. A fleshy worm, 54/6 with a length of 20 cm or more, bearing bushy gills on the anterior region of the body. It lives in U-shaped burrows, being particularly common in intertidal muddy sand. The posterior opening of its burrow is ringed by characteristic casts of faeces whilst the other is marked by a shallow depression in the sediment surface. It is used extensively as bait by anglers. Found in the North Atlantic and Mediterranean.

### Order Opheliida

Burrowing worms with rather short bodies and a prostomium without appendages. There is an eversible thin-walled proboscis. The parapodia are poorly developed and bear finely tapering setae on all but the first segment. The gills, when present, are often cylindrical projections distributed along the sides of the body. They live in sand or mud, ingesting large quantities of the bottom deposit from which they extract organic matter.

Families: Opheliidae, Scalibregmidae.

### Family Opheliidae

*Ophelina acuminata* Oersted. A torpedo-shaped worm up 54/1 to 5 cm long with a pointed prostomium. Burrows in sub-tidal muddy sand and gravel. This species has a cos-mopolitan distribution.

*Travisia forbesi* Johnston. A thicker-bodied species than 53/10 *O. acuminata*, occasionally reaching 5 cm in length. The red gills can be clearly seen along the sides of the specimens in the photograph. Burrows in intertidal and subtidal muddy sand and gravel in the Atlantic and North Pacific.

### Family Scalibregmidae

*Scalibregma inflatum* Rathke. A distinctive species, 54/5 which reaches a length of 6 cm or more and has a body which is inflated anteriorly. Burrows in subtidal muddy sand. This species has a cosmopolitan distribution.

### Order Phyllodocida

In terms of species this is by far the largest order in the polychaetes. Its members are mostly active predators and scavengers. They have many similar segments and a distinct prostomium which normally bears sensory appendages (antennae), none of which are modified as food-collecting organs. Food is captured by a cylindrical eversible proboscis which may have jaws. Locomotion is aided by well developed parapodia, which bear both simple and compound setae.

## Suborder Phyllodociformia

Members of this group have at least two pairs of antennae on the prostomium, in addition to eyes. Two or more pairs of feelers arise from the peristomium. The proboscis has no jaws. All the parapodia are directed laterally. The worms in three of the four families in this suborder (Alciopidae, Lopadorhynchidae, Pontodoridae) are exclusively planktonic, passing through the whole of their life cycle swimming in the water. The fourth family, the Phyllodocidae, live on the bottom, mainly in shallow waters, where they are more commonly associated with rocky habitats than with soft deposits.

Families: Phyllodocidae, Alciopidae, Lopadorhynchidae, Pontodoridae.

### Family Phyllodocidae [paddle-worms]

54/2 *Eulalia viridis* (Linnaeus). A characteristically green-coloured worm, up to 15 cm in length, that is common under stones, in rock crevices and amongst undergrowth, both intertidally and subtidally. It is known to scavenge on dead and dying barnacles and mussels. The specimen in the photograph has its proboscis everted and its large paddle-like parapodia clearly visible. This species has a cosmopolitan distribution.

### Family Alciopidae

54/3 An unknown species of alciopid, revealing the specialist features of members of this planktonic family. These include transparency of the body, very long foliaceous parapodia and enormous eyes, presumably for seeking their prey.

## Suborder Aphroditiformia

The prostomium of members of this suborder may have up to three antennae, and the peristomium one pair of feelers. A pair of long sensory appendages (palps) flank the mouth. The proboscis, when armed, has two or four jaws. The parapodia of the first few segments are directed forward on either side of the head. Many families in this group have part of the notopodial lobe of at least some segments modified to form scales over the back of the worm. These families are collectively called the scale-worms. They occur most commonly under stones and amongst undergrowth on rocky bottoms, although a few are found on open sand and mud.

Scale-worm families: Aphroditidae, Polynoidae, Polyodontidae, Pholoididae, Eulepethidae, Sigalionidae.

Other families: Chrysopetalidae, Palmyridae, Pisionidae.

### Family Aphroditidae

54/7, 54/9 *Aphrodita aculeata* Linnaeus [sea-mouse]. A bulky flattened worm, up to 20 cm in length, with a felt of hairs masking the dorsal scales. It cannot be readily mistaken for any other species. The sides of the body bear iridescent setae (Plate 54/9) and large brown bristles which cause severe irritation if they puncture the skin. It ploughs through soft sediments below tide level in search of its prey. It has a cosmopolitan distribution.

### Family Polynoidae

54/8 *Alentia gelatinosa* (Sars). A species which reaches a length of 10 cm; its distinctive coloration can be seen through the transparent gelatinous scales. Found under stones and in crevices from low-water mark downwards in the Mediterranean and North-east Atlantic.

*Lepidonotus clava* (Montagu). A broad flattened worm, 55/1 up to 3 cm long, whose scales do not completely cover the upper surface. Commonly found under stones and in seaweed undergrowth on the shore and subtidally. Found in the North-east Atlantic and Mediterranean.

## Suborder Nereidiformia

The prostomium has at least one pair of antennae, and the peristomium never less than one pair of feelers. The palps flanking the mouth are short and often articulated. The proboscis is sometimes armed with a pair of lateral jaws and with small hardened teeth. All the parapodia are directed laterally. The Syllidae, the largest family within this suborder, are small inconspicuous members of the invertebrate fauna, but nevertheless are voracious carnivores on colonial hydroids and sea mats in shallow-water rocky areas. They far outnumber any other polychaete family in such areas. The best-known family within this group is undoubtedly the Nereidae, which contains the ragworms so familiar to anglers.

Families: Hesionidae, Pilargiidae, Syllidae, Nereidae.

### Family Hesionidae

*Kefersteinia cirrata* (Keferstein). This fragile worm, which 55/6 may reach a length of 4 cm, feeds on a variety of small prey animals. These are enveloped by its proboscis which can be seen partially everted in the photograph. In common with most predators it has well developed eyes. Abundant in silt-free undergrowth, particularly subtidally in the East Atlantic and Mediterranean.

### Family Syllidae

*Autolytus prolifera* (Müller). A small delicate species, 55/9 not much more than 1 cm in length, that lives amongst the undergrowth in rocky subtidal areas in the Atlantic and Mediterranean. The specimen in the photograph is a male reproductive phase, which swims up into the plankton from the bottom during the breeding season.

*Trypanosyllis zebra* (Grube). A broad flattened species 56/1 that may reach 5 cm in length. The jointed appearance of the dorsal extensions of the notopodium (cirri) is typical of many syllids. The striped back is characteristic of this species. It has a cosmopolitan distribution.

### Family Nereidae

*Neanthes virens* (Sars) (= *Nereis virens*) [King ragworm]. 55/7 This ragworm, which may occasionally reach 50 cm or more in length, has an impressive pair of large jaws on its eversible proboscis. The jaws are not always used as an offensive weapon, but can be used for pulling seaweed into the mouth. Found primarily in sand in the intertidal zone of the North Atlantic.

*Nereis pelagica* (Linnaeus). This species, which may grow 55/5 to over 10 cm in length, is found under stones on muddy sand and gravel, as well as amongst seaweed undergrowth. The photograph shows the eyes and antennae on the prostomium, the feelers on the peristomium, and the two large palps around the mouth. It has a cosmopolitan distribution.

## Suborder Glyceriformia

The prostomium bears two pairs of antennae. The peristomium has no feelers, and no palps are present around the mouth. The proboscis may be unarmed (Lacydoniidae), have four evenly spaced jaws

(Glyceridae) or have a circlet of jaws (Goniadidae). All the parapodia are directed laterally.

Families: Glyceridae, Goniadidae, Lacydoniidae.

### Family Glyceridae

8, 55/10 *Glycera alba* (Müller). This species may grow to 7 cm or more in length. A side view of the anterior end of the worm (Plate 55/8) shows the characteristically conical prostomium and the papillated proboscis partly everted. Plate 55/10 reveals its four jaws arranged in a square at the end of the proboscis. Distributed throughout the North-east Atlantic.

## Suborder 'uncertain'

Within the Phyllodocida five families exist which cannot easily be placed in any of the four suborders. Three of these contain planktonic representatives (Iospilidae, Tomopteridae, Typhloscolecidae) whilst members of the other two families are bottom-dwellers.

Families of uncertain affiliation: Iospilidae, Nephtyidae, Sphaerodoridae, Tomopteridae, Typhloscolecidae.

### Family Nephtyidae

56/2 *Nephtys cirrosa* Ehlers. The body of this worm is square in cross-section and about 10 cm long when fully grown. It has a pearly sheen on the ventral surface. The worm in the photograph has its proboscis extended, revealing the characteristic longitudinal rows of papillae. It is found most commonly in clean intertidal sand in the East Atlantic.

### Family Tomopteridae

56/5 *Tomopteris helgolandica* Greeff. A planktonic worm about 1 cm long. The body and parapodia are flattened and translucent. There are no setae. Internal chitinous rods support a pair of long feelers. Found in the Atlantic.

## Order Amphinomida

The body segments are all more or less alike and bear well developed parapodia. The prostomium is distinct and lies between the anterior body segments which encroach on either side of it. The proboscis is a rasping ventral pad used for food collection. Amphinomids are most commonly found in shallow warm-water habitats.

Families: Amphinomidae, Euphrosynidae.

### Family Amphinomidae [fire-worms]

2, 55/3, 55/4 *Hermodice carunculata* (Pallas). A worm which reaches 25 cm in length and which has a distinctly square cross-section. It feeds on corals (Plate 55/3) using its rasping proboscis. It is protected from predators by bundles of white setae, composed of calcium carbonate, which are readily shed when the worm is attacked (Plate 55/4). These worms should not be handled without thick gloves, since the setae can cause severe irritation which may last for many hours. The worms are often seen crawling over rocks and coral in warm shallow waters where they sometimes aggregate to spawn (Plate 55/2). Cosmopolitan in warm waters.

## Order Eunicida

Members of this order have body segments which are all more or less alike, and a distinct prostomium which may

or may not bear appendages. They have an eversible proboscis directed ventrally, with two to five pairs of jaws, often forming a complex jaw apparatus. There are distinct parapodia along the sides of the body. Several parasitic families exist within the order and are not listed below. The remaining families contain tubiculous or burrowing forms which are predators or scavengers.

Families: Onuphidae, Eunicidae, Lumbrineridae, Arabellidae, Lysaretidae, Dorvilleidae.

### Family Onuphidae

*Hyalinoecia tubicola* (Müller). This species resides in a   56/3, 56/4
horny translucent tube which resembles the quill of a feather. The worm, which may reach a length of 10 cm, lives with its tube lying horizontally on the surface of off-shore deposits, where it scavenges for food by protruding the front part of its body from the tube (Plate 56/3). When disturbed it withdraws into the middle of the tube and both ends are closed by V-shaped valves (Plate 56/4). It is able to move over the bottom by dragging its tube behind it. Found in the Atlantic and Indo-Pacific.

## Order Sternaspida

Short, broad, worms not more than 2 or 3 cm long and consisting of only a few indistinct segments. The first three segments bear rows of stout pointed setae. They have a characteristic stiff horny shield at the rear end on the ventral side. Gill filaments are also situated at the rear end. They feed on buried organic matter and are normally found burrowing head-first into sediment using the anterior setae. The rear of the burrow is protected by the horny shield and the gill filaments extend into the water above to enable the animal to respire.

Family: Sternaspidae.

## Order Oweniida

Members of this order are bottom-dwellers and have their bodies, which are composed of a few long segments, encased in well constructed sandy tubes. The prostomium and peristomium are fused together. The proboscis is a ventral muscular pad at the entrance to the pharynx.

Family: Oweniidae.

### Family Oweniidae

*Owenia fusiformis* Delle Chiaje. The body of this species,   56/6
which reaches a length of 10 cm, is enclosed in a tough membranous tube strengthened by overlapping sand grains or shell fragments. The tube projects a short way above the sediment in which it is buried, and the animal feeds on food particles caught on short frilled tentacles that project into the water. It has a cosmopolitan distribution.

## Order Flabelligerida

The anterior end of the body in the two families making up the order is retractable into the body. Both families have green blood. The majority of forms within the bottom-living Flabelligeridae have the body surface covered by papillae that secrete mucus and consequently are heavily impregnated with sand or mud. The other family, the Poeobiidae, contains only a single planktonic genus, members of which have a sac-like body without any sign of external segmentation or setae.

Families: Flabelligeridae, Poeobiidae.

## Order Fauveliopsida

Both the prostomium and peristomium are without appendages in these small, smooth-bodied, deep-water polychaetes. The proboscis is a simple ventral muscular pad. The notopodia and neuropodia are reduced and bear blade-like setae only.

Family: Fauveliopsidae.

## Order Terebellida

Basically tube-dwelling polychaetes with the body divided into two (Ampharetidae, Terebellidae, Trichobranchidae) or three (Sabellariidae, Pectinariidae) regions. The prostomium does not have appendages, and the peristomium bears feeding tentacles that can be withdrawn into the mouth in the Ampharetidae but not in the Terebellidae. The terebellids usually have very long, highly contractile tentacles compared with those of the ampharetids. Members of this order have a proboscis consisting of a ventral muscular pad. At least one pair of gills are normally present at the anterior end of the body. The setae in the first segments of the body in the sabellariids are modified to form a plug at the entrance to the tube, and those at the anterior end in the pectinariids are modified for digging. These strong digging setae and the short tusk-shaped tube of cemented sand grains make the pectinariids difficult to confuse with other families.

Families: Sabellariidae, Pectinariidae, Ampharetidae, Terebellidae, Trichobranchidae.

### Family Sabellariidae

57/1 *Sabellaria alveolata* (Linnaeus) [honeycomb worm]. The common name of this 3–4 cm long worm derives from the fact that it constructs reefs of numerous tubes built with cemented sand grains. Colonies are usually found attached to rocks in intertidal sandy areas where there is strong wave action. The porches surrounding the entrances to the tubes can be seen clearly in the photograph. Found in the North-east Atlantic and Mediterranean.

### Family Terebellidae

56/8 *Amphitrite johnstoni* Malmgren. A large species, up to 25 cm long, with a conspicuous patch of light-coloured glandular tissue on the underside at the anterior end. The bushy gills situated just behind the tentacles are red in colour. The specimens in the photograph have been removed from their mucous tubes which are normally covered with sand and pieces of shell. Common in intertidal sand and gravel and in shallow water in the North Atlantic and adjacent Mediterranean.

56/7, 56/9 *Eupolymnia nebulosa* (Montagu). The body, which reaches 15 cm in length, is orange with characteristic white spots (Plate 56/7). The highly contractile tentacles extend over the substratum in search of food particles (Plate 56/9) which are caught in mucus and passed along a ciliated groove to the mouth. Occurs in mucous tubes buried in sediment or attached to stones and shells in shallow water. It has a cosmopolitan distribution.

57/6, 57/9 *Lanice conchilega* (Pallas) [sand mason]. These worms, which may reach a length of 20 cm or more, live in characteristic fringed sand- and gravel-covered tubes that protrude from the sediment. The fringes trap suspended food particles that are then removed by its tentacles. The rows of setae seen in Plate 57/9 are used to hold the worm securely in its tube. Found intertidally and in shallow water. It has a cosmopolitan distribution.

*Loimia medusa* (Savigny). The red gills of this 25 cm-  57/2 long worm can be seen behind the tentacles. It lives in a debris-covered mucous tube in or on the bottom in shallow water. It has a cosmopolitan distribution.

### Family Trichobranchidae

*Terebellides stroemi* Sars. This highly characteristic  57/7 worm, with well developed gills at the anterior end, reaches a length of 7 cm. It lives in membranous sand-covered tubes in muddy sand and gravel. It has a cosmopolitan distribution.

## Order Sabellida

The prostomium is reduced in size and is fused with the peristomium from which arises a rather stiff fan of tentacles (strictly gills) in all but one aberrant family (Sabellongidae). The tentacles of these tube-dwellers are thrust out into the water for respiratory purposes as well as to collect plankton and organic debris from the water. Those forms with a calcareous tube (Serpulidae, Spirorbidae) often have a calcareous plug (operculum) which blocks the entrance to the tube after the tentacles have been withdrawn. The cylindrical body is divided into a thorax and abdomen, the point of interchange being recognised sometimes only by a reversal of the setal types on the notopodium and neuropodium. Setal types include blade-like forms, setae with a distinct knee, and uncini.

Families: Sabellidae, Sabellongidae, Serpulidae, Spirorbidae.

### Family Sabellidae [fan-worms]

*Bispira volutacornis* (Montagu). A sturdy sabellid up to  57/3 15 cm long with the tentacles arranged in two spirals. Black eye-spots occur on the back of the tentacles, which, as in all sabellids, can be rapidly withdrawn in times of danger. The membranous mud-encrusted tubes protrude from rock crevices in subtidal shaded areas. Found in the North-east Atlantic and Mediterranean.

*Fabricia sabella* (Ehrenberg). A minute species, no more  57/8 than 3–4 mm long, with a pair of eyes at the posterior end. It lives in sand- or mud-covered mucus tubes, commonly attached to seaweeds or amongst mussel byssus threads, in the intertidal zone and in shallow waters. Found in the Mediterranean, North Atlantic and North-east Pacific.

*Myxicola infundibulum* (Renier). A species with a stout  57/10, 5? body up to 15 cm long with the fan of tentacles united by a web, except at their tips. The tips of the tentacles taper to a point and are not covered with feathery side-branches as they are lower down (Plate 57/10). They live in thick gelatinous tubes which are normally almost completely buried in subtidal sand or mud. Recorded from all seas other than the Indo-Pacific.

*Sabella penicillus* Linnaeus [peacock fan-worm]. The  57/12 body may be 20 cm or more in length and the fan of tentacles variously patterned. The membranous mud-encrusted tubes may be found protruding often for several centimetres above the bottom sediment. Found intertidally and in the shallow subtidal zone of the North Atlantic and Mediterranean.

*Sabella* sp. A form living in clusters of membranous  58/1

sand- and mud-encrusted tubes protruding from crevices between rocks and on reefs in the Caribbean.

57/4, 57/5   *Sabellastarte magnifica* (Shaw). Normally the tentacles are banded with various shades of red, brown, or white (Plate 57/5) but occasionally they are one colour (Plate 57/4). The worm, which grows to a length of 15 cm or more, is found in shallow waters of tropical seas.

58/4, 58/6   *Sabellastarte sanctijosephi* (Gravier) (=*S. indica*). A sturdy species up to 10 cm long with a large number of tentacles which may occasionally be banded. The fine feathery tentacles can be clearly seen in Plate 58/6. Found in the Indo-Pacific.

58/3, 58/7   *Spirographis spallanzani* Viviani. This species, which may reach a length of 20 cm, has a spiral of tentacles extending from a mud-encrusted membranous tube. The tentacles are variously patterned with orange, brown and white bands. Found in soft subtidal sediments in the Mediterranean and adjacent Atlantic.

### Family Serpulidae [calcareous tube-worms]

58/2, 58/5   *Filograna implexa* Berkeley. A small worm, less than 5 mm long, living in aggregations of thin calcareous tubes winding around each other (Plate 58/2). Each worm typically has eight tentacles. The colonies are found attached to hard objects below tide level in all seas.

58/8   *Hydroides norvegica* (Gunnerus). The entrance to the calcareous tube of this species, which may reach a length of 3 cm or more, is protected by an elaborate calcareous operculum when the worm withdraws its tentacles. Tubes are found attached to any hard objects, including ships' bottoms. It has a cosmopolitan distribution.

8/9, 58/10   *Pomatoceros triqueter* (Linnaeus) [keel-worm]. The calcareous tubes are triangular in cross-section, with a well-defined dorsal keel which extends as a rostrum above the tube entrance. Large numbers of tubes are often found encrusting stones and shells in the intertidal and subtidal zone (Plate 58/9). When removed from its tube, the stout worm can be seen to have a fan of short feathery tentacles and a large cone-shaped operculum, sometimes crowned with two or three spines (Plate 58/10). Found in the North-east Atlantic and Mediterranean.

59/1   *Serpula vermicularis* Linnaeus. A cosmopolitan species which grows to more than 5 cm in length. It has a trumpet-shaped operculum with a crenulated edge. The tube, which is circular in cross-section, has noticeable growth rings and is usually only attached by its base to the substratum. Tubes often grow in clusters.

59/2, 59/6   *Spirobranchus giganteus* (Pallas). A distinctive worm with two branches of spiralling tentacles emerging from a calcareous tube which is often buried deep within a coral head. The colour of the tentacles is very variable. A common species on reefs in the Caribbean and Indo-Pacific.

### Family Spirorbidae

59/3   *Janua (Janua) pagenstecheri* (Quatrefages). This common species has a longitudinally ridged calcareous tube which is coiled in an anticlockwise direction. The coil diameter rarely exceeds 2 mm. It is found on stones, shells of living and dead animals, and on a variety of seaweeds from the intertidal zone to depths in excess of 100 m. It has

a world-wide distribution but has not been recorded from the North-west Atlantic or North-west Pacific.

*Spirorbis spirorbis* (Linnaeus). The tube of this species is   59/4 smooth and coiled in a clockwise direction. There is often a flange at the entrance to the tube which increases the area of attachment. It is typically attached to fucoid seaweeds in the intertidal zone but is also found on kelp. Occurs in the North Atlantic.

## Order 'Archiannelida'

The five families comprising this group were at one time thought to be primitive forms from which the rest of the polychaetes had evolved. It is, however, now generally recognised that the anatomically simple form of these small worms is an adaptation to living between sand grains in soft sediments and not a primitive feature. It is quite possible that the families are not closely related to one another but should be re-assigned within the known orders of polychaetes.

Families: Dinophilidae, Nerillidae, Polygordiidae, Protodrilidae, Saccocirridae.

## Class Clitellata
approx. 6000 species

This class consists of annelids that are without parapodia and have few or no setae. Its members are exclusively hermaphrodite and when mature possess a glandular area of skin a few segments long (clitellum) which secretes an egg cocoon during the breeding season.

## Subclass Oligochaeta

The body segments are externally all more or less alike and the few setae they possess arise directly from the body wall. The small prostomium always lacks appendages. Only a few segments of the body possess gonads, the testes invariably occurring anterior to the ovaries.

The subclass contains the earthworms as well as many freshwater and a few marine forms.

Marine oligochaetes are small, bottom-living forms which feed on organic debris. They may occur in very high numbers in areas where there is organic pollution.

## Order Plesiopora

The orders of the oligochaetes are distinguished from one another by the position of the openings of the ducts leading from the testes to the body surface. In the Plesiopora, containing the marine representatives, the male pores open on the segment immediately behind that containing the testes.

Families: Naididae, Tubificidae, Enchytraeidae.

## Family Enchytraeidae

*Enchytraeus albidus* Henle. A colourless species, 2–3 cm   59/5 long, which is found on the shore amongst living and decaying seaweed and under stones. The clitellum of the specimen in the photograph is clearly visible. This species has a cosmopolitan distribution.

# Arthropods

Detailed anatomical studies of the animals previously grouped together within the phylum Arthropoda have led Manton (1973, 1977) to the conclusion that arthropods are not derived from a common ancestor, as had previously been supposed. It seems likely that arthropod features developed independently at least three times, producing the Crustacea, Chelicerata and Uniramia. Hence the old concept of the phylum Arthropoda has been abandoned here and the Crustacea, Chelicerata and Uniramia raised to phylum status. Of these phyla the Uniramia have no truly marine representatives.

The Crustacea and Chelicerata have several features in common. These include the bilateral symmetry of the body which is protected by a hard external skeleton that is periodically moulted to allow growth to occur. The body is made up of a series of similar segments, many of which bear pairs of jointed appendages. The body cavity is formed from enlarged blood spaces. The highly developed nervous system and accompanying sensory organs have contributed greatly to the evolutionary success of these two phyla.

# Phylum Crustacea

approx. 30 500 species

Most crustaceans are marine, but a proportion occur in freshwater and a few have ventured on to land. There are several parasitic species.

The majority of planktonic animals throughout the oceans of the world are crustaceans, and they are also well represented on the sea-bottom from the shore to abyssal depths. They range in size from microscopic planktonic forms to the large bottom-living spider crabs, whose leg span may be over 3.5 m. Some of the more massive lobsters may weigh over 10 kg. Crustaceans are of great economic importance, for not only are many species eaten directly by man, but they form an important food source for many commercial fish. Some species destroy timber structures by their boring activities, others foul the bottoms of ships.

The crustacean body is usually divided into a head, thorax, abdomen and a tail-piece (telson) but there is a tendency for the anterior thoracic segments to fuse with the head. In many species a skin-fold (the carapace) extends backwards from the hind part of the head to enclose the important anterior regions of the body. The chitinous external skeleton is strengthened by the addition of calcium salts. The front part of the head possesses two pairs of sensory appendages (antennae) and at least three pairs of mouth-parts (mandibles, maxillules, maxillae) occur behind the mouth. The paired appendages on the thorax and abdomen are usually two-branched (biramous) limbs, which perform a variety of functions. Gills, typically associated with the limbs, are present in many crustaceans, but the smaller forms usually respire through the general body surface. A pair of many-faceted (compound) eyes is present in the adult of many species.

With few exceptions, sexes are separate and a series of free-swimming planktonic larval stages is a common feature of the life-cycle.

## Class Cephalocarida
8 species

A primitive group of shrimp-like crustaceans less than 4 mm long, which was discovered as recently as 1955. The horseshoe-shaped head bears two pairs of short antennae. Neither eyes nor a carapace are present. Appendages occur on the head and thorax but not on the abdomen. The biramous thoracic limbs are all similar, each possessing an external leaf-like lobe at its base, a structure unique to the class. The telson bears a pair of processes, the caudal furca, each of which has a bristle sometimes half as long as the animal (Figure 16).

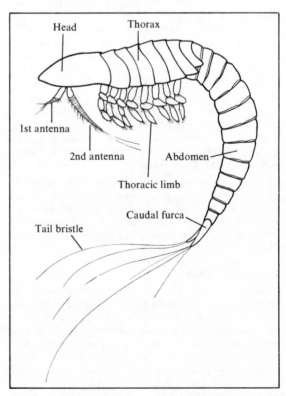

**Figure 16** *The main external features of the cephalocarid,* Hutchinsoniella. [*After Waterman & Chace.*]

**Phylum Crustacea**

| Phylum | Class | Superorder | Order | Suborder | Infraorder |
|--------|-------|------------|-------|----------|------------|

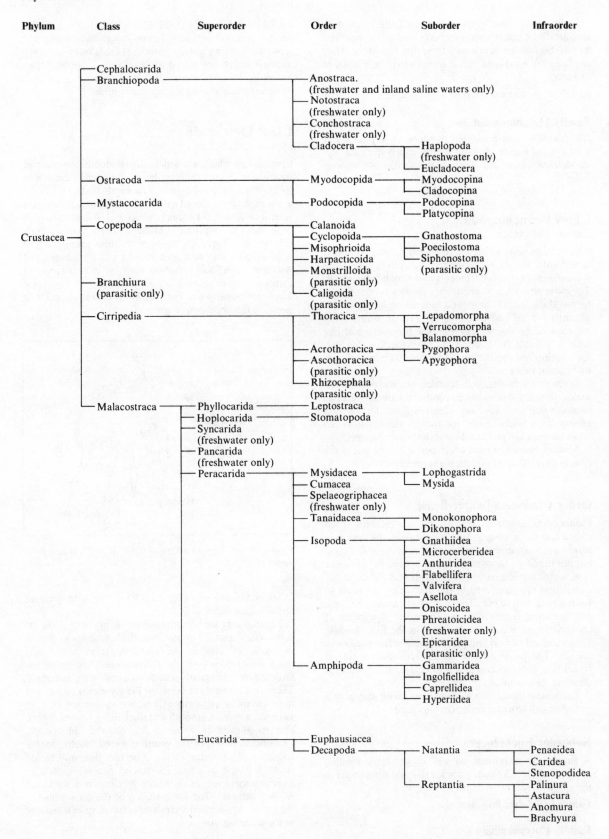

Classification follows Kaestner (1970) supplemented by Moore (1961) for the Ostracoda, Newman & Ross (1976) for the Cirripedia, Naylor (1972) and Birstein (1973) for the Isopoda and Barnard (1969) for the gammarid Amphipoda.

Plate 59
Phylum Crustacea

Unlike most free-living crustaceans, they are hermaphrodite. Cephalocarids are bottom-dwellers and feed by filtering bottom debris with their thoracic limbs. They are found from the intertidal zone to depths of 1500 m or more.

Families: Hutchinsoniellidae, Lightiellidae.

### Family Hutchinsoniellidae

59/9 *Hutchinsoniella macracantha* Sanders. A cephalocarid with a body length of about 3 mm, which lives on and in shallow-water soft sediments of the North-west Atlantic.

## Class Branchiopoda

approx. 800 species

Small crustaceans which are restricted mainly to inland waters with only a few marine representatives. The body is composed of distinct segments and may be shortened. The carapace, when present, may form a bivalved protective shell. Paired compound eyes are usually present. The first pair of antennae and one pair of mouth-parts are reduced or absent. The thoracic limbs are leaf-like with a gill at the base, and are used for feeding, locomotion and respiration. The limbless abdomen ends in a caudal furca.

Sexes are separate, but females greatly outnumber males. In many species, when conditions are favourable, eggs develop into new individuals without fertilisation taking place. When conditions are adverse, thick-walled resistant eggs are produced by sexual reproduction.

Marine representatives, which occur in only one of the four orders (Cladocera), are planktonic.

### Order Cladocera [water-fleas]

Cladocerans are small transparent crustaceans, often only a few millimetres in length. Normally the carapace forms a bivalved shell enclosing the body and limbs, but not the head. In some predatory species the carapace is reduced, serving only as a brood-pouch. The pair of compound eyes are fused in the midline of the body to form a single large eye. The second antennae are large and are used for swimming. There are 4–6 pairs of thoracic limbs, which are leaf-like in the filter-feeding forms and cylindrical and grasping in the few predatory species. The posterior end of the abdomen is turned forward in most species and bears claws and spines for cleaning the carapace.

The young usually develop without larval stages in a brood-pouch situated under the carapace.

### Suborder Eucladocera

A simple eye is present, as well as the large median compound eye. The young always develop directly from the egg without intermediate larval stages.

Families: Sididae, Polyphemidae.

### Family Polyphemidae

59/7 *Evadne nordmanni* Lovén. A small cladoceran about 1 mm long, in which the carapace is reduced leaving the head and limbs uncovered. The oval body of this egg-carrying specimen bears a small terminal spine. A predatory species common in the plankton of the North Atlantic and Mediterranean during the summer months.

59/10 *Podon polyphemoides* Leuckart. This cladoceran, which reaches a length of about 1 mm, has its head separated from the body by a deep groove. It feeds on other small animals which are seized by the thoracic limbs. Often found in the plankton of the North-east Atlantic and Mediterranean during spring and summer.

## Class Ostracoda

approx. 2000 species

Crustaceans which are widely distributed in the sea and freshwater. Adults range in length from less than 1 mm to 25 mm in *Gigantocypris*. The head, body and limbs are completely enclosed within a bivalved carapace which resembles a shell. The short thorax and abdomen appear unsegmented externally. Most animals possess only simple eyes. There are never more than seven pairs of appendages. The first and second pairs of antennae, which are used for locomotion, are large in the swimming forms and shorter, with spines, in the burrowing types. There are usually two pairs of limbs on the thorax and none on the abdomen (Figure 17).

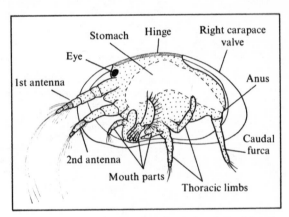

**Figure 17** *Side view of a typical ostracod with the left valve removed.*

The sexes are separate and there are a number of larval stages in the life cycle.

The majority are bottom-dwellers living in or upon the substratum, and are found from shallow areas to depths in excess of 2000 m. In addition, there are a few planktonic forms. Ostracods have diverse feeding habits, filter-feeders, predators and scavengers being common. There is a correlation between life style of animals and their carapace structure. Carapace valves are thin in swimming species, smooth and thick in burrowing forms, and roughened in those living on the substratum. Classification within the group is based chiefly on the structure of carapace valves. The fact that oil-bearing strata are rich in certain ostracod fossils has led to their extensive use as indicator species when searching for new oilfields. The fossil record of the class stretches back over 500 million years and is the most extensive of any crustacean group.

### Order Myodocopida

An entirely marine order, with convex edges to the carapace valves. Compound eyes are often present and the second antennae usually have two branches. The caudal furca bears strong spines.

## Suborder Myodocopina

The anterior end of each carapace valve has a notch, through which an antenna can extend. Compound eyes are present. There are two pairs of thoracic limbs.

Families: Thaumatocypridae, Cypridinidae, Cylindroleberididae, Sarsiellidae, Halocyprididae, Asteropidae, Conchoeciidae.

### Family Cypridinidae

59/12   *Cypridina* sp. When disturbed, species of this genus produce luminescent clouds of material from glands near the mouth. Some of the specimens in the photograph are carrying eggs in their brood pouches. *Cypridina*, which is rarely more than 5 mm long, is found in the plankton of most seas.

59/11   *Gigantocypris* sp. One of the largest known ostracods, with soft, almost spherical, red shell-valves up to 25 mm in length. It is found at depths of 200–4000 m in the Atlantic, where it preys upon passing planktonic animals, which are seized by its antennae.

### Family Conchoeciidae

59/8   *Conchoecia valdiviae* (Müller). A bright-red planktonic ostracod up to 6 mm long which normally lives at depths of 800–1200 m. It swims with the aid of antennae which are protruded through notches in the shell valves, and feeds mainly on other small crustaceans. Found in the equatorial Atlantic.

## Suborder Cladocopina

The carapace valves have no notches. There are no eyes and no thoracic limbs. All species live between sand grains on the sea bottom.

Family: Polycopidae.

## Order Podocopida

Marine, freshwater and a few terrestrial representatives occur. The ventral edges of the, sometimes sculptured, carapace valves are straight or concave. No compound eyes are present and the second antennae have one branch reduced or absent. The caudal furca is frequently absent.

## Suborder Podocopina

The carapace valves have a concave ventral edge, and the second antennae have only one branch. They possess two pairs of thoracic limbs, the second pair being used for walking and grooming.

Families: Bairdiidae, Macrocyprididae, Cyprididae, Paracyprididae, Pontocyprididae, Cytheridae, Bythocytheridae, Cytherettidae, Cytherideidae, Cytheruridae, Hemicytheridae, Leguminocythereididae, Leptocytheridae, Loxoconchidae, Paradoxostomatidae, Pectocytheridae, Progonocytheridae, Psammocytheridae, Schizocytheridae, Trachyleberididae, Xestoleberididae.

## Suborder Platycopina

These marine animals have a straight ventral edge to the carapace valves. A pair of thoracic limbs is present: these limbs are leaf-like in females and pincer-like in males.

Family: Cytherellidae.

## Class Mystacocarida
10 species

Minute marine crustaceans up to 0.5 mm in length. They have an elongate cylindrical body with a head divided into two regions, a short anterior section and a much longer posterior part. Only simple eyes are present. Both pairs of antennae are prominent, the second pair together with the first pair of mouth-parts being used for locomotion. Toothed furrows, of unknown function, are situated on the posterior part of the head and on the body. There are four pairs of small thoracic limbs with only one branch (Figure 18).

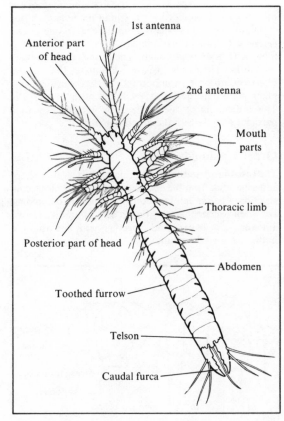

**Figure 18** *External features of the mystacocarid,* Derocheilocharis. [*After Delamare & Chappuis.*]

The sexes are separate and there are several larval stages in the life cycle.

The animals live between sand grains in the intertidal and subtidal zones, where they feed by collecting organic debris and micro-organisms with long bristles on their mouth-parts.

Family: Derocheilocaridae.

## Class Copepoda
approx. 7500 species

Mainly marine crustaceans, ranging from less than 1 mm to several millimetres in length. Copepods are free-living as well as parasitic. The bodies of some parasitic forms are often so highly modified that they are almost unrecognisable as crustaceans. The body of free-living forms is usually short and cylindrical and is composed of a head, thorax and abdomen. The posterior part of the head is fused with one or more thoracic segments. Both

**Plate 60**

**Phylum Crustacea**

compound eyes and a carapace are absent, although simple eyes are better-developed than in other crustacean groups. The anterior part of the head often projects forward as a small beak. The antennae are well developed, the first pair being large with many bristles. Biramous thoracic limbs, sometimes with the aid of the antennae, are used for locomotion. The thin abdomen lacks appendages apart from the last segment, which bears a caudal furca with bristles. In tropical planktonic forms these bristles are often long and elaborate to aid flotation. Although copepods are usually transparent, they may have a red or blue coloration due to oil droplets in the body. Luminescent organs are present in some forms.

The sexes are separate, and the first antennae of the male are often modified for grasping the female during copulation. Females usually carry the eggs in one or two egg sacs. The majority of planktonic copepods are filter-feeding herbivores, although some of the larger ones are predators. Copepods usually dominate the animal plankton, forming an important link between the small plant plankton and larger predators such as the fish. The bottom-living species feed by picking up food particles directly with the mouth-parts.

## Order Calanoida

The fused head and thorax of calanoids is oval in shape and separated from the abdomen by a conspicuous constriction (Figure 19a). The multi-segmented first antennae are often longer than the body. The biramous second antennae are shorter and, together with the thoracic limbs, are used for swimming.

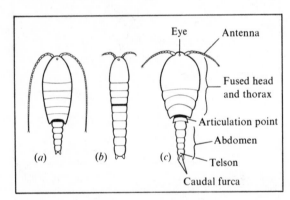

**Figure 19** *Body regions of free-living copepod groups:* (a) *calanoid*, (b) *harpacticoid*, (c) *cyclopoid.*

In males, one of the first pair of antennae is frequently modified for grasping the female during pairing. Most females carry fertilised eggs in a single egg sac, but some species shed their eggs directly into the water after fertilisation.

The majority of calanoids live in the surface waters of the oceans but a few can be found as deep as 4000 m.

Families: Calanidae, Megacalanidae, Eucalanidae, Paracalanidae, Calocalanidae, Pseudocalanidae, Stephidae, Spinocalanidae, Aetideidae, Euchaetidae, Phaennidae, Scolecithricidae, Tharybidae, Temoridae, Metridiidae, Diaixidae, Centropagidae, Pseudodiaptomidae, Lucicutiidae, Heterorhabdidae, Augaptilidae, Arietellidae, Pseudocylopidae, Ridgewayiidae, Candaciidae, Pontellidae, Bathypontiidae, Acartiidae, Tortanidae, Mormonillidae.

### Family Megacalanidae

*Bathycalanus* sp. Members of this genus are amongst the largest of the free-living copepods, females of 18 mm in length having been found. They are deep-water forms, ranging from 400 m to depths in excess of 2000 m.    60/1

*Megacalanus princeps* Wolfenden. A deep-water species reaching a length of about 10 mm. It has been encountered most frequently at depths below 400 m in the Atlantic.    60/2

### Family Aetideidae

*Valdiviella brevicornis* Sars. A deep-water species, distinguished from other species in the genus by its relatively short antennae. The female specimen in the photograph is carrying an egg sac on her abdomen. Recorded at depths below 500 m in the Atlantic and Caribbean.    60/4

### Family Heterorhabdidae

*Hemirhabdus* sp. Members of this genus have been recorded at depths between 250 m and 1000 m in the Atlantic.    60/3

## Order Cyclopoida

The body of cyclopoids is pear-shaped with an obvious constriction about two-thirds of the way along the body, between the fourth and fifth free thoracic segments (Figure 19c). The first antennae are short compared with those in calanoids, and the second antennae usually have only one branch. The first antennae of males are sometimes modified for grasping the female, who carries two egg sacs.

The order includes many marine planktonic and bottom-dwelling forms as well as freshwater representatives.

### Suborder Gnathostoma

In this suborder, eyes are not prominent. Males have the first antennae modified for grasping. In both sexes the first two pairs of mouth-parts are modified for chewing, and the first pair of thoracic limbs are unmodified.

Families: Oithonidae, Cyclopinidae, Cyclopidae.

### Suborder Poecilostoma

Prominent eyes are common in this group. In males of many species the second antennae have a grasping function whilst the first antennae are unmodified. The first pair of mouth-parts are not used for chewing, and the second pair of mouth-parts are reduced. The first pair of thoracic limbs often have a grasping function.

Families: Oncaeidae, Corycaeidae, Sapphirinidae.

### Family Sapphirinidae

*Sapphirina* sp. A flattened, leaf-like, cyclopoid with prominent eyes situated close together at the anterior end. The beautiful interference colours shown in the photograph result from light reflecting on guanine crystals in the skin. Representatives of this genus are common in the surface waters of tropical seas.    60/9

## Order Misophrioida

The bottom-dwelling members of this order possess both calanoid and cyclopoid characteristics. Externally they are similar to the cyclopoids in body segmentation but the biramous antennae have a very large outer branch, as found in the calanoids. The female carries a single egg sac.

Family: Misophriidae.

## Order Harpacticoida

The body, which rarely exceeds 2 mm in length, is normally widest at its anterior end and tapers to the posterior without any marked external constrictions (Figure 19b). The first antennae are very short and the second antennae are biramous. Often in males both of the first antennae are modified for grasping the female, who commonly carries her eggs in a single egg sac.

A few harpacticoids are planktonic, but most live on or in both mud and sand where they feed on organic debris and algae.

Families: Longipediidae, Canuellidae, Aegisthidae, Cerviniidae, Pontostratiotidae, Ectinosomidae, Neobradyidae, Phyllognathopodidae, Darcythompsoniidae, Chappuisiidae, Tachidiidae, Harpactididae, Tisbidae, Porcellidiidae, Peltidiidae, Pseudopeltidiidae, Tegastidae, Thalestridae, Balaenophilidae, Parastenhelliidae, Diosaccidae, Miracidae, Metidae, Ameiridae, Paramesochridae, Tetragonicepsidae, Canthocamptidae, Cylindropsyllidae, Louriniidae, Parastenocaridae, Cletodidae, Laophontidae, Ancorabolidae, Ismardiidae.

### Family Laophontidae

6, 60/10    *Microlaophonte trisetosa* Boxshall. The female, which reaches a length of about 0.4 mm, is seen carrying an egg sac beneath her abdomen (Plate 60/10). The smaller male, which lacks red pigmentation, has antennae modified for grasping the female during copulation (Plate 60/6). The species occurs in organically rich sediments amongst mangroves in the Caribbean.

## Class Cirripedia [barnacles]
approx. 900 species

The barnacles are an entirely marine group whose members range in size from minute parasitic forms to the larger goose barnacles, which may reach 80 cm in length. The majority of non-parasitic forms are attached to rocks and other solid objects by a cement-like secretion from glands on the first pair of antennae, and literally 'stand on their heads'. The second antennae and compound eyes are absent in adults. The body is indistinctly segmented and the abdomen much reduced. The carapace, which encloses the body and the limbs, is usually strengthened with calcareous plates. The number and shape of these plates aids the identification of many species. There are normally six pairs of biramous thoracic limbs which are protruded between the carapace valves to 'comb' food particles from the water.

Although many barnacles are hermaphrodite, cross-fertilisation occurs by means of a long penis which can be extended from one barnacle to another. Sexes remain separate in some forms. Eggs hatch into planktonic larvae which settle on the bottom after a series of larval stages.

Barnacles occurring on rocky shores are extremely resistant to wave damage, exposure to the air, and to extremes of temperature. Some cirripedes are of economic importance since they foul the bottom of ships, buoys, pilings and the legs of oil-rigs.

## Order Thoracica

Barnacles that are permanently attached to rocks and other solid objects by the anterior region of the head. The attachment region may be elongated or flattened into a disc. The carapace is usually covered by calcareous plates. There are six pairs of limbs evenly distributed along the thorax.

Most are hermaphrodite, but a few species have separate sexes with males much smaller than the females. In some instances dwarf males accompany hermaphrodite forms.

## Suborder Lepadomorpha [goose barnacles]

The anterior part of the head is elongated to form a stalk which bears the rest of the head and thorax (Figure 20a). The flexible stalk often lacks calcareous plates and carries the remains of the first antennae and the cement glands.

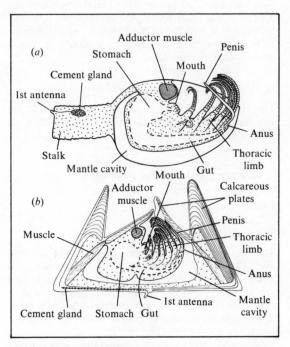

**Figure 20** *Sections through barnacles to reveal their basic structure;* (a) Lepas, (b) Balanus.

The majority are hermaphrodite, but in a few forms the sexes are separate.

Many members of this group attach themselves to floating objects such as driftwood and buoys.

Families: Scalpellidae, Heteralepadidae, Iblidae, Koleolepadidae, Lepadidae, Malacolepadidae, Poecilasmatidae, Oxynaspididae, Trilasmidae.

### Family Scalpellidae

*Pollicipes polymerus* Sowerby. A colonial species of goose    60/7 barnacle up to 5 cm high, with a large number of separate calcareous plates. The stalk is protected by scales. Colonies are found attached to rocks intertidally in regions of strong water movement. It differs from most other barnacles in holding its feeding fan of thoracic limbs erect without combing movements until sufficient food has accumulated on them, only then are they withdrawn. Occurs in the North-east Pacific.

60/5    *Scalpellum scalpellum* (Linnaeus). A slender species which may reach a length of 5 cm. As in other members of the family, the stalk is covered with scales. It is found attached to colonial hydroids and sea-mats below tide level in the North-east Atlantic and Mediterranean.

### Family Lepadidae

60/8    *Lepas anatifera* (Linnaeus). A species with 5 translucent white plates protecting the body, which is situated on a stalk reaching a length of 25 cm when fully extended. It is normally found attached to floating objects such as buoys, boats and driftwood. It has a cosmopolitan distribution.

61/1    *Lepas anserifera* (Linnaeus). This species differs from *L. anatifera* in having a short stalk and plates which are distinctly furrowed. Attached to floating objects in warm waters.

61/3    *Lepas fascicularis* Ellis & Solander. A species with five thin translucent plates protecting a body between 3 and 4 cm long when fully grown. After initial attachment to a drifting object (a tar ball in the photograph) a spongy bubble raft is secreted to aid flotation. The specimens have their limbs in the feeding position. It has a cosmopolitan distribution.

## Suborder Verrucomorpha

The anterior part of the head is flattened into a broad attachment disc. The body is enclosed within an asymmetrical box-like wall made up of four plates. Two other plates form a movable lid.

    All species in this suborder are hermaphrodite.

    Most are deep-sea forms, although a few are found intertidally.

    Family: Verrucidae.

## Suborder Balanomorpha

The anterior part of the head forms a flattened attachment disc. The body is enclosed within a symmetrical wall which is derived from eight calcareous plates. In several species these plates may become fused or lost, and in one genus, *Pyrgoma*, the wall comprises a single plate. A movable lid, formed originally from four plates, is present in many forms (Figure 20*b*).

    Nearly all species are hermaphrodite, a few forms being accompanied by dwarf males.

    Members of the group are common on rocky shores.

    Families: Catophragmidae, Chthamalidae, Coronulidae, Bathylasmatidae, Tetraclitidae, Archaeobalanidae, Pyrgomatidae, Balanidae.

### Family Chthamalidae

61/4    *Chthamalus stellatus* (Poli) [star barnacle]. This species with a diameter of up to 1–2 cm has six similar-sized shell plates surrounding an oval opening. The arrangement of the joints between the plates forming the movable lid within the opening is characteristic of the species (Figure 21*a*). A common rock-dwelling species of intertidal barnacle in the North-east Atlantic and Mediterranean, but is also found in the Caribbean.

### Family Tetraclitidae

61/5    *Tetraclita squamosa* (Bruguière). This species, which reaches a diameter of 3 cm, has four markedly striated shell plates arranged in a cone. Common on intertidal rocks in the Indo-Pacific.

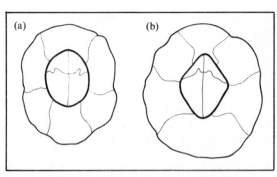

**Figure 21** *The differences in form between* (a) Chthamalus stellatus, *and* (b) Balanus balanoides. [*After Southward.*]

### Family Balanidae

*Acasta spongites* Darwin. A characteristic species with six plates and a rounded base, which is usually found embedded in a sponge. The specimen in the photograph, which has its feeding limbs extended, has been removed from a sponge. Found in shallow waters of the Mediterranean and North-east Atlantic.    61/6

*Balanus balanoides* (Linnaeus) [acorn barnacle]. One of the six plates surrounding the diamond-shaped opening is larger than the others and overlaps the adjacent plates (Figure 21*b*). This species, which reaches a diameter of 15 mm, is found in large numbers on intertidal rocks (Plate 61/9). The biramous thoracic feeding limbs can be clearly seen as they comb the water for food (Plate 61/2). Occurs in the North-east Atlantic.    61/2, 61▮

*Balanus crenatus* Bruguière. A rock-dwelling species with six plates, each of which decreases in width towards the diamond-shaped opening. The base can reach a diameter of 2 cm. Primarily a subtidal species, but sometimes found in crevices and under stones low on the shore in the North Atlantic and North Pacific.    61/7

*Balanus perforatus* Bruguière. The six strongly built shell plates of this species are often striated and have a purplish tinge. It is found on the lower shore and in shallow water in the Mediterranean and North-east Atlantic.    61/8

## Order Acrothoracica

Small barnacles that burrow into calcareous rocks, corals, and mollusc shells. The anterior part of the head forms an attachment disc. The carapace is not strenthened by calcareous plates. There are usually less than six pairs of thoracic limbs, the first pair being situated near the mouth and separated from the remainder, which are attached towards the rear of the thorax.

    The sexes are separate, the males being very much smaller and usually attached to the females.

## Suborder Pygophora

Barnacles having a complete gut, and which commonly bore into corals.

    Families: Lithoglyptidae, Cryptophialidae, Zapfellidae.

## Suborder Apygophora

Barnacles having an incomplete gut, and which are usually found boring into mollusc shells.

    Family: Trypetesidae.

# Class Malacostraca
approx. 19 100 species

This class is the largest and most varied group of Crustacea, its members occurring in the marine, freshwater and terrestrial environments. Sizes range from a few millimetres to a limb-span of over 3 m in the case of the giant spider crabs. The majority of malacostracans are free-living but there are some parasitic forms.

The bodies of most species have a heavily calcified external skeleton. Both pairs of antennae are well developed, the first pair usually being biramous. Stalked compound eyes are frequently present, as is the carapace, although both are lacking in some forms. The thorax, which is often fused with the head, has eight segments, each bearing a pair of appendages. The first three pairs are usually used for food gathering whilst the others are mainly locomotory. Gills are normally situated at the bases of the thoracic limbs. The abdomen typically has six segments (rarely seven), each with a pair of appendages which are variously modified for swimming and reproduction, occasionally for respiration. The last pair are broadened and with the last abdominal segment (telson) may form a tail fan; a caudal furca is rarely present (Figure 22).

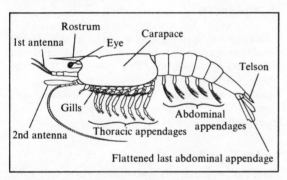

Figure 22 *External features of a generalised malacostracan.*

The sexes are usually separate, although a few hermaphrodite species exist. The female often carries her fertilised eggs in special thoracic brood-chambers or on the abdominal limbs. The life history is frequently complex, involving a series of larval stages, but direct development occurs in some species.

Free-living malacostracans are found in a wide range of marine habitats, some are planktonic filter-feeders, whilst others are bottom-dwelling scavengers.

Many species in this class play an important role in marine food webs, whilst lobsters, crabs, shrimps and prawns are harvested directly by man.

## Superorder Phyllocarida

Members of this primitive superorder have the thorax and part of the abdomen covered with a large bivalved carapace which, at the anterior end, bears a hinged plate covering the head. The antennae are long, and protrude from the carapace. Although the compound eyes are reduced in some forms, in many they are situated on the ends of movable stalks. The eight pairs of leaf-like thoracic limbs are frequently used for filter-feeding. The abdomen comprises seven segments, and ends in a long caudal furca.

The sexes are separate, and the female carries her eggs on the thoracic limbs. The eggs usually develop directly into a young adult.

The group has a long fossil record, extending back over 300 million years.

## Order Leptostraca

This is the only phyllocarid order which contains living representatives. All are small marine forms ranging in length from a few millimetres to 4 cm. They are mainly bottom-dwellers living in mud or amongst seaweed.

Family: Nebaliidae.

### Family Nebaliidae

*Nebaliopsis typica* Sars. A species which is atypical of    61/10
the Leptostraca in that it is probably a permanent member of the plankton, whereas the rest are bottom-living. It is thought to feed on eggs of planktonic animals. Records of this species are sparse but it is probably cosmopolitan.

## Superorder Hoplocarida

Members of this group range in length from 4 to 35 cm. The body is dorsoventrally flattened and is often covered with ridges and spines. There is a short carapace covering the head and only part of the thorax. At its anterior end the carapace forms a movable plate overhanging the large eye stalks. The prominent first antennae have three branches. The first five pairs of thoracic limbs are uniramous and modified for grasping. The second pair, which are greatly enlarged and bear spines, are used to capture prey in a manner similar to that employed by the praying mantis. The broad, well developed, abdomen has five pairs of large biramous swimming limbs, the sixth pair forming a tail fan with the telson. Gills are present on the abdominal limbs.

The sexes are separate and the eggs, which are laid in clusters by the female, are held securely amongst the thoracic limbs. Larvae hatch from the eggs and pass through a series of stages before reaching adulthood.

## Order Stomatopoda [mantis shrimps]

This is the only hoplocarid order with living representatives. They are mainly coastal bottom-dwellers living in burrows. Many of the tropical species are brilliantly coloured. All mantis shrimps are predators, feeding on other crustaceans, worms, molluscs and fish. Some species are themselves eaten by man, particularly in Japan and the Mediterranean countries.

Families: Squillidae, Bathysquillidae, Gonodactylidae, Lysiosquillidae.

### Family Squillidae

*Squilla mantis* Fabricius [mantis shrimp]. A predatory    62/4
bottom-living species which may reach a length of 20 cm or more. It has two characteristic black spots on the tail. Found on soft sediments, usually below 10 m, in the Mediterranean and adjacent Atlantic.

## Superorder Peracarida

Members of this mainly marine group are usually small, but some of the deep sea isopods may attain lengths of over 30 cm. The eyes may be stalked or unstalked and

Plate 62

Phylum Crustacea

the jaws (mandibles) bear a characteristic movable tooth. A carapace is present in primitive shrimp-like forms but is absent in the more specialised animals. The thoracic limbs flex between the fifth and sixth joints.

The female has a brood-pouch under the thorax, formed from special plates situated at the base of the thoracic limbs. Eggs develop directly into juveniles within the pouch.

Filter-feeding is common in the group, but there is a tendency for the larger forms to adopt a more direct method of feeding.

## Order Mysidacea [opossum shrimps]

Although some deep-sea species can reach lengths of over 20 cm, the majority are small elongate shrimp-like forms. Movable stalked eyes are usually present and are covered by an extension of the carapace (rostrum). The carapace is fused with the first three thoracic segments only. Gills may be present on the biramous thoracic limbs. Some species swim using their thoracic limbs, but others use their abdominal appendages. The last pair of abdominal limbs form a leaf-like tail fan with the telson.

Filter-feeding is common in the group, but a few of the deep-sea forms are scavengers. Mysids often live in large swarms which are an important food for some fish.

Almost all the bottom- and mid-water mysids swim up towards the water surface at night and sink back down again during the day. This behaviour, which relates to light intensity, is particularly noticeable during the spawning period.

### Suborder Lophogastrida

The thoracic limbs have branched gills, and the biramous abdominal appendages, which are used for swimming, are well developed in both males and females. The brood pouch of females is formed from seven pairs of plates.

Families: Lophogastridae, Eucopiidae.

### Family Lophogastridae

62/1 *Gnathophausia* sp. Members of this planktonic genus, which may reach a length of 25 cm, secrete a luminescent material from a gland near the mouth parts. They have been recorded from depths below 500 m.

### Suborder Mysida

The thoracic limbs do not have branched gills, and the abdominal appendages are usually reduced in females. The brood-pouch of females is formed from only two or three pairs of plates.

Families: Petalophthalamidae, Mysidae.

### Family Mysidae

62/3 A sample of mysids from an oceanic plankton haul. The large eyes are a prominent feature, as are the brood-pouches in several specimens.

## Order Cumacea

Small marine and brackish-water crustaceans with a large head and thorax and a narrow abdomen. The carapace is joined dorsally to three or four thoracic segments. It completely encloses the head and thorax, and forms a large respiratory chamber on either side of the body; anteriorly it extends forward beyond the head. Although

the antennae are often large in the male they are very small in the female. The immovable eyes are poorly developed and often fused together. Animals swim using their thoracic limbs assisted by flexing of the abdomen. The short first pair of thoracic limbs bear gills which extend into the respiratory chambers. The male bears biramous cylindrical appendages on the abdomen, but these are absent in females. There is no tail fan.

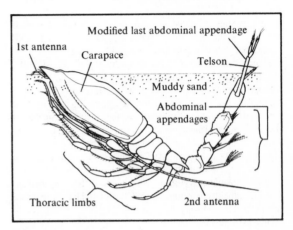

**Figure 23** *An adult cumacean,* Diastylis, *in its natural position within the sediment.*

Cumaceans live in burrows in sand and mud (Figure 23). Many species filter-feed but some pick up debris and algae. Large swarms often occur near the water surface during the breeding season.

Families: Bodotriidae, Leuconiidae, Nannastacidae, Lampropidae, Pseudocumidae, Diastylidae, Ceratocumidae.

## Order Tanaidacea

Tanaidaceans are mainly marine and usually only a few millimetres long. The carapace, which covers only the first two thoracic segments, forms a respiratory chamber on either side of the body. The antennae have one or two branches, and the eyes, when present, are situated on short immovable stalks. The first pair of thoracic limbs have gills which project into the respiratory chambers, and the second pair of limbs are modified into large grasping organs. The short abdomen has biramous appendages in the male, but these are usually absent in females. The last pair of appendages are slender and do not form a tail fan (Figure 24).

Sexes are usually separate but a few hermaphrodite forms occur.

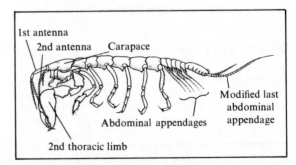

**Figure 24** *A typical tanaidacean* Apseudes, *showing the main external features.*

Tanaidaceans live in burrows or tubes in mud or amongst plants. They are present in both shallow and deep waters, some occurring at depths as great as 8000 m. Filter-feeding is common, but some pick up food with the grasping thoracic limbs.

## Suborder Monokonophora

The antennae have two branches. The brood-pouch of females is formed from four pairs of plates. In the male the paired reproductive openings lie on a central cone on the ventral side of the eighth thoracic segment.

Families: Apseudidae, Kalliapseudidae.

## Suborder Dikonophora

The antennae have only one branch. The brood-pouch of females is formed from one to four pairs of plates. In the male the paired reproductive openings lie on a pair of cones on the eighth thoracic segment.

Families: Neotanaidae, Paratanaidae, Tanaidae.

## Order Isopoda

A large crustacean group with free-living representatives in the marine, freshwater and terrestrial environments. A few highly modified parasitic forms also occur. The body, which is usually flattened from top to bottom, consists of a head which incorporates the first (occasionally the second) thoracic segment, a thorax, and a short abdomen. There is no carapace. The first pair of antennae are much smaller than the second pair and the eyes are unstalked. The first pair of thoracic limbs are used as accessory mouth-parts, whilst the remainder are usually cylindrical walking legs. The biramous abdominal appendages are flattened and often function as gills. In the male the second pair (or first and second pair) of these appendages are rod-like and are used to transfer sperm to the female. The young remain in the brood pouch of the female for some time after hatching.

Most marine isopods are bottom-dwellers, their dull-coloured leathery external skeleton providing excellent camouflage. Many are found in soft sediments and in crevices from the shore to depths in excess of 10 000 m. Most feed on organic debris, whilst others browse on algae or are predators on small animals. One family, the Limnoriidae, is of economic importance as its members bore into marine timbers and submarine cables.

## Suborder Gnathiidea

Small marine forms, typically 2–5 mm long, living in burrows in mud, in rock crevices, or in corals, tubes of worms, or in sponges. The first two thoracic segments are fused with the head, whilst the last thoracic segment is much reduced and lacks appendages. Jaws (mandibles) are absent in the female but in the male are large hook-like structures projecting in front of the head. The abdomen is small and much narrower than the thorax. Adults do not feed but there is a larval stage which is a temporary fish parasite.

Family: Gnathiidae.

## Suborder Microcerberidea

Minute forms living between sand grains, sometimes in underwater caves. The first pair of limbs of the long thorax bears claws.

Family: Microcerberidae.

## Suborder Anthuridea

The marine forms often live amongst kelp holdfasts or in worm tubes. The body is long and cylindrical with the well developed first pair of thoracic limbs bearing claws. The first abdominal appendages are broad and cover the remaining appendages.

Family: Anthuridae.

## Suborder Flabellifera

The body is noticeably flattened, with some of the abdominal segments fused together. The bases of the thoracic limbs are expanded into large plates and the last pair of abdominal appendages are fan-shaped. The wood-borers (gribbles) belong to this suborder.

Families: Cirolanidae, Sphaeromatidae, Serolidae, Anuropidae, Limnoriidae, Excorallanidae.

### Family Cirolanidae

*Eurydice pulchra* Leach. A species up to 0.8 mm long, with antennae that almost reach to the end of the thorax. The oval thorax and large compound eye can be seen clearly in the photograph. It is said to feed on injured animals in sand and mud on the shore and in shallow water. Found in the North-east Atlantic. 62/6

### Family Sphaeromatidae

*Dynamena* sp. In common with other members of this family, it has a domed appearance when viewed from the side and is capable of rolling itself up into a ball when disturbed. They are herbivorous scavengers which occur in crevices, under stones and amongst seaweed on the shore. 62/7

### Family Limnoriidae

*Limnoria quadripunctata* Holt [gribble]. An isopod less than 4 mm long, with short antennae and four characteristic small protruberances on the anterior part of the telson (Plate 62/5). The species of *Limnoria* are very difficult to distinguish but they all bore into submerged timber, often in great numbers, riddling it with burrows (Plate 62/2). Plate 62/8 shows a wooden pile exposed at low water which has been ravaged by the gribble, the scourge of harbour authorities before the advent of concrete jetty piles. This species has a cosmopolitan distribution. 62/2, 62/5, 62/8

## Suborder Valvifera

Many members of this group live amongst algae or in hydroid and bryozoan undergrowth. A few species burrow in sand and mud. The body may be flattened or almost cylindrical, with most of the abdominal segments fused together. The last pair of abdominal appendages are long and leaf-like and cover the other appendages.

Families: Idoteidae, Arcturidae.

### Family Idoteidae

*Idotea baltica* (Pallas). A large idoteid, the males may reach a length of over 3 cm but the females rarely reach 2 cm in length. The disparity in size, which is com- 62/9, 63/1

**Plate 63**                                                                                    **Phylum Crustacea**

mon in most isopods, is seen clearly in the photograph of a copulating pair (Plate 63/1). The species is characterised by the presence of a large central spine and two smaller lateral spines on the rear of the telson (Plate 62/9). The colour of most idoteids is very variable and should not be used to try to distinguish between species. Commonly found scavenging amongst seaweed undergrowth and seaweed debris below tide level. It has a cosmopolitan distribution.

63/3    *Idotea emarginata* (Fabricius). Males of this species may reach 3 cm in length. The telson is broad and without spines on the posterior margin. A specimen is seen on the left of another species, *I. granulosa*, in the photograph. Common amongst seaweed undergrowth and seaweed debris below tide level in the North-east Atlantic.

63/3    *Idotea granulosa* Rathke. A species rather similar to *I. baltica*, but it is generally smaller and has only one spine at the end of the telson. Common amongst seaweed undergrowth in the intertidal zone, where its colour often matches that of the sheltering seaweed. Found in the North-east Atlantic.

## Suborder Asellota

Although occurring in a wide range of habitats, representatives of this suborder are often found on the shore amongst encrusting plants and animals. The flattened body has the abdominal segments fused into a single unit. Most of the abdominal appendages function as gills and are enclosed within a chamber formed by the anterior appendages. The last pair of appendages are cylindrical.

Families: Desomosomatidae, Munnopsidae, Abyssianiridae, Echinothambernidae, Thambernidae, Microparasellidae, Microjaniridae, Iaeropsidae, Ianiridae, Ianirellidae, Haploniscidae, Munnidae, Ischnomesidae, Macrostylidae, Nannoniscidae.

## Suborder Oniscoidea

This group contains the familiar land-dwelling woodlice, but a few species occur on the seashore. The flattened body has distinct abdominal segments, and the bases of the thoracic limbs are expanded into large plates.

Families: Ligiidae, Tylidae.

### Family Ligiidae

63/4    *Ligia oceanica* (Linnaeus) [sea-slater]. This species, which has an oval flattened body up to 3 cm in length, bears a superficial resemblance to a woodlouse. During the day it hides in crevices, under stones and amongst seaweeds on the upper shore and emerges to feed on decaying seaweed and other organic matter at night. This species, although capable of living for long periods of time totally submerged in seawater, prefers to avoid submersion. An inhabitant of North Atlantic shores.

## Order Amphipoda

A group of mainly free-living marine crustaceans, although a few are external parasites of whales and fish, and some species live in freshwater. The body is generally curved and flattened from side to side and has the first (occasionally the second) thoracic segment fused with the head. There is no carapace. The antennae vary in size and shape and the eyes are unstalked. The first pair of thoracic limbs are used as accessory mouth-parts, whilst the remainder are walking legs. Typically the first and second pairs of walking legs have pincers or claws for grasping and often the last three pairs are directed backwards. Gills are present at the bases of some thoracic limbs. The abdomen is not distinct from the thorax and has swimming appendages and backward-pointing limbs which can be used for pushing or jumping.

The male is frequently larger than the female and often rides on her back for several days prior to mating.

Amphipods are common bottom-dwellers in intertidal and shallow-water areas, although there are many deep-water forms also. There are some planktonic species. Most species scavenge plant and animal debris but some are filter-feeders and predators.

## Suborder Gammaridea

The majority of amphipods belong to this suborder. Only the first thoracic segment is fused with the head, which usually has small eyes. There are large plates at the bases of the thoracic limbs and the abdomen and its appendages are well developed.

Families: Acanthonotozomatidae, Ampeliscidae, Amphilocidae, Ampithoidae, Anamixidae, Aoridae, Argissidae, Astyridae, Atylidae, Bateidae, Beaudettiidae, Calliopiidae, Cheluridae, Colomastigidae, Corophiidae, Cressidae, Dexaminidae, Dogielinotidae, Eophliantidae, Eusiridae, Gammaridae, Haustoriidae, Hyalellidae, Hyalidae, Hyperiopsidae, Isaeidae, Ischyroceridae, Kuriidae, Lafystiidae, Laphystiopsidae, Lepechinellidae, Leucothoidae, Liljeborgiidae, Lysianassidae, Melphidippidae, Ochlesidae, Oedicerotidae, Pagetinidae, Paramphithoidae, Pardaliscidae, Phliantidae, Phoxocephalidae, Pleustidae, Podoceridae, Prophliantidae, Sebidae, Stegocephalidae, Stenothoidae, Stilipedidae, Synopiidae, Talitridae, Thaumatelsonidae, Vitjazianidae.

### Family Ampithoidae

*Ampithoe ramondi* Audouin. A species up to 2 cm long    63/5 with the two pairs of antennae of approximately the same length. The second pair of walking legs are larger and different in shape from the first pair. Its coloration varies depending on the seaweed to which it is attached. Feeds on algae and other organic material in the Mediterranean.

*Ampithoe rubricata* (Montagu). A species of variable    63/7 colour and about 2 cm long, which lives in a tube constructed from pieces of seaweed. Found amongst undergrowth on the shore and in shallow water in the North Atlantic.

### Family Aoridae

*Microdeutopus gryllotalpa* (Costa). A slender species up    63/8, 63/ to 1 cm long, which shows several of the family characteristics. The first pair of antennae are longer than the second pair. The last pair of walking legs are longer than the others. The first pair of walking legs of this species have a large claw. It lives in tubes amongst seaweeds in the North Atlantic. A male (Plate 63/8) and a female specimen with eggs (Plate 63/9) can be seen in the photographs.

### Family Calliopiidae

*Apherusa jurinei* (Milne Edwards). A slim-bodied species    63/10 the female of which rarely exceeds 5 mm in length. The colour is very variable in this species, which is found amongst intertidal seaweeds in the North-east Atlantic.

## Family Corophiidae

63/11   *Corophium volutator* (Pallas). The body, which is not compressed laterally as in most amphipods, may reach 1 cm in length and is dominated by a very large pair of second antennae. It occurs, often in large numbers, in U-shaped burrows in intertidal muddy sand in the North Atlantic.

## Family Eusiridae

63/6   *Meteusiroides* sp. Members of this genus are unusual amongst the amphipods in that they are planktonic. This specimen was taken in a haul from the tropical Atlantic.

## Family Gammaridae

63/2   *Elasmopus pocillimanus* (Bate). A typical curved gammarid with a body length of up to 1 cm. The first antennae are larger than the second and have a small branch halfway along; a characteristic feature of this family. It is often found in harbours living amongst the sea-lettuce, *Ulva lactuca*, and other intertidal algae in the Mediterranean and adjacent Atlantic.

63/12   *Gammarus locusta* (Linnaeus). Males of this species can reach lengths of 2 cm, although females are usually smaller. The second antennae are only slightly shorter than the first, and the posterior margins of the telson and last three abdominal segments bear small spines. Found under stones and seaweeds in the intertidal zone and in shallow water in the Mediterranean, North Atlantic, Arctic, and North Pacific.

63/13   *Marinogammarus marinus* (Leach). A species with a body length of up to 2 cm which is similar to *G. locusta*, differing mainly in the shorter length of the branches of the last abdominal appendages. Occurs in the Atlantic intertidally under stones and is often found in estuaries, since it is tolerant of low salinities. The female specimen in the photograph is brooding eggs.

64/1   *Melita palmata* (Montagu). Male individuals such as this may reach a length of 1 cm. Specimens, which are usually of a reddish brown hue, are found under stones on the shores of the North Atlantic.

## Family Hyalidae

64/2   *Hyale nilssoni* (Rathke). An amphipod with a body less than 1 cm long and bearing short first antennae. The second pair of thoracic legs are large with well developed claws. Commonly found amongst seaweeds and in crevices on rocky shores in the North Atlantic.

64/10   *Hyale schmidti* Heller. This species, seen on red algae in the photograph, is found amongst algae in clean shallow-water areas of the Mediterranean and adjacent Atlantic.

## Family Ischyroceridae

64/3   *Jassa falcata* (Montagu). Specimens are less than 1 cm in length and have the last segment of the first and second pairs of thoracic legs bent back to form a pincer. They live in tubes constructed of fine sediments and amongst kelp holdfasts and other seaweeds. This species has a cosmopolitan distribution.

64/4   *Parajassa pelagica* (Leach). A species whose body length is less than 1 cm, with characteristic tufts of bristles on the antennae. It is often found amongst seaweed undergrowth in shallow waters of the North-east Atlantic and Arctic.

## Family Leucothoidae

*Leucothoe spinicarpa* (Abildegaard). This species has a   64/5 body often over 1 cm long and the first and second antennae are of almost equal length. The first pair of thoracic legs have well developed pincers. Often found living in the cavities of sponges and tunicates or in kelp holdfasts. This species has a cosmopolitan distribution.

## Family Podoceridae

*Podocerus variegatus* Leach. This 5 mm long, spider-like   64/11 amphipod is one of the few forms which is not flattened from side to side. The large second antennae, characteristic of the genus, are clearly seen in the photograph. The first pair of thoracic legs have well developed claws, and the abdomen is usually folded under the body. Common in wave-exposed algae, where it moves around rather like a spider. Found in the North-east Atlantic and Mediterranean.

## Family Talitridae

*Orchestia gammarella* (Pallas) [sand-hopper]. A species   64/6, 64/7 with a body length of up to 2 cm, and the first antennae much shorter than the second (Plate 64/6). The second thoracic leg of the male terminates in a large pincer (Plate 64/7). The species is often found amongst decaying seaweed and stones on the upper and middle shore, and jumps strongly when disturbed. Both the male and the female (Plate 64/6) have strongly developed propulsive limbs. Found in the North Atlantic and Mediterranean.

## Suborder Ingolfiellidea

Very small (1–3 mm long) elongate worm-like forms living between sand grains. Only the first thoracic segment is fused with the head, which has no eyes. The abdominal appendages are greatly reduced.

Family: Ingolfiellidae.

## Suborder Caprellidea [skeleton shrimps]

The body is not compressed from side to side and the first two thoracic segments are fused with the head. The first pair of antennae are larger than the second and the thoracic limbs have large pincers or claws for grasping. The abdomen is very short and has no appendages in the female and only one or two reduced pairs in the male.

Family: Caprellidae.

## Family Caprellidae

*Caprella linearis* (Linnaeus) [ghost shrimp]. A slender-   65/2 bodied amphipod up to 2 cm long, which is often found attached to algae, hydroids or sea-mats by the hind legs. The body is held in a posture similar to that of a praying mantis and food is captured by the large claws of the first two pairs of thoracic limbs. The abdomen is reduced. The photograph shows a female carrying eggs in her brood-pouch. This species has a cosmopolitan distribution.

## Suborder Hyperiidea

This suborder contains transparent oceanic forms. The first thoracic segment is incorporated in the swollen head, which often has very large eyes. The plates at the bases of the thoracic limbs are small.

Families: Mimonectidae, Proscinidae, Scinidae, Lanceolidae, Vibiliidae, Cystisomatidae, Paraphronimidae, Hyperiidae, Phronimidae, Phrosinidae, Lycaeopsidae, Lycaeidae, Pronoidae, Oxycephalidae, Platyscelidae, Parascelidae.

### Family Scinidae

64/8   *Scina* sp. A common genus with a dorsoventrally flattened body usually less than 1 cm long. The head has long antennae and small eyes (an unusual feature amongst the hyperiids). Members of this genus, which contains many bioluminescent species, are found at all depths in the open ocean.

### Family Lanceolidae

64/9   *Megalanceola* sp. Members of this genus are large, reaching 2 cm in length. They have a short head with small oval eyes and a poorly developed rostrum. The first antennae are shorter and more robust than the second pair.

65/1   *Scypholanceola* sp. Individuals, which may reach a length of 6 cm, are blind and have a soft reflective external skeleton. This specimen was taken in a plankton haul from a depth of 1000 m.

### Family Vibiliidae

65/7   *Vibilia stebbingi* Behning & Woltereck. A species up to 1 cm in length, with basal segments of the first pair of antennae rigid and leaf-like. It has been captured from depths between 40 and 1000 m. It has a cosmopolitan distribution.

### Family Cystisomatidae

65/6   *Cystisoma* sp. A large characteristic hyperiid, which can reach a length of 12 cm. It is almost completely transparent, apart from some pigmentation in the large compound eyes and in the anterior part of the gut. Found in plankton hauls from 200–1000 m.

### Family Hyperiidae

65/9   *Parathemisto gaudichaudii* (Guérin). This species may reach 25 mm in length and is most frequently found in the top 100 m of colder northern and Antarctic waters.

65/8   *Pegohyperia princeps* Barnard. A very rare species 25 mm or more in length, with a metallic sheen on the external surface. This specimen was taken from a depth of about 1000 m and is unusual in that most species from this depth have a strong coloration.

### Family Phronimidae

65/4   *Phronima* sp. An oceanic hyperiid up to 4 cm long, with large compound eyes divided into upper and lower portions. The female attacks tunicates (particularly salps and *Pyrosoma*), or the swimming bells of siphonophores, and eats away the interior to provide a shelter for herself and her young. The photograph shows the head and limbs of a female facing out of a 'barrel house'. The white collar around the interior of the barrel is a collection of young larvae.

### Family Phrosinidae

65/5   *Primno macropa* Guérin. Members of this genus have a globular head with large eyes and short antennae. The clawed thoracic limbs are a prominent feature of this species, which may reach 1 cm in length. A cosmopolitan species (absent only from the Arctic) usually taken at depths of 200–500 m.

### Family Lycaeidae

*Lycaea bovallioides* Stephensen. The anterior part of the    65/3 body is globular in this species, as in most other members of the family. It is found in oceanic surface waters.

*Lycaea nasuta* Claus. A species similar to *L. bovallioides*    65/10 in form and behaviour. Both species are sometimes found living in salps. The specimen in the photograph is seen from above.

### Family Oxycephalidae

*Streetsia* sp. A characteristic oceanic genus with the head    66/3 prolonged into a pointed rostrum. Found most commonly in surface waters.

## Superorder Eucarida

This group contains some of the largest crustaceans, and includes the familiar shrimps, lobsters and crabs. The eyes are always borne on stalks and the jaws (mandibles) lack a movable tooth in the adult. The thoracic limbs flex between the fourth and fifth joints. They have a large carapace, which is fused to the dorsal side of all thoracic segments.

The female lacks a thoracic brood-pouch and the eggs are carried instead on the abdominal limbs. The life cycle has a series of larval stages.

Many eucarids are predators or scavengers, but filter-feeding occurs in some forms.

## Order Euphausiacea

Shrimp-like crustaceans which are flattened from side to side. The short carapace does not cover the gills and extends forward forming a roof over the head. There are large stalked eyes and long biramous antennae. The first thoracic limbs are not used as accessory mouth-parts. The first six pairs are long and fringed with hairs, and often are used for filter-feeding. In several species the second and/or third thoracic limbs are modified for grasping. Branched gills are situated at the bases of the thoracic limbs. The leaf-like abdominal appendages are fringed with long hairs and are used for swimming, the last pair forming a tail fan with the telson.

In the male the first two pairs of abdominal appendages are modified for transferring sperm to the female. Fertilised eggs may be liberated into the sea or may be retained between the thoracic limbs of the female. Light-producing organs are present in the eye stalks and along the body in both sexes, and appear to be of special significance during mating and spawning.

Euphausids are permanent swimmers living mainly in the open ocean, although a few are found along coasts. Filter-feeding is common, but there are a few predators. Some species aggregate in large swarms and form an important source of food for fish and some whales.

Families: Bentheuphausiidae, Euphausiidae.

### Family Euphausiidae

*Euphausia sanzoi* Torelli. An oceanic species which    66/1 reaches a length of 3 cm. The fringed hairs of the front thoracic limbs, which can be clearly seen in the photograph, are used for filter-feeding. The green coloration in the fore-gut is from phytoplankton recently ingested. Found in surface waters in the tropics.

*Thysanopoda* sp. Species in this genus regularly reach a    66/2 length of 8 cm or more when adult. The red coloration is typical of deep-living species.

## Order Decapoda

This predominantly marine order includes the largest crustaceans. The head and thorax are fused together dorsally and are covered by the carapace. The carapace extends down on either side of the body, enclosing the gills. There are usually several rows of gills on the thorax, forming a large surface area for respiration. The first three pairs of thoracic limbs are modified as mouthparts, whilst the remaining five pairs are leg-like. In some groups the abdomen is well developed with swimming appendages and a tail fan, whilst in others it is very much reduced.

In the male the first two pairs of abdominal appendages are rod-like and are used for transferring sperm to the female. The female usually carries the eggs on her abdominal appendages.

The decapods have invaded nearly every type of bottom habitat in the sea, from the shore to the depths of the ocean. There are also several permanently planktonic species. Most species are scavengers on plant and animal remains, but there are also predatory and filter-feeding forms. Some decapods are an important food source for man and play a major role in the fish-farming projects of several countries.

## Suborder Natantia [shrimps and prawns]

Swimming decapods with light external skeletons whose bodies frequently are flattened from side to side. There is often a well developed rostrum which may be toothed. The second antenna has a large flat lobe arising from its base. The thoracic legs are long and slender and the first three pairs may carry pincer-like claws. The well developed abdomen has a full complement of swimming appendages. The first abdominal segment is not markedly smaller than the rest.

## Infraorder Penaeidea

The first three pairs of thoracic legs are of more or less equal length and possess pincer-like claws. The thoracic gills are tree-like. The side plates of the second abdominal segment do not overlap those of the first. Fertilised eggs are shed directly into the sea in most forms.

Families: Penaeidae, Sergestidae.

### Family Penaeidae

66/5 *Funchalia villosa* (Bouvier). This large, 10 cm long, oceanic prawn is a predator, detecting its prey by means of pressure-sensors situated on its long antennae. The kink in the antenna is a natural feature. It is found in surface waters of the ocean at night but migrates into deeper water during the day.

66/4 *Parapenaeus longirostris* (Lucas). A deep water species, which can reach a length of 13 cm, often caught over sand and mud at depths of 200–500 m in the Atlantic and adjacent Mediterranean. Sometimes sold in the fish-markets of Spain and Portugal.

66/7 *Plesiopenaeus edwardsianus* (Johnson). This species of red prawn, which can reach a length of 30 cm, has conspicuous spines on the dorsal side of the abdomen. It is rarely found in less than 500 m of water in the Atlantic and adjacent Mediterranean. Occasionally sold in Mediterranean fish-markets.

### Family Sergestidae

*Sergestes armatus* (Kroyer). Members of this genus reach a length of 10 cm or more, and have long whip-like antennae. This species is found at depths of 250–500 m in the open ocean. The coloration is typical of prawns taken at these depths.                                                      66/6

## Infraorder Caridea

Only the first two pairs of thoracic legs have pincers. The thoracic gills have a central shaft with plate-like branches on either side. The side plates of the second abdominal segment overlap those of the first (Figure 25*a* and Plate 67/3). The abdomen flexes down strongly from the third segment in some species. Fertilised eggs are carried on the abdominal appendages of the female.

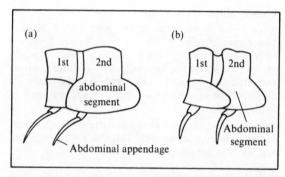

**Figure 25** *Overlapping side plates of the 1st and 2nd abdominal segment in* (a) *Caridea, and* (b) *Stenopodidea.*

Families: Procaridae, Oplophoridae, Nematocarcinidae, Stylodactylidae, Pasiphaeidae, Bresiliidae, Disciadidae, Eugonatonotidae, Rhynchocinetidae, Campylonotidae, Palaemonidae, Gnathophyllidae, Psalidopodidae, Alpheidae, Ogyrididae, Hippolytidae, Processidae, Pandalidae, Thalassocarididae, Physetocarididae, Glyphocrangonidae, Crangonidae.

### Family Oplophoridae

*Acanthephyra purpurea* Milne Edwards. An oceanic species reaching 15 cm or more in length, with a spiny rostrum and smooth dorsal carapace. It usually occurs at depths below 500 m, but has been taken in plankton hauls from depths of as little as 200 m at the entrance to the English Channel.                                                          66/8

*Notostomus gibbosus* Milne Edwards. This oceanic prawn is readily distinguished from *N. longirostris* by the short rostrum and high domed carapace.                                66/9

*Notostomus longirostris* Bate. The spines on the long rostrum of this 12 cm long species extend back on the dorsal side of the carapace (cf. *A. purpurea*). It is frequently encountered in plankton hauls from depths of over 700 m.                                                                     67/1

*Oplophorus spinosus* (Brullé). This species is characterised by the long backward-pointing spines on the dorsal side of the abdomen. The specimen in the photograph is brooding eggs between the abdominal appendages.        67/4

*Systellaspis debilis* Milne Edwards. A deep-water species reaching 8 cm in length, with a long spiny rostrum and a telson terminating in a point. The black spots on the sides of the body are light-producing organs. This specimen, taken from a depth of 600–700 m, is devouring a fish, *Cyclothone braueri*.                                               67/2

## Family Pasiphaeidae

67/5   *Parapasiphaea cristata* Smith. The reduced rostrum of this species, taken in a plankton haul from a depth of 800–1000 m, is characteristic of all species in this family. The red pigment concentrated in the external skeleton camouflages the prawns, since it does not show up in the blue-green daylight penetrating to these depths, nor in the blue-green light produced by the light organs of predators.

67/3   *Pasiphaea hoplocerca* Chace. The small pointed rostrum of the 10 cm long specimen in the photograph is clearly visible, as is the large side plate of the second abdominal segment which characteristically overlaps that of the first abdominal segment in all species in the infraorder Caridea. This specimen was taken in a plankton haul between 400 and 700 m.

## Family Rhynchocinetidae

67/11   *Rhynchocinetes rugulosus* Kubo [hinged-beak prawn]. The common name of this species derives from the large hinged rostrum which can be directed upwards at right angles to the carapace. This feature, together with the pronounced body and leg markings, is characteristic of the genus. Found on shallow reefs in the Indo-Pacific.

67/7   *Rhynchocinetes* n.sp. A group of large-eyed specimens on red colonial sea-squirts. This, as yet undescribed species, is found on reefs in the Indo-Pacific.

## Family Palaemonidae

67/6   *Leander tenuicornis* (Say). A stout-bodied species, up to 5 cm long, which is usually found in floating *Sargassum* weed. The colour of these animals matches that of the weed. Found in the North-west Atlantic.

67/9   *Palaemon elegans* Rathke (=*Leander squilla*). A semi-transparent prawn up to 5 cm long. The rostrum, which terminates in a point, has saw-like teeth along the upper and lower edges. Found in rock-pools and in rocky subtidal areas in the North-east Atlantic and Mediterranean.

67/12   *Palaemon serratus* (Pennant) (=*Leander serratus*) [common prawn]. The common prawn can grow to a length of up to 10 cm. The toothed rostrum curves upwards and ends in a double point. Occurs in rock-pools and in shallow rocky areas. This species is often collected for food. Found in the North-east Atlantic and Mediterranean.

The genus *Periclimenes* contains many species of small shrimps up to 2 cm long. They live in association with larger animals such as sea anemones, jellyfish, sea slugs and sea cucumbers. In return for the shelter provided by the host they often clean debris and parasites from its surface. Many are strikingly patterned, whilst others closely match the colours of the host species.

67/8   *Periclimenes amethysteus* (Risso). A species often found in association with the sea anemone *Aiptasia* in the Mediterranean. The photographed specimen is in an aggressive posture.

67/10   *Periclimenes brevicarpalis* Schenkel. This female specimen with a distinctive 'wart' on its head is seen living in association with the giant sea anemone, *Cryptodendrum adhesivum*, in the Red Sea.

68/1   *Periclimenes holthuisi* Bruce. A translucent species which is often found amongst the tentacles of the upside-down jellyfish *Cassiopeia*. Found in the Indo-Pacific.

*Periclimenes imperator* Bruce. This variably marked species is frequently found on sea cucumbers such as *Stichopus* and *Synapta* (Plate 68/3) and occasionally on the sea slug, *Hexabranchus sanguineus* (Plate 68/4). Found in the Indo-Pacific.   68/3, 68

*Periclimenes sagittifer* (Norman). A translucent Mediterranean species associated with sea anemones. The female specimen in the photograph is amongst the tentacles of *Condylactis aurantiaca*.   68/2

*Stegopontonia commensalis* Nobili. A species up to 2 cm in length which is adapted for living on the spines of the sea urchin, *Echinothrix diadema*. The larger of the two specimens in the photograph is a female. Found in the Indo-Pacific.   68/6

## Family Gnathophyllidae

*Drimo elegans* (Risso). A stout-bodied shrimp up to 3 cm long, with a characteristically spotted surface. Often found associated with starfish in the Mediterranean.   68/7

*Gnathophyllum americanum* Guérin [bumblebee shrimp]. The stout body of this 3 cm long species is characteristically patterned. The foreshortened head is also rather distinctive. Found associated with starfish and sea urchins in the Caribbean and Indo-Pacific.   68/9

## Family Alpheidae

*Alpheus macrocheles* (Hailstone) [snapping shrimp]. The body of this species reaches 3 cm or more in length, and has the carapace extending forward to a point over each eye. The claws of the first pair of walking legs are greatly enlarged. The snapping sound produced when the claws are suddenly closed is a sound often heard by divers underwater. Found on soft sediments usually below 30 m in the Mediterranean and adjacent Atlantic.   68/5

*Synalpheus* sp. [snapping shrimp]. Several species occur in this genus and can sometimes be differentiated from one another by colour. They live in the water canals of sponges, amongst anemones and in crevices. This specimen was photographed in the Caribbean.   68/10

## Family Hippolytidae

*Hippolyte inermis* Leach (=*H. prideauxiana*) [chameleon prawn]. This 3–4 cm long species is capable of changing its colour depending on the type of bottom on which it is resident. It commonly occurs in sea-grass meadows together with several other closely related species in the Mediterranean and North-east Atlantic.   68/8

*Lysmata seticaudata* Risso. A species up to 3 cm long characterised by the continuation of the rostral spines onto the carapace. It is sometimes found in groups in caves as well as amongst rocks in more open subtidal areas in the Mediterranean.   69/1

*Saron marmoratus* (Olivier). A species about 5 cm long, with characteristic bunches of bristles along its back. The legs are banded and the male is distinguished from the female, which often occurs with it, by its greatly elongated first pair of walking legs. Found on reefs and coral rubble in the Indo-Pacific.   68/11

*Saron* sp. Two unknown species from the Red Sea, one which occurs on corals (Plate 69/2) and the other a night-active sand-dwelling species (Plate 69/5).   69/2, 69/

*Thor amboinensis* (de Man). A small shrimp about   69/3

2 cm in length, usually found associated with giant anemones such as *Condylactis gigantea* and *Stoichactis helianthus*. Found in the Caribbean.

### Family Pandalidae

69/9 *Heterocarpus grimaldii* Edwards & Bouvier. A deep-water oceanic species showing the poorly developed first pair of pincered walking legs, a feature which is typical of the family. It is easily recognised as belonging to the infraorder Caridea since the side plate of the second abdominal segment overlaps that of the first.

69/4 *Parapandalus narval* (Fabricius). This bottom-dwelling night-active species is often located in groups in caves during the daytime. The animals in the photograph are facing into the prevailing current. Found in the Mediterranean and adjacent Atlantic.

69/8 *Parapandalus richardi* (Coutière). This 5 cm long species is found in the open ocean usually at depths between 250 and 750 m. The female in the photograph is carrying eggs between her abdominal limbs.

69/10 *Plesionika acanthonotus* (Smith). An oceanic species with a short rostrum which is smooth on the upper surface and toothed underneath. The photographed specimen is carrying eggs beneath the abdomen.

### Family Physetocarididae

69/7 *Physetocaris* sp. A beautifully patterned oceanic mid-water specimen with large leaf-like lobes at the base of the second antennae.

69/11 Many of the crustaceans living in the top few centimetres of the open ocean have a blue coloration. The prawn in the photograph is one such animal which is often associated with the floating *Velella*.

## Infraorder Stenopodidea

The third pair of thoracic legs are much longer than the first two pairs, although all three pairs bear pincers. The thoracic gills have a central shaft with unbranched filaments arising from it. The side plates of the first abdominal segment overlap those of the second (Figure 25*b*, p. 75). Fertilised eggs are carried on the abdominal appendages of the female.

Family: Stenopodidae.

### Family Stenopodidae

70/3 *Stenopus hispidus* (Olivier) [banded coral shrimp]. A beautifully patterned shrimp with a spiny surface, reaching 5 cm or more in length. It commonly feeds on the external parasites of fish. The shrimp waves its long white antennae to attract the fish to its cleaning station. The fish remains entirely passive during the cleaning operation, which is beneficial to both parties. Found on reefs in the Caribbean and Indo-Pacific.

69/6 *Stenopus spinosus* Risso. A cave-dwelling species of cleaner-shrimp which reaches 8 cm in length. As with other members of this genus, the body surface is covered with spines and the long antennae are white. Found in the Mediterranean.

## Suborder Reptantia [lobsters, hermit crabs and true crabs]

Walking decapods with a robust external skeleton whose body is frequently flattened from top to bottom.

Often they have no rostrum or flattened antennal lobes. The thoracic legs are well developed, with the first pair usually much larger than the others and commonly bearing heavy pincer-like claws which are used for catching food and for defence. The abdominal appendages are not primarily used for swimming and are often poorly developed. The first abdominal segment is smaller than the rest.

## Infraorder Palinura [spiny lobsters]

Robust forms with a spiny carapace which is either flattened from top to bottom with a keel at the sides, or is sub-cylindrical. The sides of the carapace are fused with the ventral plate in front of the mouth (Figure 26*a*). The first four pairs of walking legs usually end in claws not pincers. The thorax has many gills consisting of unbranched filaments. The long armoured abdomen, with well developed keels at the sides in some species, ends in a broad tail fan.

Families: Polychelidae, Palinuridae, Scyllaridae, Eryonidae.

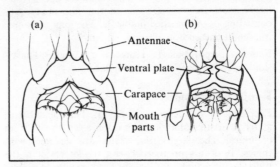

**Figure 26** *Ventral view of the anterior end of* (a) Panulirus *(Infraorder Palinura), and* (b) Nephrops *(Infraorder Astacura) showing the arrangement of ventral plate and carapace.*

### Family Palinuridae

*Palinurus elephas* (Fabricius) (= *P. vulgaris*) [crawfish or 70/1 spiny lobster]. A large species up to 50 cm in length, without pincers on the walking legs, except on the fifth leg in females. The body surface and the antennae are spiny. It is found usually below 20 m on rocky and sandy areas. During the winter months the adults migrate offshore to breed. Juveniles remain offshore for several years and do not come into shallower water until they are mature. This crustacean is prized for eating. Found in the North-east Atlantic and Mediterranean.

*Panulirus argus* (Latreille) [crawfish or spiny lobster]. 70/4 This species reaches a length of 50 cm or more, and has a few large spots on the upper surface of the abdomen. The young, in contrast to those of *Palinurus elephas*, are found in shallow water. A common species amongst reefs and rocks in the Caribbean. On occasions they migrate at night in long nose-to-tail lines for distances of several kilometres.

*Panulirus versicolor* Latreille [painted crayfish]. A strongly 70/6 patterned crawfish with a body length of 40 cm or more. The black and white stripes across the green abdomen are characteristic. Found on Indo-Pacific reefs.

### Family Scyllaridae

*Parribacus antarcticus* (Lund) [Antarctic slipper lobster]. 70/2 As with all members of this family, the knobbly carapace is flattened from top to bottom and the second pair of antennae are modified into shield-shaped structures. The body of this species may reach 20 cm in length. This nocturnal lobster is found on rocks and reefs in the Indo-Pacific and Caribbean.

70/8    *Scyllarides nodifer* (Stimpson) [slipper lobster]. This species, which reaches 15 cm in length, is found in sea-grass beds and on reefs in the Caribbean.

70/7    *Scyllarides* sp. This slipper lobster, which is about 30 cm long, was found in the rocky subtidal area of Sydney, Australia.

70/9    *Scyllarus arctus* (Linnaeus). A species rarely exceeding 15 cm in length that is found on rocky ground below about 10 m in the Mediterranean and adjacent Atlantic.

## Infraorder Astacura

Robust forms with a cylindrical carapace which is not fused with the ventral plate in front of the mouth (Figure 26*b*). The first three pairs of walking legs end in pincers, but the first pair of legs are much heavier and larger than the others. The thorax has many gills consisting of unbranched filaments. The long abdomen is straight and ends in a broad tail fan.

Families: Nephropidae (= Homaridae), Axiidae, Thalassinidae, Callianassidae, Laomediidae.

### Family Nephropidae

70/5    *Enoplometopus occidentalis* (Randall). This lobster with a characteristic coloration can reach a length of 20 cm or more. The large front claws are covered with spines. It occurs on the deeper parts of reefs in the Indo-Pacific where it forages for food, mainly at night.

71/3    *Homarus gammarus* (Linnaeus) [common lobster]. Large specimens of this big-clawed species may reach 50 cm in length. Unlike the crawfish, it is rarely found foraging in the open during the day but prefers to remain in crevices amongst rocks with only its antennae and claws protruding. It feeds on a variety of shellfish which are crushed by the powerful claws. This species, which occurs in the Mediterranean and North-east Atlantic, is highly prized as a food.

70/10    *Nephropsis atlantica* Norman. A deep-water oceanic species with a characteristically bristled dorsal surface and long lateral spines on the abdominal plates. The long rostrum bears lateral spines and the small eyes are not pigmented. Found in the Atlantic.

## Infraorder Anomura

A group of decapods of variable form which includes the squat lobsters, hermit and porcelain crabs. The carapace is not fused with the ventral plate in front of the mouth. The first pair of thoracic legs often end in large pincers, and the fifth pair (sometimes also the fourth) are smaller than the rest and often concealed by the carapace. The thoracic gills are variable in structure. The abdomen, which is well developed and flexed ventrally in some species, may be soft and asymmetrical. The last pair of abdominal appendages are usually present and may form a tail fan.

Families: Pylochelidae, Diogenidae, Coenobitidae, Lomisidae, Paguridae, Lithodidae, Galatheidae, Chirostylidae, Porcellanidae, Hippidae, Albuneidae.

### Family Diogenidae

*Dardanus lagopodes* (Forskål). The left claw is distinctly larger than the right in this reef-dwelling Indo-Pacific hermit crab.    71/1

*Dardanus megistos* (Herbst). A species which may reach a length of 20 cm and is characteristically marked all over its body. Like most other hermit crabs it is a scavenger but it will break open mollusc shells when the opportunity arises. Inhabits the deeper parts of Indo-Pacific reefs.    71/6

### Family Coenobitidae

All hermit crabs protect their soft abdomen within old sea shells or any other suitable hollow object. When threatened they withdraw completely within the shell and block the entrance with their front claws. Coenobitids are unusual amongst hermit crabs in that they are capable of living out of the sea as adults for varying lengths of time, some only needing to return to the sea to breed.

*Coenobita clypeatus* (Herbst) [land hermit crab]. This species with a carapace length of up to 3 cm is able to live permanently out of the sea as an adult. It retains moisture by living in burrows in the sand and by foraging mostly at night or in shaded areas. Common in the Caribbean.    71/2

*Coenobita jousseaumei* Bouvier. An Indo-Pacific species that lives high on the shore and burrows into sand during the day. It needs to return to the sea at least once a day to replenish the water within its protective shell.    71/4

### Family Paguridae

*Aniculus maximus* Edmondson. This highly coloured hermit crab has its front end covered with bristles which provide additional protection from predators. During the period required to transfer to a larger shell the soft abdomen of the crab is very vulnerable. Found on deeper parts of Indo-Pacific reefs.    71/5

*Paguritta harmsi* (Gordon) [coral hermit crab]. This small Indo-Pacific hermit crab inhabits old worm tubes in coral. It has feathery antennae which are held out in the water to filter out plankton. When retracted the claws completely block the entrance to the tube.    71/7

*Pagurus bernhardus* (Linnaeus). A species up to 10 cm long, commonly found inhabiting whelk shells when adult. Various animals live in association with the crab including the polychaete *Nereis fucata* which lives inside the whelk shell and protrudes its head from the entrance to pick up scraps left by the crab. The anemone, *Calliactis parasitica*, is often found clinging to the outside of the whelk shell. Occurs on sand and gravel bottoms from the intertidal zone downwards in the North-east Atlantic and Mediterranean.    71/8

*Pagurus calidus* Risso. A rare species up to 8 cm long which has spines and bristles on the walking legs. It normally occurs below 50 m but is found in caves at shallow depths in the Mediterranean.    72/1

*Petrochirus diogenes* Linnaeus. One of the largest species of hermit crab known, reaching a length of 30 cm or more. It frequently inhabits conch shells. The specimen in the photograph is breaking open a mollusc with its strong claws. Found on open sand and in sea-grass beds in the Caribbean.    72/5

## Family Galatheidae [squat lobsters]

72/3   *Galathea strigosa* (Linnaeus). The bright blue pattern on the back of this 10 cm-long species is very striking, particularly when seen underwater. The rostrum characteristically bears three pairs of laterally placed spines. It is found amongst seaweed on rocky bottoms in shallow water, in the North-east Atlantic and Mediterranean.

72/4   *Galathea* sp. An unknown species of squat lobster from deep Atlantic waters.

72/2   *Munida bamffica* (Pennant) (= *M. rugosa*). The first pair of walking legs bear pincers and are at least twice as long as the body, which can reach a length of 5 cm. As well as a central pointed rostrum there are characteristic spines over the eyes. Found on rocky and sandy bottoms, usually below 10 m in the North-east Atlantic and Mediterranean.

## Family Porcellanidae [porcelain crabs]

The porcelain crabs, although superficially similar to crabs, are more closely related to squat lobsters. The presence of long antennae, the small last pair of walking legs and the presence of a tail distinguishes them from true crabs.

72/10   *Neopetrolisthes ohshimai* Miyake. A small species with a carapace length of about 1 cm which lives amongst the tentacles of the giant anemone *Stoichactis kenti*. The complicated filter-feeding apparatus can be seen clearly in front of the head. Found in the Indo-Pacific.

72/8   *Pisidia longicornis* (Pennant) (= *Porcellana longicornis*). The long-clawed porcelain crab seen here in company with a sea-urchin, *Psammechinus miliaris*, has a body length of less than 1 cm. It is a filter-feeder that is found under stones and amongst seaweed on the shore and in shallow waters of the North-east Atlantic and Mediterranean.

## Family Hippidae

72/6   *Hippa* sp. [mole crab]. Members of this genus have an almost cylindrical body up to 2 cm long, with the first pair of walking legs lacking pincers. They burrow in sand and gravel in search of organic debris. Found in the Indo-Pacific.

# Infraorder Brachyura [true crabs]

A large and specialised group, with the body flattened from top to bottom. The carapace, which has a keel at its margin, is fused with the ventral plate in front of the mouth. The antennae are often very small. The first pair of thoracic legs are well developed with strong pincers. The other four pairs are variously developed, ending normally in a claw-like projection. The thoracic gills have plate-like branches on either side of a central shaft. The abdomen, which is small and tightly folded under the thorax, is not used in locomotion. The last pair of abdominal appendages are absent. In the male the first two pairs of abdominal appendages are used to transfer sperm to the female and the third to fifth pair of appendages are absent.

Families: Homolodromiidae, Dromiidae, Homolidae, Raninidae, Dorippidae, Calappidae, Leucosiidae, Majidae, Parthenopidae, Corystidae, Atelecyclidae, Thiidae, Cancridae, Pirimelidae, Portunidae, Xanthidae, Geryonidae, Goneplacidae, Pinnotheridae, Ocypodidae, Palicidae, Mictyridae, Grapsidae, Gecarcinidae, Hapalocarcinidae.

## Family Dromiidae

*Dromia personata* (Linnaeus) (= *D. vulgaris*) [sponge crab].   73/1 The 'furry' body and high domed back of this species are characteristic, as are the pink front claws. The body, which may be over 10 cm across, is often concealed by pieces of sponge, sea-squirts or even sea-weeds which are held over the carapace with its modified rear thoracic legs. Found amongst rocks in the intertidal zone and in shallow water in the Mediterranean and North-east Atlantic.

*Dromidia antillensis* Stimpson [lesser sponge crab]. A   72/7 species with similar shape and habits to *Dromia personata* but which rarely reaches more than 3 cm across the carapace. The photograph reveals the means by which the crab holds objects down on its back (in this case a sponge). The sponge has grown to conform to the outline of the crab carapace. Found in the Caribbean.

## Family Raninidae

*Ranina ranina* Linnaeus [spanner crab]. The carapace of   72/9 members of this family is elongated, but does not cover the abdomen. The common name of this crab derives from the pincer arrangement at the tip of the first pair of thoracic legs. The movable claw is bent at right angles to the flattened end of the immovable joint, rather like the end of an adjustable spanner. Found in coral sand around reefs in the Indo-Pacific, usually with only the front part of its body exposed.

## Family Calappidae [box crabs]

*Calappa flammea* (Herbst). The domed carapace of this   73/2 species reaches a width of over 10 cm. The common family name relates to the fact that the powerful front claws and the more slender rear limbs fit into depressions in the body surface, leaving a smooth outline similar to that of a box. This feature, together with their ability to bury themselves in sand, makes them a difficult target for predators. They break open and eat molluscs, particularly gastropods. Common in shallow waters of the Caribbean.

*Calappa* sp. Specimen with a more knobbly appearance   73/3 than *C. flammea* which was photographed on coral sand in the Indo-Pacific.

## Family Leucosiidae

*Ilia nucleus* (Herbst) [nut crab]. This species has a   73/4 characteristic nut-shaped body with the eyes and antennae close together on an anterior protruberance of the carapace. It is found amongst stones and in sea-grass beds in subtidal waters of the Mediterranean.

## Family Majidae [spider crabs]

*Achaeus cranchi* Leach. A species with a body of less than   73/5 1 cm in length, coated with hairs on the body and legs. It has a short rostrum which is bifurcated at the tip. The species lives amongst seaweeds and in sea-grass beds in shallow waters of the Mediterranean.

*Macropodia longirostris* (Fabricius). A delicately built   73/6 species with long thin limbs and a triangular body not more than 1.5 cm long. The rostrum is composed of two long parallel spines. It is found in a variety of subtidal habitats, where it feeds on small crustaceans as well as on hydroids and sea-mats, in the Mediterranean and adjacent Atlantic.

*Macropodia* sp. An unknown species photographed on   73/8 the soft coral, *Dendronephthya*, in the Red Sea.

73/7, 73/9   *Maja squinado* (Herbst) [spiny spider crab]. A large species with a spine-covered strawberry-shaped body up to 20 cm long (Plate 73/7). At certain times of the year large numbers of both sexes may aggregate together (Plate 73/9). Males copulate with recently moulted females before the breeding aggregations disperse. Young specimens decorate the carapace with seaweeds, sponges, and colonial animals. Found subtidally on rocks and sand in the North-east Atlantic and Mediterranean, where they graze on algae and colonial organisms.

73/10   *Mithrax spinosissimus* (Lamarck). An oval-bodied crab up to 20 cm long. The body and legs all bear spiny protuberances, those of the claws being in a single row along the outer edge. A night-active species occurring in shallow waters of the Caribbean.

74/1   *Pisa tetraodon* (Pennant). The triangular hairy carapace of this species is rounded at the rear end and about 3 cm long. The three or four spines on the edge of the carapace distinguish this species from others in the genus. This specimen is heavily decorated with pieces of sponge. Occurs in subtidal rocky areas of the Mediterranean.

74/3   *Stenorhynchus seticornis* (Herbst) [arrow crab]. The carapace of this species rarely exceeds 1 cm in width, but has a characteristic long spiny rostrum and very long thin legs. It moves slowly around on undergrowth feeding on crustaceans and small colonial animals. Found in the tropical Atlantic and Caribbean.

## Family Cancridae

74/2, 74/4   *Cancer pagurus* Linnaeus [edible crab]. A distinctive crab with a pie-crust edge to the carapace, which reaches a width of 25 cm or more. The claws of the first pair of thoracic legs are massive (Plate 74/2), the larger of the two being used to break open mollusc shells. The other claw has sharper teeth inside the pincer, and is used more usually for tearing and cutting soft tissue. Small specimens are found under stones on the shore, but larger animals occur in rocky subtidal areas, where, like the lobster, they take up more or less permanent residence in a particular rock crevice (Plate 74/4). This species is highly prized as a food. Occurs in the North-east Atlantic and Mediterranean.

74/6   *Cancer productus* Randall [red crab]. The carapace width of this species rarely exceeds 15 cm and its claws are not so massive as those of *C. pagurus*. It occurs commonly in both intertidal and shallow subtidal areas of the North-east Pacific, where it can find sufficient cover.

## Family Portunidae

74/7   *Callinectes sapidus* Rathbun [blue crab]. A large omnivorous species up to 20 cm across the carapace which, because of its size when adult, cannot be readily confused with other species in this genus. It is found on muddy sand bottoms in shallow water along the eastern seaboard of North America, where it is fished extensively for food.

74/8   *Carcinus maenas* (Linnaeus) [common shore crab]. A species with an oval to diamond-shaped carapace up to 8 cm across, which bears three blunt protruberances between the eyes and five protruberances on each side. The colour of the body varies from light brown to dark green. Common on rocky and sandy shores as well as below the tide line in the North Atlantic and Mediterranean.

74/5   *Lissocarcinus orbicularis* Dana. The striking pattern of this 3–4 cm wide Indo-Pacific species may occasionally be reversed to white on a brown background. The fifth pair of legs are not broadly flattened as in most portunids. It lives on the body surface of sea-cucumbers.

*Macropipus corrugatus* (Pennant) (= *Portunus corrugatus*). This species is characteristically red with a wrinkled carapace up to 5 cm across. The last pair of legs are flattened and have a low dorsal keel. Found amongst rocks in shallow waters of the Mediterranean and North-east Atlantic.   75/5

*Macropipus depurator* (Linnaeus). On occasions this species can be confused with *Carcinus maenas* because of the similarity of the carapace. However, *Macropipus* always has the last joint of the fifth leg flattened to aid with swimming. The specimen in the photograph is eating a lugworm. Usually occurs on soft sediments below tide level in the North-east Atlantic and Mediterranean.   75/7

*Macropipus puber* (Linnaeus) [velvet swimming crab]. The red eyes, velvety upper surface to the carapace and distinctive black pattern of lines on the limbs distinguishes this species from others in the genus. It is a large form whose carapace may reach 8 cm across and has up to ten small teeth between the eyes. It is omnivorous and defends itself ferociously when attacked. Found amongst seaweeds and in crevices on rocky bottoms in shallow water in the North-east Atlantic.   75/6

## Family Xanthidae

*Carpilius corallinus* (Herbst). A large crab which may reach 15 cm across the characteristically marked smooth carapace. The specimen in the photograph has lost two of its limbs, probably as a result of attack by a predator. This crab is greatly prized as a food by man. It occurs on reefs and amongst coral rubble in shallow waters of the Caribbean.   75/1

*Eriphia sebana* (Shaw & Nodder). The bulky body of this aggressive species reaches about 5 cm across. The red eyes are characteristic of this species from Indo-Pacific reefs.   74/9

*Etisus splendidus* Rathbun. This Indo-Pacific species has an irregularly toothed edge to the carapace and first pair of legs. The blunt pincers are used for breaking open mollusc shells.   75/8

*Lophozozymus pictor* (Fabricius). A distinctive Indo-Pacific species of reef-dwelling crab whose carapace rarely exceeds a width of 10 cm.   75/2

*Pilumnus hirtellus* (Linnaeus) [hairy crab]. A crab about 2 cm across the carapace; it is difficult to confuse with others because of the long bristles covering its body and limbs. Found under stones and in crevices on the shore and amongst seaweed in rocky subtidal areas of the North-east Atlantic and Mediterranean.   75/3

*Xantho poressa* (Olivi). The carapace of this xanthid is about 2 cm across at its widest point and tapers sharply to the rear. Its colour is very variable and may be yellow, brown, green or blue. Found on sand or amongst rocks in shallow waters of the Mediterranean.   75/4

## Family Pinnotheridae

*Pinnotheres pinnotheres* (Linnaeus) [pea crab]. The round membranous carapace of this species is up to 1.8 cm across. At a certain stage of their life cycle, adult females live within the shells of mussels (and other bivalves), feeding on the organic debris filtered from the water by the host. The smaller males are said to pass from mussel to mussel in order to fertilise the females. The red   76/1 coloration seen through the carapace of the female is that of the gonad, sometimes infected by a small crustacean parasite, *Pinnotherion vermiforme*. Found in the North-east Atlantic and Mediterranean.

### Family Ocypodidae

Members of the genus, *Ocypode* [ghost crabs], live in deep burrows above the water line. At low tide, particularly at night, they leave their burrows and scavenge for organic debris on the shore. They enter the water only to moisten the gills or when disturbed.

76/2 *Ocypode ceratopthalma* (Pallas). This Indo-Pacific species, with a carapace width of 4–5 cm, is easily recognised by the fact that the eyestalks project well beyond the eyes. The eyes of all ocypodids surround the eyestalk and are able to view through 360° horizontally.

76/4 *Ocypode cordimana* Desmarest. A smaller species than *O. ceratopthalma* whose eyestalks do not project beyond the eyes. The specimens in the photograph are running with the typical sideways action along the water's edge in search of organic debris deposited by the water. Found in the Indo-Pacific.

76/5 *Ocypode quadrata* (Fabricius). The species name of this animal alludes to the square carapace which reaches about 5 cm across. The large sockets in which the eyes are laid for protection are clearly visible at the front of the carapace of the photographed specimen. Found on Caribbean beaches.

76/3 *Ocypode saratan* (Forskål). In the mating season, the male of this species builds special pyramids of sand up to 15 cm high, and about 50 cm from its burrow, to attract females within the area of its burrow. The female follows a path of compacted sand towards the burrow of the male who entices her inside by waving his large yellowish-pink claw. Found in the Indo-Pacific.

*Uca* sp. [fiddler crab]. The species of *Uca* are very    76/6
difficult to distinguish. They build burrows in sheltered muddy areas both above and below the high-tide line. They scavenge for food in much the same way as *Ocypode*. The typically large-clawed male in the photograph is surrounded by balls of mud removed when making its burrow. Occurs in the Indo-Pacific.

### Family Grapsidae

*Grapsus albolineatus* Lamarck [swift-footed crab]. Fast-    76/7
moving greenish coloured crabs with a carapace up to 5 cm across. They live in crevices on Indo-Pacific rocky shores just above the surf line and forage for organic debris amongst the surf, using their spine-tipped legs for hanging on.

*Percnon planissimum* (Herbst). The 3-cm wide body and    76/8
the limbs of this crab are much flattened. The body shape enables them to live in narrow crevices on rocky shores and to flatten themselves against the rock when washed by surf whilst searching for food. Occurs in the Indo-Pacific.

### Family Gecarcinidae

*Gecarcinus ruricola* (Linnaeus). The characteristically    76/9
black-topped carapace of this omnivorous species is domed and has sides that converge posteriorly. The first pair of walking legs are enlarged but are not massive. This Caribbean species and its relatives live in deep burrows well above the high tide level and do not necessarily forage on the shore. They only need to return to the sea to reproduce, and are able to use freshwater as well as seawater to keep their gills moist.

# Phylum Chelicerata

approx. 31 000 species

The King crabs, the sea spiders and a few mites are the marine representatives of this large phylum, which is dominated by the land-dwelling spiders and the economically important mites and ticks.

The body is divided into two regions, an anterior prosoma, which bears the mouth parts and locomotory appendages, and a posterior opisthosoma, which often has reduced or modified appendages. There are no antennae, and the only pair of appendages situated in front of the mouth (the chelicerae) have become modified for feeding. The second pair of appendages (the pedipalps), which are located behind the mouth, are used for a variety of functions in the different chelicerate groups. Four pairs (sometimes more) of swimming or walking legs are usually present.

## Class Merostomata 4 species

Marine chelicerates which have a long tail spine at the end of the body, and five (or six) of the opisthosomal appendages modified as gills.

## Order Xiphosura [King or horseshoe crabs]

King crabs, which can reach a length of 60 cm, have a heavily armoured body showing little sign of segmentation when viewed from above. The prosoma is covered by a horseshoe-shaped carapace and is separated by a hinge from the opisthosoma. Two compound eyes, situated laterally, and one pair of median simple eyes are present. The chelicerae are small, and the remaining appendages, including the pedipalps, are used for walking or swimming. Four pairs of these limbs bear pincer-like claws, but the fifth pair has leaf-like processes which sweep away silt during burrowing. The mouth is surrounded by the spiny bases of limbs, which help to crush the food. Short spines are situated on the edge of the opisthosoma. Its long mobile tail spine aids forward movement and can be used for righting the animal should it get turned over. The opisthosoma has six pairs of appendages, the first pair forming a protective flap over the two reproductive openings, while the other five pairs with their leaf-like folds, function as gills.

Plate 77

Phylum Chelicerata

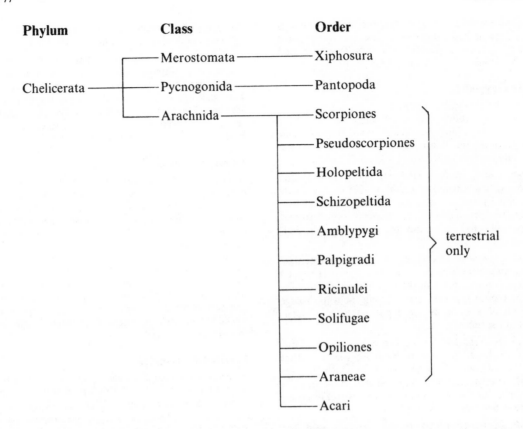

| Phylum | Class | Order |
|--------|-------|-------|
| Chelicerata | Merostomata | Xiphosura |
| | Pycnogonida | Pantopoda |
| | Arachnida | Scorpiones |
| | | Pseudoscorpiones |
| | | Holopeltida |
| | | Schizopeltida |
| | | Amblypygi |
| | | Palpigradi |
| | | Ricinulei |
| | | Solifugae |
| | | Opiliones |
| | | Araneae |
| | | Acari |

terrestrial only

Classification as used by Webb, Wallwork & Elgood (1978) supplemented by Hedgpeth (1954) for the Pycnogonida.

The sexes are separate and the smaller males have their pedipalps modified for grasping the female. They move into the intertidal zone to breed, and eggs are fertilised by the male as they are deposited in the sand by the female. A larva emerges from the egg and undergoes a series of moults before the adult form is reached.

King crabs are bottom dwellers, living in sandy and muddy regions of coastal waters. One species occurs along the North American Atlantic coastline whilst the other three are found in the West Pacific.

They are omnivores, feeding mainly on worms, molluscs and algae. Food is picked up by the chelicerae and passed to the mouth where it is crushed by the limb bases. These few species are the sole surviving members of a larger group which flourished some 200 million years ago.

Family: Limulidae.

### Family Limulidae

77/1   *Limulus polyphemus* (Linnaeus). This species, which may reach a length of 50 cm, is found commonly in the shallow subtidal regions of the North-west Atlantic from Nova Scotia to Florida.

## Class Pycnogonida [sea-spiders]
approx. 600 species

Pycnogonids are exclusively marine spider-like animals, usually only a few millimetres long, although some of the deep-sea forms may have a leg span of over 50 cm. The body is long and narrow with the opisthosoma reduced to a small stump. The prosoma is divided into a 'head' and a segmented trunk region. The head bears a cylindrical proboscis, which is sometimes as long as the body, and two pairs of eyes situated on a rounded protruberance. Three pairs of appendages, chelicerae, pedipalps and special egg-carrying legs, are often present in the head region. The trunk has from four to six pairs of walking legs, very long in some species, which end in claws. Pycnogonids are mainly crawling animals, but some are able to swim by flapping movements of the legs.

The sexes are separate, and the male fertilises the eggs as they are laid by the female. The eggs, which normally are carried by the male, hatch into larvae which moult several times before the adult form is reached.

Pycnogonids are cosmopolitan, although they are apparently more abundant in colder waters. They are mainly bottom-dwellers, ranging from the intertidal zone to depths of over 6000 m. Many species feed on attached coelenterates and bryozoans. The young of a few species attach to planktonic medusae.

## Order Pantopoda

Characters similar to those of the class.

Families: Nymphonidae, Ammotheidae, Tanystylidae, Phoxichilidiidae, Endeidae, Pycnogonidae, Pallenidae, Colossendeidae.

### Family Ammotheidae

*Achelia* sp. Members of this genus, which are widely distributed in shallow waters, have both chelicerae and pedipalps present. This 0.3 mm long specimen was located amongst hydroids and bryozoans overgrowing algae and sea-grasses in the Mediterranean.   77/2

### Family Endeidae

77/4 *Endeis pauciporosa* Stock. The slender body and lack of chelicerae and pedipalps is typical of the family. This Indo-Pacific species is well camouflaged and difficult to see when moving slowly amongst undergrowth in shallow water. The photograph shows a male carrying eggs, and a female of the same species.

### Family Colossendeidae

77/3 *Colossendeis colossea* Wilson. The lack of chelicerae and the slender body of this species are characteristic of the family. This gigantic deep-water species has a leg span which occasionally reaches 50 cm. The photograph taken from a submersible shows a specimen patrolling across the muddy bottom at 1900 m depth off the Atlantic coast of North America. The species has a world-wide distribution.

77/5 *Colossendeis* sp. The long proboscis and pedipalps typical of the family can be seen clearly in the photograph. Members of this genus often have pairs of limbs additional to the four large pairs normally present. These are used for egg carrying.

### Class Arachnida

approx. 30 000 species

A large group of mainly terrestrial animals with the only truly marine species belonging to the order Acari (mites and ticks).

In the arachnids the prosoma is undivided and covered dorsally by a solid carapace. The opisthosoma is usually well developed and clearly segmented, although it is fused with the prosoma in some forms. Six pairs of appendages are situated on the prosoma: the chelicerae and pedipalps which may perform a variety of functions including seizing of prey, and four pairs of walking legs. Several kinds of respiratory structures occur on the opisthosoma but gills are never present.

The sexes are usually separate and the transfer of sperm to the female by the male is usually accompanied by elaborate courtship behaviour. The fertilised egg frequently develops directly into a miniature adult, but a 'larval stage' with fewer appendages than the adult sometimes occurs.

### Order Acari [mites and ticks]

A large group of arachnids, which contains marine mites as well as many terrestrial and parasitic forms. The round, oval or flattened body usually lacks divisions of any kind, the prosoma and opisthosoma being completely fused together. The chelicerae and pedipalps, which are often adapted for piercing and sucking, are carried on a projection at the anterior end of the body.

Body form is very variable in members of this order which have colonised many different habitats.

The body of marine species is typically flattened, occasionally circular in cross-section, with a length ranging from 0.2 mm to 2 mm. Eyes are usually present. They do not swim, the four pairs of legs, which are often long with sensory hairs or bristles, being used for crawling (Figure 27).

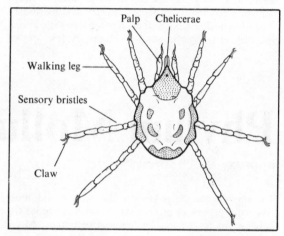

**Figure 27** *External features of a typical marine halacarid.*

In marine representatives the sexes are separate and the fertilised egg hatches into a larva with three pairs of legs, which, after several moults, develops into an adult.

They occur intertidally and in deeper waters of all seas where they can be found crawling over algae and other attached organisms. The majority of marine species belong to a single family, the Halacaridae, which contains both algal-feeding and predatory forms.

Main marine family: Halacaridae.

# Phylum Tardigrada

[water-bears] approx. 400 species

| Phylum | Order |
|--------|-------|
| Tardigrada | Heterotardigrada |
| | Mesotardigrada (freshwater only) |
| | Eutardigrada |

Classification follows Morgan & King (1976). N.B. The tardigrades have been tentatively placed by Manton (1977) in the phylum Uniramia (see p. 62).

A group of minute animals less than 1 mm long which live mainly in the water films surrounding mosses, liverworts and lichens. Only about 25 species are marine. The body is short and plump with no well-defined head. Eye spots and sensory appendages are frequently present. The skin is thick and often forms a series of plates on the dorsal surface giving the animal a segmented appearance. The skin is moulted at intervals to allow growth to occur. There are four pairs of short stubby legs, ending in claws or 'toes' with adhesive discs, that are used for creeping over the substratum (Figure 28).

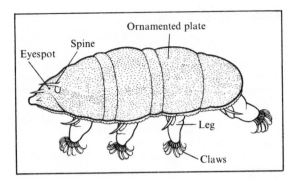

**Figure 28** *Generalised marine tardigrade.*

The sexes are separate and since males are normally scarce it is likely that the egg develops directly into a miniature adult without being fertilised in many cases.

Marine tardigrades dwell mainly between sand grains and amongst algae in shallow waters. A few species live in close association with other animals such as mussels, barnacles and the wood-boring isopods. The majority feed on the fluid contents of plant cells which are pierced by sharp mouth spines. A few seem to be carnivorous.

## Order Heterotardigrada

Tardigrades with sensory appendages at the sides of the head region.

Families: Batillipedidae, Oreellidae, Stygarctidae, Halechiniscidae.

## Order Eutardigrada

Tardigrades with no sensory appendages at the sides of the head region. Only very few marine species occur within this group.

Family: Macrobiotidae.

# Phylum Mollusca

## approx. 75 000 species

The phylum, which is one of the largest and most successful groups of invertebrate animals, includes the familiar snails, mussels, oysters and octopuses. Many molluscs are found in the sea, but they have also invaded freshwaters and the land.

Although the body form is very variable in the seven classes which make up the group, there are certain features which are common to nearly all molluscs. Apart from a primitive deep-sea form, *Neopilina*, molluscs show little evidence of segmentation. The body, which is basically bilaterally symmetrical, usually has a head, a muscular foot and a hump of tissue (visceral mass) which contains the gut and other organs. A sheet of skin, the mantle, extends from the dorsal body wall to cover part or all of the body. Between the mantle and the visceral mass is a cavity, the mantle cavity, which houses the gills (when present), the anus and openings of the excretory and reproductive organs. In most molluscs the mantle secretes a protective calcareous shell into which the body can retract, but in others the shell may be internal or absent altogether.

Sexes are often separate, but hermaphrodite forms also occur. Eggs and sperm may be shed into the water where fertilisation takes place, or copulation may occur. Planktonic larvae are frequently present in the life cycle.

Adult molluscs are mainly bottom-dwellers but planktonic and actively swimming forms also occur. Molluscs are particularly abundant in the intertidal zone and in shallow waters, where many feed by scraping bacterial, fungal and algal films off the bottom by means of a toothed tongue-like organ, the radula, a structure unique to the phylum. Carnivorous forms often use the radula for attacking their prey. Many species lacking a radula are filter-feeders and strain food particles from the water or from the surface of the sediment in which they live.

Molluscs are an important food source for fish, aquatic birds and mammals including man. Shellfish such as oysters, mussels and scallops are farmed in many areas of the world. Shell-collecting by both amateurs and professionals occurs on a large scale and certain shells are used for making jewellery and buttons.

## Class Monoplacophora
approx. 10 species

The few animals in this class are small, with conical limpet-like shells up to 4 cm in diameter. The body has an apparently segmented arrangement of gills, excretory organs, muscles and nerves. There is a definite head which, however, lacks eyes and tentacles. The mouth, which is surrounded by sensory structures, lies in front of the broad foot which dominates the undersurface of the body. A median terminal anus is present. A groove containing the gills separates the edge of the foot from the mantle which encircles the body at the edge of the shell (Figure 29).

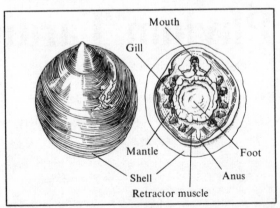

**Figure 29** *Dorsal and ventral view of the monoplacophoran,* Neopilina. [*After Lemche & Wingstrand.*]

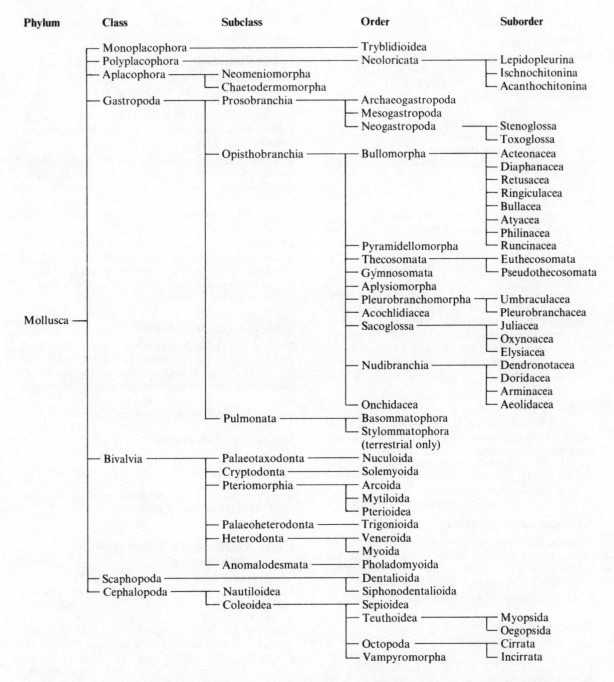

| Phylum | Class | Subclass | Order | Suborder |
|--------|-------|----------|-------|----------|

No generally acceptable overall classification of the Mollusca exists at the moment. The Monoplacophora and Polyplacophora are classified according to Moore (1960). The Aplacophora have been arranged in the manner suggested by A. Scheltema (personal communication). The classification of the prosobranch and pulmonate Gastropoda follows Taylor & Sohl (1962) and opisthobranch Gastropoda, Thompson (1976). The Bivalvia are arranged as in Moore (1969) and the Scaphopoda according to Palmer (1974). The classification of Voss (1977) has been adopted for the Cephalopoda.

The sexes are separate, and fertilisation of eggs by sperm probably takes place in the surrounding water. The species so far described in this class belong to the single genus *Neopilina*, the first specimens of which were discovered in dredgings from deep sea trenches off the Pacific coast of Costa Rica in 1952. Since then other species have been dredged from trenches in other parts of the Pacific and in the Atlantic and Indian oceans. They apparently feed on organic matter in the bottom mud.

## Order Tryblidioidea

Characters identical to those of the class.

Family: Tryblidiidae.

# Class Polyplacophora [chitons]
approx. 600 species

An entirely marine group of molluscs which range from a few millimetres to 35 cm in length. The oval flattened body is covered by a shell consisting of eight overlapping transverse plates. The plates are encircled by, and partially embedded in, an area of mantle called the girdle, which usually bears spines or scales. The ventral surface of the body is dominated by a broad muscular foot which is separated by a groove from the head and mantle. A variable number of gills are situated in this groove. The head, which lacks eyes and tentacles, bears a central mouth, whilst the anus is situated in the midline just behind the foot.

In almost all chitons the sexes are separate and eggs and sperm are usually shed into the water. The fertilised egg develops into a larva which leads a short planktonic life before changing directly into an adult. In a few species the young are brooded between the gills of the parent.

Chitons are cosmopolitan and are commonly found clinging to rocks in intertidal and shallow water areas, although a few species have been found at great depths. Many remain concealed under rocks or in crevices during the day and emerge to feed at night. They are mainly plant-feeders, scraping algae from rocks with a well developed radula.

## Order Neoloricata

The shell plates in this order, which contains all living chitons, are arranged like overlapping tiles on a roof. Each plate (apart from the first) has two processes extending from it which fit beneath the plate in front. The outer edges of the plates commonly have teeth that aid attachment of the plates to the mantle.

## Suborder Lepidopleurina

A group of fairly primitive chitons which lack toothed outer edges to the shell plates. The mantle does not extend over the shell. There are only a few pairs of gills in the mantle groove and these are near the anus.

Families: Lepidopleuridae, Hanleyidae, Choriplacidae.

### Family Lepidopleuridae

77/6    *Lepidopleurus cajetanus* (Poli). The concentric ridges on the outer edges of the plates of this 3 cm long species enable it to be readily identified. Found on rocks and shells in subtidal waters of the Mediterranean.

## Suborder Ischnochitonina

Small to quite large chitons whose shell plates always have toothed outer edges. The mantle, which is very variable in width and ornamentation, does not extend over the shell. Gills occupy most of the mantle groove but they are not usually situated close to the anus.

Families: Subterenochitonidae, Ischnochitonidae, Schizoplacidae, Callochitonidae, Callistoplacidae, Chaetopleuridae, Mopaliidae, Schizochitonidae, Chitonidae.

### Family Ischnochitonidae

77/9    *Tonicella lineata* (Wood). This species, which reaches a length of 5 cm, often has a pattern of dark brown lines on the plates and white spots on the surrounding smooth girdle. It is found in the intertidal zone, where it browses particularly on coralline algae such as *Lithothamnion*. Occurs in the North-east Pacific.

### Family Chitonidae

*Acanthopleura gemmata* Blainville (= *Acanthozostera gemmata*). A large chiton that may reach a length of 15 cm. The plates are surrounded by a broad girdle covered with spines. It is an intertidal species which hides in shaded crevices when exposed during the day. Found in the Indo-Pacific.    77/7

*Acanthopleura haddoni* Winckworth. A smaller species than the above, rarely reaching more than 5 cm in length. The characteristically notched plates are surrounded by a spiny girdle. Like many species of chiton it tends to return to the same position on the rock after feeding at high tide. Found in the Indo-Pacific.    77/8

*Chiton olivaceus* Spengler. This characteristically patterned species, which reaches a length of 4 cm has its plates surrounded by a narrow girdle covered with scales. Found on rocks and shells on the shore and in shallow waters of the Mediterranean.    77/10

## Suborder Acanthochitonina

Contains some of the largest known chitons, which have well developed teeth at the edges of the shell plates. The shell is covered or partially covered by the mantle. A limited number of gills occupy only part of the mantle groove.

Family: Acanthochitonidae.

### Family Acanthochitonidae

*Acanthochitona communis* (Risso). A distinctive species up to 5 cm long in which the girdle partly obscures the plates. Tufts of bristles are embedded in the girdle between adjacent plates. Found on hard bottoms below tide level in the Mediterranean.    78/1

# Class Aplacophora [solenogasters]
approx. 250 species

An exclusively marine group of small worm-like molluscs. The mantle may completely enclose the body, which is not divided into specific regions. There is no shell but the mantle surface is covered with a cuticle in which calcareous spicules are embedded, often giving the animal a silvery appearance. The foot is either reduced or absent altogether. Surrounding and partly projecting over the slit-like mouth is an area free of spicules (oral shield) and at the posterior end of the body the anus opens into a cavity which often contains gills. The majority of solenogasters are hermaphrodite but some species have separate sexes. Little is known of the breeding habits, but in some species fertilisation is internal and the young are retained in the body of the parent. In others the fertilised egg develops into a free-swimming larva.

These bottom-living animals have a world-wide distribution and are found from shallow waters to depths of 4000 m. Some species live in burrows in soft sediments, where they feed on organic debris and microscopic organisms. Others live amongst algae, hydroids or corals, on which they apparently feed.

## Subclass Neomeniomorpha

Members of this subclass possess a median longitudinal

groove on the ventral surface which contains the reduced foot. Gills may be present in the anal cavity. They are hermaphrodite forms which live mainly amongst algae, hydroids and corals.

Families: Lepidomeniidae, Neomeniidae, Proneomeniidae, Gymnomeniidae.

## Subclass Chaetodermomorpha

The ventral groove and foot are absent, but there is a pair of gills in the anal cavity. The sexes are separate and they live in vertical burrows in soft sediments with the posterior end (and gills) situated at the entrance to the burrow.

Families: Limifossoridae, Prochaetodermatidae, Chaetodermatidae.

### Family Chaetodermatidae

78/2 *Falcidens gutterosus* Salvini-Plawen. These small atypical worm-like molluscs may reach a length of 15 mm or more. Backward-pointing calcareous spicules are embedded in the skin, giving the body of the specimen in the photograph a silvery appearance at all but the rear end. Found burrowing in mud, usually at depths below 40 m, in the Mediterranean.

## Class Gastropoda

approx. 64 500 species

This is by far the largest and most diverse class of the Mollusca and it contains marine, freshwater and land-dwelling forms. There are also some parasitic species.

The body is asymmetrical, having a well defined head with eyes, tentacles and a radula. There is a prominent ventral foot, mantle and visceral mass. The shell, when present, is in one piece and is spirally coiled at least in the young stages. The body can be retracted into the shell. The shell, however, may be internal or absent altogether in some species. The large visceral mass, which contains most of the gut, the heart, excretory and reproductive organs, is normally covered by the mantle, and is often permanently contained in the larger coils of the shell. In gastropods, the visceral mass and mantle have rotated 180° on the foot so that the mantle cavity containing the gills and anus has become situated anteriorly above and behind the head. As a result of this rotation (torsion) the gut and nervous system have become twisted into a U-shape (Figure 30). Torsion has occurred independently of shell spiralling. All living gastropods at some stage of their development, commonly in the larval stage, undergo torsion. In a number of species a partial reversal of the torsion process has occurred (detorsion) and in these animals the mantle cavity and shell are reduced or absent. The mantle often forms a tube, the siphon, through which a current of water is drawn into the mantle cavity. In some gastropods another siphon carrying the outgoing water current may be present.

The ventral foot is usually a flat creeping sole, but this is often modified in swimming and burrowing forms. Primitive gastropods possess a pair of gills in the mantle cavity, each consisting of a series of leaf-like plates on a central shaft. In more advanced forms there is a tendency for one or both gills to be lost. In the latter case, gas exchange either occurs through the body surface or through specially developed respiratory structures.

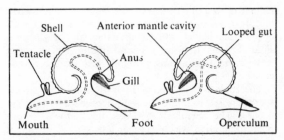

**Figure 30** *A hypothetical pre-torsion gastropod* (left), *and one which has undergone torsion* (right) *to produce an anterior mantle cavity.*

Both species with separate sexes and hermaphrodite forms occur within the class. Eggs and sperm may be shed into the sea, but more frequently copulation, with resultant internal fertilisation, takes place. The eggs, which are then laid in gelatinous masses, or in capsules, may hatch into a fairly advanced type of free-swimming larva or into a small replica of the adult.

Gastropods have a world-wide distribution and are found from the intertidal zone to abyssal depths. They have become adapted to living on all types of sea bottom, but several swimming and floating species also occur. Feeding habits are very diverse; some forms rasp algae from rocks with the radula, others are scavengers, particle feeders or carnivores.

Gastropods are important economically in that many are food for fish and for man. Some species, e.g. the predatory muricids, can cause extensive damage to oyster beds, whilst others act as hosts for parasites of fish and shore-birds.

## Subclass Prosobranchia

Gastropods which show pronounced torsion, with the mantle cavity, gills and anus situated in front of the visceral mass. The shells of most species are coiled in a spiral. The head has a single pair of tentacles and usually has a mobile projection, the snout, which bears the mouth anteriorly. A pair of eyes are situated close to the tentacles and a well-developed radula is present. In many prosobranchs the foot carries a rounded horny or calcareous plate, the operculum, which closes the aperture when the animal withdraws into its shell. A pair of gills, or more commonly a single gill, is situated in the mantle cavity, and in some species the mantle forms a respiratory siphon.

The sexes are usually separate. In some species eggs and sperm are shed into the water, but in others internal fertilisation occurs. There is often a free-swimming larval stage, but direct development into a tiny adult is found in some species.

Although several freshwater and a few terrestrial species occur, the majority of prosobranchs are marine. They are particularly common on rocky shores and in shallow waters.

### Order Archaeogastropoda

A mainly marine group of primitive gastropods, which contains the familiar limpets, abalones and top shells. The shell, which is usually lined with mother-of-pearl, may be spirally coiled with an operculum or cone-shaped with no operculum. Holes or notches frequently occur in the shell. The radula usually has numerous teeth arranged in transverse rows. There is either a pair of gills or a single gill present, and they are composed of plate-like leaflets situated on either side of a central shaft. The mantle usually has no siphon, the incoming water current being drawn directly through the mantle edges.

The male has no penis. In most species eggs and sperm are released into the seawater where fertilisation occurs. The fertilised egg develops into a simple planktonic larva which changes into a more advanced type typical of the gastropods.

Many archaeogastropods are found on rocky shores where they rasp algae from rocks with the radula.

Superfamily: Pleurotomariacea.
Families: Pleurotomariidae, Scissurellidae, Haliotidae.

Superfamily: Fissurellacea.
Family: Fissurellidae.

Superfamily: Patellacea.
Families: Acmaeidae, Patellidae, Lepetidae.

Superfamily: Cocculinacea.
Families: Cocculinidae, Lepetellidae.

Superfamily: Trochacea.
Families: Trochidae, Stomatellidae, Turbinidae, Skeneidae, Phasianellidae, Orbitestellidae.

Superfamily: Neritacea.
Family: Neritidae.

## Family Haliotidae

This family contains only one genus. Members have an ear-shaped shell with a small spiral at the apex. The shell has a series of holes around the whorl through which water currents and waste matter pass out. The inner side of the shell has a mother-of-pearl layer which is used for costume jewellery. They are algal grazers which cling to rocks with a large muscular foot. The foot of some species is eaten by man.

78/5    *Haliotis cracherodii* Leach [abalone]. This species, which may reach 15 cm in length, is found under stones and in crevices in the intertidal and shallow subtidal regions of the North-east Pacific.

78/3    *Haliotis lamellosa* Lamarck [ormer]. The ridged surface of the shell, which may reach 8 cm in length, is often encrusted with calcareous algae. Found under stones and in crevices in shallow subtidal areas of the Mediterranean.

## Family Fissurellidae [keyhole limpets]

78/4    *Diodora gibberula* (Lamarck). The 1.5 cm long shell of this species is typical of the family, and bears a hole at its apex through which water is ejected in a constant stream. This species is thought to filter plankton from the water. Found on rocks in shallow Mediterranean waters.

## Family Patellidae [limpets]

78/10   *Patella aspera* Lamarck (= *P. athletica*). The shell of this species is up to 5 cm long and pale in colour inside and out. The foot is coloured yellow or orange. Often found in rock pools encrusted with calcareous algae high on exposed shores of the North-east Atlantic.

78/8    *Patella intermedia* Jeffreys (= *P. depressa*). The shell is shorter than that of *P. aspera*, not usually exceeding 3–4 cm in length, and bears light and dark bands on the exterior and interior of the shell. The foot is dark coloured. Found on rocks of the middle shore on exposed coasts of the North-east Atlantic.

*Patella vulgata* Linnaeus [common limpet]. A taller    78/9, 78/ limpet than the two previous species, with a light-colour shell exteriorly and interiorly. The foot is coloured yellow or orange. By far the commonest species on all rocky shores where it grazes on young algae. The limpets in Plate 78/11 have been able to graze all the algae except for that growing on their own shells. Occurs in the North-east Atlantic.

*Patina pellucida* Linnaeus [blue-rayed limpet]. This    78/6, 78/ small limpet, usually less than 2 cm long, bears prominent blue spots on its dorsal side when young (Plate 78/6). As the animals mature, the spots join to form continuous lines. They are found in large numbers on the seaweed *Laminaria* in the North-east Atlantic.

## Family Trochidae [topshells]

The topshells are readily distinguished by their conical shape and rounded base. The aperture of the shell is covered by an operculum when the animal withdraws into its shell. All are grazers.

*Calliostoma ligatum* (Gould). The beautifully coloured    79/1 shell of this topshell reaches a height of 2 cm or more. It is found both intertidally and in shallow subtidal rocky areas of the North-east Atlantic.

*Calliostoma zizyphinum* (Linnaeus) [painted topshell].    79/2, 79/ The height of the shell may reach 2.5 cm. The species cannot readily be confused with others due to the characteristic purple and lilac banding around the shell (Plate 79/2). Old specimens may have the pattern obscured by growth of encrusting calcareous algae (Plate 79/3). Found on rocks in shallow waters of the North-east Atlantic and Mediterranean.

*Clanculus pharaonis* Linnaeus. The bright red 2 cm high    79/4 shell of this species has a characteristic black and white pattern running in bands around the many-whorled shell. The aggregated specimens seen here are in the act of spawning. Found on subtidal rocks in the Indo-Pacific.

*Gibbula umbilicalis* (da Costa). The shell is a compressed    79/5 cone less than 1.5 cm, high with characteristic purple stripes running vertically down the shell. It is a common inhabitant of rocky shores in the North-east Atlantic.

*Monodonta lineata* (da Costa). The 2.5 cm high shell of    79/8 this species has a distinctive zig-zag pattern of purple lines. It is sometimes confused with the winkle *Littorina littorea* but can be distinguished by the tooth that occurs on the shell aperture. Often found in very large numbers in rock pools high on the shore accompanied by *G. umbilicalis*. Occurs in the North-east Atlantic.

## Order Mesogastropoda

A large order of gastropods which contains many marine and several freshwater and terrestrial species. The periwinkles, slipper and worm shells, conchs and cowries all belong to this group.

The shell, which is usually spirally coiled with a horny operculum, lacks mother-of-pearl on the inner surface. The radula characteristically has seven teeth in each transverse row. In algal-feeders the snout is short and mobile, but in carnivorous forms it is elongated into a proboscis which is used to capture the prey. A single gill, with leaflets on one side of the central shaft only, is present in the mantle cavity. The mantle has a siphon, which is often situated in a special groove on the anterior edge of the shell aperture.

A penis is generally present in the male, and internal fertilisation occurs. The eggs, which are laid either in jelly-like masses or in capsules, usually develop directly into a fairly advanced planktonic larva.

Mesogastropods, which have a world-wide distribution, are mainly bottom-dwellers although a few free-swimming forms exist. Predators, algae and debris feeders occur within the order.

Superfamily: Littorinacea.
Families: Lacunidae, Littorinidae.

Superfamily: Rissoacea.
Families: Hydrobiidae, Truncatellidae, Rissoidae, Assimineidae, Vitrinellidae, Skeneopsidae, Omalogyridae, Rissoellidae, Cingulopsidae.

Superfamily: Architectonicacea.
Family: Architectonicidae.

Superfamily: Cerithiacea.
Families: Turritellidae, Vermetidae, Caecidae, Planaxidae, Potamididae, Cerithiidae, Cerithiopsidae, Triphoridae.

Superfamily: Epitoniacea.
Families: Epitoniidae, Janthinidae.

Superfamily: Strombacea.
Family: Struthiolariidae, Aporrhaidae, Strombidae.

Superfamily: Hipponicacea.
Families: Hipponicidae, Fossaridae, Vanikoridae.

Superfamily: Calyptraeacea.
Families: Trichotropididae, Capulidae, Calyptraeidae.

Superfamily: Lamellariacea.
Families: Lamellariidae, Eratoidae.

Superfamily: Cypraeacea.
Families: Cypraeidae, Ovulidae.

Superfamily: Atlantacea.
Families: Atlantidae, Carinariidae, Pterotracheidae.

Superfamily: Naticacea.
Family: Naticidae.

Superfamily: Tonnacea.
Families: Cassididae, Cymatiidae, Bursidae, Tonnidae, Ficidae.

## Family Littorinidae

79/9    *Littorina littorea* (Linnaeus) [edible periwinkle]. The shell of this herbivorous species is not flattened and reaches a height of over 2 cm. The brown or black shell is usually sculptured and banded. It is seen, often in very large numbers, in crevices and in shallow rock-pools on the shore. Although a gill-breather it is able to withstand more dessication than *L. obtusata*. Found in the North Atlantic and Mediterranean, where it is eaten extensively by man.

79/6    *Littorina obtusata* (Linnaeus) [flat periwinkle]. The spire of the shell in this 1 cm high herbivorous species is characteristically flattened. The colour of the shell is very variable ranging from orange-yellow to dull brown. Banded specimens are not uncommon. It lives in the intertidal zone, often in large numbers amongst seaweeds. Since it is a gill-breather it benefits from the humid atmosphere around the weeds when exposed at low tides. Found in the North-east Atlantic.

79/7    *Littorina rudis* Maton [rough periwinkle]. This small grazing species has a distinctly whorled ridged shell, less than 1 cm in height. It is found in crevices and on open rock higher on the shore than the preceding *Littorina*

species, as a result of the mantle cavity being modified for air breathing. Occurs in the North-east Atlantic.

## Family Rissoidae

*Rissoa variabilis* Mühlfeld. A small species with an   79/10 elongated ribbed shell less than 1 cm in length. It grazes on microscopic algae covering seaweed and sea-grasses in shallow Mediterranean waters.

## Family Skeneopsidae

*Skeneopsis planorbis* (Fabricius). A species with a   80/7 translucent shell barely 2 mm in diameter, bearing a superficial resemblance to that of a freshwater planorbid snail. These little-known tiny molluscs are found intertidally amongst undergrowth, particularly in the calcareous alga, *Corallina* in the North Atlantic.

## Family Vermetidae

Members of this family live in worm-like shell tubes, usually attached to rocks and other hard objects. The early tube whorls are recognisably those of gastropods, but in many species, later tube formation is relatively straight.

*Vermetus adansonii* Daudin. This small Indo-Pacific   80/5 species, with a tube diameter of less than 5 mm, often develops in clusters which are overgrown by soft corals and other colonial animals. The entrance to the tube is plugged by a concave operculum.

*Vermetus maximus* Sowerby. The tube diameter of this   80/9 Indo-Pacific species is often in excess of 2 cm and its length up to 15 cm. The bizarrely marked concave operculum can be seen plugging most of the tube entrance but similarly marked tentacles protrude from above.

*Vermetus* sp. This Indo-Pacific vermetid, and many   80/6 other members of the family, secretes fine mucous threads from the foot into the surrounding water to trap plankton. The threads and trapped food are subsequently pulled down into the mouth.

## Family Potamididae

*Velacumantis australis* Quoy & Gaimard [Australian   80/10 mud whelk]. The long conical shell of this species reaches a length of 4 cm. This common inhabitant of mud flats is a secondary host for a parasite which infects wading birds. It also causes 'swimmers itch' when the larvae burrow into human skin. Found in the Zealandic region.

## Family Cerithiidae

*Cerithium echinatum* Lamarck. The knobbly shell of this   80/8 cerith, which reaches 6 cm in height, is frequently encrusted by algae and small worm tubes. This specimen is grazing over algal-covered coral. Abundant on Indo-Pacific reef flats.

## Family Triphoridae

*Triphora perversa* (Linnaeus). The narrow many-whorled   80/1 shell of this species is up to 3 cm long and bears two or three regular bands of bumps in each whorl. Found on sand, gravel and broken ground below tide level in the Mediterranean and North-east Atlantic.

## Family Janthinidae

*Janthina* sp. The violet shell of this oceanic species is   80/2 up to 2 cm high. It maintains itself at the water surface by means of a bubble to which its eggs are attached. This specimen is accompanied by a nudibranch and by an anemone which is attached to its shell. *Janthina* itself attacks and eats velellids (Plate 11/3).

### Family Aporrhaidae

80/3   *Aporrhais pespelicani* (Linnaeus) [pelican's-foot shell]. A characteristic species with the last whorl of the shell flared out into a 'foot' covering the head. In order to reveal its shape the living specimen in the photograph has been removed from the muddy gravel in which it normally burrows. Found in the North-east Atlantic and Mediterranean.

### Family Strombidae

The shells of strombids are characterised by a deep notch at the anterior end of the outer lip, through which one of the eyestalks normally protrudes. The operculum is long and narrow and can be used as a lever to right itself. They have a large muscular foot. Strombids are herbivores, feeding on algae and organic debris in shallow sandy areas.

80/4   *Lambis crocata* Link. Members of this genus are commonly known as spider shells because of the long curved projections on the outer lip of the adult shell aperture. A ventral view of this 15 cm long Indo-Pacific specimen reveals the curved narrow operculum and the strombid notch at the anterior end of the outer lip.

81/1, 81/2   *Strombus gigas* Linnaeus [Queen conch]. This species has a heavy shell up to 30 cm in length which is often covered with encrusting organisms (Plate 81/1). When inverted, the beautiful pink coloration of the inside of the shell aperture can be seen (Plate 81/2). This edible species is found on sand and in sea-grass beds in the Caribbean.

81/4   *Strombus vomer* Röding. The shell of this Indo-Pacific strombid rarely reaches 10 cm in length. An eyestalk can be seen protruding through the strombid notch in the anterior part of the shell.

81/5   *Tibia insulaechorab* Röding. This species, which reaches 15 cm in length, has a characteristic shell shape at the anterior end. Found on sand in the Indo-Pacific.

### Family Calyptraeidae

81/3   *Calyptraea chinensis* (Linnaeus) [Chinaman's hat]. The translucent white, low, limpet-shaped shell of this species may reach 2.5 cm in diameter. The living specimen in the photograph is seen from below, revealing the thin curving partition above which the bulk of the body is situated. Found attached to stones and shells on soft bottoms in the Mediterranean and North-east Atlantic.

81/6   *Crepidula fornicata* (Linnaeus) [slipper limpet]. A distinctive oval species up to 5 cm long, with the ventral partition characteristic of the family covering about half the underside of the shell. Individuals are normally found grouped in chains with the oldest animals at the bottom. The animals change sex from males to females as they get older. Hence the animals at the bottom are females and the younger animals at the top, males. These filter-feeding animals are major pests on oyster beds where they not only compete with the oysters for food but tend also to smother them. Found in the North Atlantic.

### Family Eratoidae

81/7   *Trivia adriatica* Monterosato. The oval domed shell with an elongated slit-like opening is characteristic of the family. When the body of the animal is extended, folds of mantle tissue partly obscure the shell. This distinctly marked 1 cm long species is found on rocky bottoms in shallow waters of the Mediterranean, where it feeds on algae, sponges and small crustaceans.

*Trivia monacha* (da Costa). A species rarely more than 81/8 1.5 cm in length, characterised by a ribbed shell which bears three black pigment spots. The shell may sometimes be obscured by mantle folds. The specimen in the photograph is seen crawling over a sea cucumber, *Cucumaria saxicola*. Found in a variety of habitats in shallow water where it feeds on colonial sea-squirts. A North-east Atlantic and Mediterranean species.

### Family Cypraeidae [true cowries]

The domed beautifully patterned glossy shells of this family are familiar objects. However, one of the most characteristic features of this group is the bilobed mantle, which can completely envelop the shell when fully extended. The mantle may be thin and translucent, or, as is the case in most cowries, be thicker and have a colour pattern completely different from that of the shell. The head, with its well developed tentacles and eyes, the broad foot and the mantle lobes protrude from the long narrow aperture on the underside of the shell. Little is known about the diet of cowries but most seem to be omnivorous. They are mainly found in shallow waters of tropical seas where they are usually nocturnal in habit.

*Cypraea arabica* Linnaeus. Shell up to 8 cm long. This 81/10 species, seen here with its papillated mantle only partially extended, is found on Indo-Pacific reefs.

*Cypraea carneola* Linnaeus. The lightweight shell rarely 81/9 exceeds 8 cm in length. The shell aperture is characteristically deep purple. Found on Indo-Pacific reef flats.

*Cypraea isabella* Linnaeus. Shell up to 5 cm long. The 82/1 totally black body of this species contrasts markedly with its light-coloured shell. Found on Indo-Pacific reefs.

*Cypraea lurida* Linnaeus. Shell up to 4 cm long. The 82/2 shell colour can be seen through the translucent mantle of the species. Found on rocky bottoms in the Mediterranean.

*Cypraea mappa* Linnaeus. Shell up to 9 cm long. When 82/4 the bilobed mantle is fully extended it does not completely cover the top of the shell. This results in an undulating mantle mark on either side of the dorsal midline which can be clearly seen in the photographed specimen. Occurs on Indo-Pacific reefs.

*Cypraea nucleus* Linnaeus. The nodular surface of the 82/3 3 cm long shell is very characteristic. The specimen in the photograph shows the long burrowing papillae of the fully extended mantle. Found on Indo-Pacific reefs.

*Cypraea pantherina* Solander. Shell up to 8 cm long. 82/5 The photographed specimen has a papillated mantle with longitudinal black and white lines which contrasts with the brown spotted shell beneath. Found in the Indo-Pacific.

*Cypraea talpa* Linnaeus. Shell up to 9 cm long. The 82/6 thick white-speckled dark mantle with stout papillae contrasts greatly with the light shell of this specimen photographed on an Indo-Pacific reef.

*Cypraea testudinaria* Linnaeus. A large shell up to 82/7 13 cm long. The dark brown mottling on the shell is just visible through the papillated mantle of this specimen which was photographed at night on an Indo-Pacific reef.

82/8    *Cypraea tigris* Linnaeus. Shell up to 15 cm long. A common species found both intertidally and subtidally in the Indo-Pacific.

## Family Ovulidae

82/9    *Calpurnus verrucosus* Linnaeus. The upper surface of the shell, which reaches 4 cm in length, has a triangular keel. The shell has characteristic button-shaped protruberances at either end. Found on sand and on soft corals in the Indo-Pacific.

83/4    *Cyphoma gibbosum* (Linnaeus) [flamingo tongue]. The shell is up to 3 cm long, with a pronounced transverse ridge across the centre of the dorsal surface. The mantle has a characteristic pattern. Found on sea-fans in the Caribbean.

83/2    *Cyphoma signatum* Pilsbry & McGinty. The 3 cm long shell is covered with a mantle with distinctive markings. Found on sea-fans in the Caribbean.

83/3    *Jenneria pustulata* (Lightfoot). This species cannot be confused with any other because of the bright orange protruberances on the upper side of the 2–3 cm long shell. The mantle lobes bear long branched papillae. Found amongst corals in the Panamanian region.

83/6    *Ovula ovum* Linnaeus [egg cowry]. The white shell, which may reach 10 cm or more in length, is distinguished from true cowries by the lack of teeth on the inner lip of the aperture. The black mantle is almost completely covering the shell in this specimen feeding on soft coral in the Indo-Pacific.

83/1    *Volva* n.sp. Long anterior and posterior extensions of the shell are characteristic of the genus. This as yet undescribed species was photographed on antipatharian coral at a depth of approximately 40 m on Lord Howe Island, South Pacific.

## Family Carinariidae

83/7    *Carinaria lamarcki* Pèron & Lesueur (= *C. mediterranea*). This species belongs to a group of molluscs which live permanently floating in the plankton of the open ocean. The shell is thin and small and has the gill protruding from it. The body is up to 10 cm long and gelatinous with a stabilising fin (probably a modified foot) opposite the shell. It is often found floating from the surface film with the shell acting as a keel. Occurs in the Atlantic and Mediterranean.

## Family Naticidae

83/5    *Natica alderi* Forbes [common necklace shell]. A characteristic glossy shell up to 3 cm in diameter, often with a brown-black pattern. It burrows in intertidal and subtidal sand in search of bivalve mollusc prey. The common name derives from the collars of eggs with embedded sand grains that are laid on the sand surface. Found in the North-east ' Atlantic and Mediterranean.

83/8    *Natica hebraea* (Martyn). The shell of this species reaches 4 cm or more across. The large foot possessed by all species in this genus is used for burrowing. Occurs on sand in the Mediterranean.

## Family Cassididae [helmet shells]

The cassids have solidly built shells which are often used as household ornaments. They are mainly tropical, many feeding on sea-urchins and sand-dollars and sometimes also on other molluscs.

*Casmaria ponderosa* Gmelin. The shell reaches 5 cm in    83/9
height. They are common on intertidal and subtidal sand in the Indo-Pacific.

*Cassis madagascariensis* Lamarck. The uniformly cream-    83/10
coloured shell of this species reaches 25 cm or more in length. It feeds particularly on the long-spined sea-urchin, *Diadema*, whose spines are crushed by the broad flat plate surrounding the aperture. The proboscis is then usually inserted into the sea-urchin through its anus. Found on sand below tide level in the Caribbean.

*Cypraecassis rufa* Linnaeus. The beautiful shell of this    83/11
Indo-Pacific species, which is 15 cm or more in length, is used extensively for carving shell cameos. Specimens are common in sea-grass beds where they feed on sea-urchins.

## Family Cymatiidae

*Cabestana spengleri* Perry. This species, with a shell up    84/1
to 15 cm in height, feeds particularly on sea-squirts and is found from shallow to deep water, particularly in harbour and estuarine areas, in the Zealandic region.

*Charonia rubicunda* Perry. The shell height of this species    84/2
may reach 20 cm or more. It is found in the Zealandic region on reefs from shallow to deep water where it feeds frequently on starfish and sea-urchins.

*Charonia tritonis* Linnaeus. A large Indo-Pacific species    84/8
with a magnificent shell 40 cm or more in height. It feeds on reef-dwelling starfish, particularly on *Linckia* and the crown-of-thorns starfish, *Acanthaster*.

## Family Tonnidae [tun shells]

*Tonna variegata* Lamarck. The shell colour of this broad-    84/7
footed species is extremely variable. Specimens up to 20 cm in height have been found in deep sandy areas around reefs. The Indo-Pacific specimen seen here is, however, gliding over coral.

## Order Neogastropoda

Marine prosobranchs which include the familiar whelks, volutes and cone shells. The spirally coiled shell, which may have an operculum, usually has an elongated groove (siphonal canal) near the shell aperture, which carries the long mantle siphon. The radula may be narrow with not more than three large teeth in each row or may have two rows of harpoon-like teeth which can inject poison into the prey. A highly extensible proboscis is usually present. The mantle cavity contains a single gill with one row of leaflets.

The male usually has a penis and internal fertilisation occurs. The eggs, which are commonly laid in capsules, often develop directly into a tiny adult.

Neogastropods are particularly common in tropical seas, although they also occur in colder waters. They are active predators feeding on other gastropods, bivalves, worms and fish.

## Suborder Stenoglossa

Prosobranchs with a narrow radula containing not    84/9
more than three large teeth in each row (Plate 84/9).

Superfamily: Muricacea.
Families: Muricidae, Magilidae.

91

Superfamily: Buccinacea.
Families: Pyrenidae (=Columbellidae), Buccinidae, Neptuneidae, Melongenidae, Nassariidae (=Nassidae), Fasciolariidae.

Superfamily: Volutacea.
Families: Olividae, Vasidae, Harpidae, Volutidae, Cancellariidae, Marginellidae.

Superfamily: Mitracea.
Families: Mitridae, Vexillidae.

### Family Muricidae

84/4 *Murex trapa* Röding. A species with prominent spines that has a body length of 11 cm. There is a horny operculum and a long straight siphonal canal bearing spines. Feeds mostly on bivalves by boring a hole through the shell. Found on sandy bottoms in the Indo-Pacific.

84/3 *Nucella lapillus* (Linnaeus) (=*Thais lapillus*) [dogwhelk]. This whelk, which has a short siphonal canal, may reach a length of 5 cm. Occurs on rocky shores, often in large numbers, amongst barnacles and mussels on which it feeds. A North Atlantic species.

### Family Pyrenidae (=Columbellidae) [dove shells]

84/5 *Pyrene rustica* (Linnaeus) (=*Columbella rustica*). This specimen with its siphon and tentacles extended has a shell length of 2–3 cm. The operculum can also be seen at the rear of the foot. Numerous on rocky bottoms in shallow waters of the Mediterranean.

### Family Buccinidae

84/6 *Euthria cornea* (Linnaeus) (=*Buccinulum cornea*). This heavy-shelled species, seen here with its head and snout extended, is up to 6 cm long. It is normally found on sand and mud deposits in the Mediterranean.

84/12 *Buccinum undatum* Linnaeus [common whelk or buckie]. A large heavy-shelled edible species up to 10 cm high, with a short siphonal canal. It is normally found on sand and mud (occasionally on rocks in muddy areas) where it feeds extensively on polychaete worms. Empty shells of this North Atlantic species are a favourite habitat for hermit crabs.

### Family Nassariidae (=Nassidae)

84/10 *Nassarius incrassatus* (Ström). This species has a variously coloured and banded shell up to 15 mm in height. Found on mud and amongst stones and algae in shallow waters of the Mediterranean and North-east Atlantic.

### Family Fasciolariidae

84/11 *Fasciolaria tulipa* Linnaeus [true tulip]. A characteristic species up to 25 cm long, with an extended spire and a long siphonal canal. The horny operculum is claw-shaped. Common in sea-grass beds in the Caribbean.

85/1 *Fusinus polygonoides* Lamarck (=*Fusus polygonoides*). The siphonal canal and spire are of approximately equal length in this species which reaches 10 cm in length. They are most active at night in coral sands, where they prey mainly on other molluscs and worms. Found in the Indo-Pacific.

85/2, 86/4 *Pleuroploca filamentosa* Röding (=*Fasciolaria filamentosa*). The shell of this brightly coloured animal reaches a height of 12 cm. Its horny furrowed operculum can be seen in the foreground of the photograph. This species attacks other gastropods (particularly trochids)

through the aperture. Common on shallow reefs in tropical waters.

### Family Olividae [olives]

*Ancillista velesiana* Iredale. This species has a thin shell up to 9 cm long with a broad foot which folds over the lower part of the shell. The foot is used for burrowing in sand. Commonly found below 20 m in the Indo-Pacific. 85/3

*Oliva flaviola* Duclos. The forepart of the foot in this Indo-Pacific species, seen here crawling over coral in search of prey, is distinctly arrow-shaped. Members of the genus are characterised by the lack of an operculum. 85/10

*Oliva porphyria* (Linnaeus) [tent olive]. The common name of this species, which reaches a length of 10 cm, derives from the characteristic pattern on the glossy shell. Found on sand from shallow to deep water in the Panamanian region. 85/5

### Family Harpidae [harp shells]

*Harpa amouretta* Röding [lesser harp shell]. The strongly ribbed shell, with characteristic brown blotches between the ribs, reaches a length of 7 cm. The foot is broad and expanded anteriorly to aid, burrowing in sand. The photographed specimen is, rather unusually, gliding over coral. An Indo-Pacific species. 85/4

### Family Volutidae [volutes]

*Cymbiola magnifica* (Gebauer) (=*Cymbiolena magnifica*) [magnificent volute]. The shell of this carnivore reaches 30 cm or more in length. Found on reefs and adjacent sand from shallow to deep water in the Indo-Pacific. 85/6

*Melo amphora* (Lightfoot) [baler shell]. One of the largest volutes, with a shell length of up to 50 cm. The shell has large shoulder spines which obscure the spire. The common name derives from the fact that Australian aboriginals used it for baling water from their canoes. Common on muddy sand from shallow to deep water, in the Indo-Pacific. 85/7

### Family Marginellidae [margin shells]

*Persicula miliaris* (Linnaeus). The shell of this species, which has a compressed spire, rarely reaches 1 cm in length. The foot, which has no operculum, extends up around the lower edge of the shell. Found on rocks and sand in the Mediterranean, where it feeds on a variety of small animals including colonial hydroids and bryozoans. 85/8

### Family Mitridae [mitre shells]

*Mitra mitra* Linnaeus. The elongated shell of this species, which reaches 15 cm or more in length, is characterised by vivid orange blotches. It has a very long proboscis which enables it to attack marine worms buried in the sediment. A night-active species which is common in shallow sandy areas of the Indo-Pacific. 85/9

## Suborder Toxoglossa

The radula has two rows of grooved harpoon-like teeth which are loosely arranged; they are used to inject poison into the prey.

Superfamily: Conacea.
Families: Turridae, Conidae, Terebridae.

## Family Turridae

86/1    *Clathurella purpurea* (Montagu) (= *Philbertia purpurea*). The purple-tinted strongly ridged shell reaches over 2 cm in length. This predatory species, which has no operculum, occurs in a variety of habitats, but is found most frequently on soft sediments. A Mediterranean and North-east Atlantic species.

## Family Conidae [cone shells]

Members of this carnivorous family are mainly confined to tropical waters and are readily recognised by their conical shape, flattened spire and long shell aperture partially protected by a narrow horny operculum. They have a venomous sting which can be dangerous to man. They feed on a variety of animals including worms, molluscs and fish.

86/2    *Conus betalinus* Linnaeus. The sparsely patterned shell of this species may reach a length of 15 cm. Occurs on intertidal and subtidal sand in the Indo-Pacific.

86/4    *Conus dalli* Stearns. A species whose shell length rarely exceeds 5 cm. The specimen has been photographed in the act of stinging a large fasciolariid, *Pleuroploca*. Found on sand and mud in the Panamanian region.

86/10   *Conus episcopus* Hwass (in Bruguière). The beautifully patterned shell of this Indo-Pacific reef dweller reaches 8 cm in length.

/3, 86/5,   *Conus geographus* Linnaeus. This Indo-Pacific species,
6/6, 86/7   which reaches a length of 12 cm or more, has a powerful sting known to have caused several human deaths (Plate 86/3). A series of three photographs shows a specimen attacking and devouring a small reef-dwelling fish (Plates 86/5–7).

87/1    *Conus marmoreus* Linnaeus. This 10 cm long species reputedly has a powerful sting dangerous to man. Found on reefs and adjacent sand in the Indo-Pacific.

86/8    *Conus striatus* Linnaeus. The shell colour and pattern of this common Indo-Pacific species is very variable. Shells may occasionally reach a length of 10 cm.

6/9, 87/2   *Conus textile* Linnaeus. The spire of this 8–10 cm long shell is less flattened than in many other species. A female specimen (Plate 87/2) has been photographed in the act of cementing its tough membranous egg capsules to a rock surface. A common Indo-Pacific species with a powerful sting.

## Family Terebridae [auger shells]

87/3    *Terebra strigata* Sowerby [zebra auger]. Like all members of this family, this species has a long narrow many-whorled shell tapering to a sharp point. The dark-brown wavy stripes are characteristic of this auger; it is found on sand intertidally and in shallow water in the Panamanian region.

# Subclass Opisthobranchia

A marine group of gastropods whose species range in length from a few millimeters to the 40 cm length of the tropical sea-hares, which may weigh as much as 2 kg. The adults show varying degrees of detorsion and the mantle cavity has moved back along the right side or is absent altogether. The shell is often reduced and is frequently covered by parts of the mantle or foot. An operculum is rarely present. In the more advanced opisthobranchs a shell is completely lacking. The head, which may be flattened and modified for burrowing, usually has one pair of eyes, a radula, and up to four pairs of tentacles. The foot is a flat creeping sole and in some forms possesses large lateral lobes (parapodia) which are used in swimming. There may be a single gill housed in the mantle cavity, but in opisthobranchs with no cavity it is lost altogether, and often new respiratory structures have developed on the body surface. Dorsal processes (cerata), which are probably used for defence, frequently occur on either side of the body in the shell-less forms.

Nearly all opisthobranchs are hermaphrodite, and there is an exchange of sperm between individuals. The eggs, which are frequently laid in jelly-like masses, normally hatch as free-swimming larvae.

They are mainly bottom-living, although a few species actively swim in the upper layers of the sea. The majority occur in shallow waters but a few are found at considerable depths. Most opisthobranchs are carnivores although there are some algal and particle-feeders. They show a remarkable ability to discriminate between food organisms, many eating only one species of prey.

## Order Bullomorpha

Members of this group, which contains the familiar bubble-shells, usually have a shell, but this may be internal or lacking altogether. An operculum is rarely present. The head often forms a characteristically flattened shield which is used for burrowing. A radula is normally present. The foot frequently bears prominent parapodial lobes which may be fin-like in some species. The mantle cavity, which has moved to a posterior position in the more advanced forms, usually contains a single gill.

The penis is unarmed and there is usually an open sperm groove running from the reproductive opening to the penis.

Many bullomorphs burrow into soft sediments and are predators feeding mainly on foraminiferans (single-celled organisms), worms and bivalves. A few species browse on algae.

## Suborder Acteonacea

The spirally coiled external shell may be thick and heavy or paper-thin with a sunken apex. A horny operculum may be present. The head shield commonly bears ear-like lobes and the radula frequently has large numbers of small teeth. Well developed parapodial lobes are sometimes present. There is no open sperm groove and the penis may be housed in the large mantle cavity.

Families: Acteonidae, Bullinidae, Hydatinidae.

## Family Bullinidae

*Bullina lineata* (Gray). The markings on this delicate    87/4
little bubble-shell, barely measuring more than 1 cm, are characteristic, as is the iridescent blue line edging the parapodial extensions of the foot. It appears seasonally in rock pools amongst algal undergrowth in the Indo-Pacific.

## Suborder Diaphanacea

Small opisthobranchs with thin shells which may be completely enclosed by the mantle. The head shield has tentacles and the anterior part of the foot usually bears wing-like tentacular processes. The posterior end of the foot may be forked. Parapodial lobes are not developed.

Families: Diaphanidae, Notodiaphanidae.

Plate 87

## Suborder Retusacea

Opisthobranchs with fragile external shells which may have a flattened spire. An operculum is rarely present. The head shield bears tentacles which commonly conceal the front of the shell. A radula and parapodial lobes are absent.

Family: Retusidae.

## Suborder Ringiculacea

The only genus, *Ringicula*, has a spirally coiled external shell. A radula is present. The head shield is broad and its middle part can be erected to form a temporary siphon when the animal is buried in the sediment. Special glands which secrete an obnoxious fluid when the animal is irritated are situated on the posterior edge of the shield.

Family: Ringiculidae.

## Suborder Bullacea

Oval or egg-shaped shells with a sunken spire, which frequently have a ridge around the elongated shell aperture. The posterior end of the head shield is forked and a radula is present. The well developed parapodial lobes frequently cover part of the shell.

Family: Bullidae.

## Suborder Atyacea

Large thin shells with a sunken spire. The head shield bears ear-shaped processes that frequently conceal the front of the shell. The radula is well developed. Large parapodial lobes often surround the shell.

Family: Atyidae.

## Suborder Philinacea

The shell in this group is either external and reduced, internal, or absent altogether. The head shield usually bears processes of some kind but these are lacking in a few species. A radula is commonly present, but is absent in the aglajids which swallow their prey whole. Well developed parapodial lobes may meet mid-dorsally and are used for swimming in some species. A single gill usually occurs in the mantle cavity.

Families: Scaphandridae, Akeridae, Gastropteridae, Aglajidae, Philinidae, Philinoglossidae.

### Family Aglajidae

87/7    *Navanax inermis* (Cooper). This species, which may reach 20 cm in length, has an internal shell. It is a voracious carnivore attacking and eating nudibranch molluscs. Found in rock pools and in sea-grass beds in the Panamanian region.

## Suborder Runcinacea

Opisthobranchs with small elongated bodies in which the shell is reduced or absent altogether. The head shield lacks processes of any kind and is fused with the mantle. A radula is present but parapodial lobes are absent. The gills are situated beneath the rear end of the mantle.

Family: Runcinidae.

## Order Pyramidellomorpha

Members of this group possess well developed spirally coiled shells which usually have an operculum. Eyes are situated between the tentacles which are grooved on the outer side. A radula and gills are absent. The characteristic feature of pyramidellomorphs is a long proboscis carrying a pointed bristle which is used to pierce the prey, particularly worms and bivalves, from which they suck body fluids.

They are most common in the warmer waters of the world although several species are found in temperate regions.

Family: Pyramidellidae.

## Order Thecosomata [shelled pteropods]

These opisthobranchs are modified for a free-swimming existence. The shell may be external and spirally coiled with an operculum, variously shaped with no operculum, or absent altogether. Species with no external shell may possess an internal transparent false shell (pseudoconch). The head bears a pair of well developed tentacles and often has a snout or proboscis. A radula is frequently present. The parapodial lobes of the foot are expanded into muscular wing-like fins that are used for swimming. The mantle cavity, which is usually well developed, is either dorsally or ventrally placed and only rarely contains a gill. An open sperm groove runs from the reproductive opening to the unarmed penis.

Shelled pteropods are found in the plankton of the open oceans, where they feed by filtering small organisms from the water with bands of cilia on the parapodial lobes or on the edge of the mantle. They show daily vertical migration in the water column, rising to the surface waters at dusk and returning to deeper layers during the day.

## Suborder Euthecosomata

Pteropods with an external shell which is either spirally coiled with an operculum or variously shaped with no operculum. The head has no snout or proboscis. The mantle cavity, which is either dorsal or ventral in position, houses a gill in a few species.

Families: Spiratellidae (=Limacinidae), Cavoliniidae.

### Family Cavoliniidae

*Cavolinia tridentata* (Forskål). [sea-butterfly]. A planktonic species with a distinctive shell shape. The two large parapodial extensions of the foot beat like wings to move this 2 cm long pteropod through the water. Mucus coating the wings traps plant plankton and organic debris. The mucous streams are then passed into the mouth by beating cilia. The species has a world-wide oceanic distribution.    87/8

## Suborder Pseudothecosomata

Pteropods with either an external spiral shell with an operculum, an internal false shell, or no shell at all. The head bears a snout or proboscis. The mantle cavity when present is ventral and occasionally houses a gill.

Families: Peraclidae, Procymbuliidae, Cymbuliidae, Desmopteridae.

## Family Cymbuliidae

87/5 *Corolla spectabilis* Dall. This planktonic pteropod is propelled through the water by two large parapodial extensions whose latticework of muscle fibres can be seen in the photograph. The conical internal shell (pseudoconch) is the only opaque part of this animal. It spins a mucous net to trap plankton. Distributed in surface waters of the ocean.

87/6 *Gleba cordata* Niebuhr. It has only recently been discovered that this species, which measures a maximum of 5 cm across its outstretched wings, spins a membrane of mucus up to 2 m across in the water. The membrane traps minute planktonic organisms. When the membrane becomes clogged it is drawn into the mouth situated at the end of a long proboscis. Distributed in surface waters of the ocean.

## Order Gymnosomata [naked pteropods]

Free-swimming pteropods which have no shell or operculum. The cylindrical body often has a conspicuous waist and the head bears two pairs of tentacles. A radula is usually present. The foot has a pair of ventral parapodial fins which are used for swimming. There is no mantle cavity but remnants of the gill are occasionally present. An open sperm groove runs from the reproductive opening to the penis which is sometimes armed with spines.

Naked pteropods are exclusively planktonic, living in the surface waters of the open ocean where they are often found in large numbers during the day. They are active predators, feeding mainly on the slower shelled pteropods which they catch with special structures situated in a muscular region of the gut, into which the mouth leads. These structures often consist of adhesive conical tentacles, suckered arms and sacs with hooks, which are protruded through the mouth by muscular action.

Families: Pneumodermatidae, Cliopsidae, Notobranchaeidae, Clionidae, Thliptodontidae, Anopsiidae, Laginiopsidae.

### Family Cliopsidae

87/9 *Cliopsis* sp. The anterior end of this planktonic species bears a reduced foot and relatively small parapodia compared with the cymbuliids. They are active predators, seizing prey by means of hooked sacs everted through the mouth. Distributed in surface waters of the ocean.

### Family Clionidae

88/2 *Clione limacina* Phipps. It has an elongated body up to 2 cm long, with small parapodia situated anteriorly. There are adhesive conical tentacles at the anterior end as well as hooked sacs. Distributed in northern oceanic waters.

## Order Aplysiomorpha [sea-hares]

Members of this order, which contains the largest living opisthobranchs, have the shell internal or absent altogether. The elongated head bears two pairs of tentacles and a well-developed radula is present. The foot, which is large with prominent parapodial lobes, can be used for creeping or swimming. The mantle is reduced, often forming just a thin covering over the shell when present. The small mantle cavity contains a gill and a gland which expels a dye (usually purplish) into the water when the animal is attacked.

An open sperm groove runs from the reproductive opening to the penis, which is often armed with spines. An exchange of sperm takes place between individuals and chains composed of several copulating sea hares frequently occur. The fertilised eggs, which are often laid in the intertidal zone, hatch into free-swimming larvae.

Sea-hares occur in shallow coastal waters where they live and feed on the larger algae (particularly green algae), cutting off pieces with the strong radula.

Family: Aplysiidae.

## Family Aplysiidae

*Aplysia dactylomela* Rang. This species, which is easily recognised by the rings and streaks of dark pigment on the body surface, may reach 30 cm or more in length. Members of this species swim frequently, using the muscular parapodial flaps of the foot. It is found amongst seaweed and sea-grass in shallow waters throughout tropical seas. 88/1

*Aplysia depilans* Gmelin. A species reaching 30 cm in length when fully grown. It occurs on seaweed-covered sand and rock in shallow waters of the Mediterranean and adjacent Atlantic. 88/3

*Aplysia punctata* Cuvier. This species is smaller than *A. depilans*, rarely reaching 20 cm in length. As with most members of the family they appear seasonally in the weed-covered intertidal zone and in shallow water to copulate and lay their characteristic egg strings. Found in the North-east Atlantic. 88/4

## Order Pleurobranchomorpha

Flattened slug-like opisthobranchs which may have an external or internal shell, or no shell at all. The head has two pairs of tentacles and a veil of skin protecting the mouth. A radula is always present. The foot, which is large and broad with no parapodial lobes, is often used for swimming. A conspicuous gland of disputable function is sometimes present at the posterior end of the foot. The large mantle often forms a skirt around the body. There is no truly enclosed mantle cavity, but the feathery gill is protected by the mantle on the right side of the body. The skin, which contains calcareous spicules in some species, can often produce a strong acid secretion when the animal is disturbed. The penis is unarmed and an open sperm groove is usually absent.

Representatives occur commonly in coastal waters where many feed on sponges and tunicates with the aid of a long extensible proboscis.

## Suborder Umbraculacea

They possess an external umbrella-like shell which frequently does not enclose the entire animal. The foot is often very broad, normally bearing warts on the dorsal surface. A large leaf-like penis is present in some species. They normally feed on sponges.

Family: Umbraculidae.

## Suborder Pleurobranchacea

The shell is internal or absent altogether. The mantle commonly encircles the body, often partly concealing

95

the head. Calcareous spicules are usually present in the skin.

Pleurobranchids feed mainly on tunicates.

Family: Pleurobranchidae.

### Family Pleurobranchidae

88/6 *Pleurobranchaea meckeli* Leue. A distinctive species, up to 10 cm long, in which the mantle is united with a veil of skin in front of the mouth. Found on sandy and stony bottoms in the Mediterranean, where it preys on worms and other molluscs.

88/5 *Pleurobranchus areolatus* Mörch. A flattened species up to 7 cm in length with a warty tessellated upper surface. The species exudes mucus in quantity. Found on reefs and reef rubble in shallow waters of the Caribbean and Panamanian regions.

88/7 *Pleurobranchus testudinarius* Cantraine. The dorsal surface of this species, which may reach 10 cm in length, is covered by large warts. It feeds mainly on colonial sea-squirts and sponges. Found in the Mediterranean and adjacent Atlantic.

88/8, 88/9 *Pleurobranchus* sp. Two unknown species from the Indo-Pacific, with the tessellated mantle obscuring the head.

## Order Acochlidiacea

Very small opisthobranchs with a body length of less than 10 mm. There is no shell, but calcareous spicules are commonly present in the skin. The head usually bears two pairs of tentacles and a narrow radula. The visceral mass is clearly separated from, and often longer than, the foot, which has no parapodial lobes. The mantle cavity and gill are absent.

Very little is known of the breeding habits of these animals, but the sexes appear to be separate in a few species and the penis is usually unarmed.

They are found living between sand grains in both tropical and temperate waters, where they apparently feed on micro-organisms in the sediments. Several species can tolerate waters of low salinity.

Families: Acochlidiidae, Hedylopsidae, Microhedylidae.

## Order Sacoglossa

Opisthobranchs whose shell can be either external with one or two valves, internal, or absent altogether. The body may be slug-like, have dorsal processes (cerata), or be flattened and leaf-like. The head normally has one pair of tentacles and a narrow radula with sharp piercing teeth. The foot, which is often long and slender and closely applied to the head and visceral mass, may have parapodial lobes. There is usually no mantle cavity or gill. The penis often possesses a strong bristle and the animal usually lacks an open sperm groove.

Sacoglossans usually occur in shallow coastal waters, where the majority feed on green algae, piercing the cell walls with the radula teeth and then sucking out the fluid contents. A few species feed on the eggs of other opisthobranchs.

### Suborder Juliacea

Small sacoglossans, often only a few millimetres in length. The body is partly enclosed within a bivalved shell, which has a small spiral at the apex of the left valve. These opisthobranchs, which in the past have been mistaken for bivalves, feed on green algae.

Family: Juliidae.

### Suborder Oxynoacea

Sacoglossans with an external shell, composed of one valve only, which does not enclose the complete animal. Some species have large fleshy parapodial lobes which may be divided into two parts. A gill is occasionally present.

Families: Volvatellidae, Oxynoidae, Cylindrobullidae.

### Suborder Elysiacea

Sacoglossans which have no shell. The body may be slug-like often bearing two rows of cerata, or leaf-shaped with flattened parapodial lobes. Both algal-feeders and egg-eating forms occur within this group.

Families: Elysiidae, Placobranchidae, Polybranchiidae, Stiligeridae, Limapontiidae, Oleidae.

### Family Elysiidae

89/6 *Elysia viridis* (Montagu). The slug-like flattened body of this species reaches 3 cm or more in length. Commonly found feeding on green seaweeds intertidally and in shallow water. Occurs in the Mediterranean and Northeast Atlantic.

89/1 *Thuridilla hopei* (Verany). A distinctive Mediterranean species up to 15 cm long, with relatively long tentacles and a conspicuous colour pattern along the back. Numerous on shallow-water algae (particularly *Cystoseira*).

89/3 *Tridachia crispata* Mörch. The mantle edge of this 5 cm long sea-slug is characteristically crinkled, and so it cannot be readily confused with any other species. The two specimens seen here are probably about to copulate in order to exchange sperm. Common in sea-grass beds and on algae-covered rocks in the Caribbean.

### Family Polybranchiidae

89/8 *Cyerce nigra* Pease. This distinctively coloured species reaches 4 cm in length and has a large number of plate-like dorsal extensions (cerata). The cerata have a respiratory function but are also used for jerky swimming movements. Found in shallow water and in pools left by the receding tide, where it feeds on green algae in particular. An Indo-Pacific species.

## Order Nudibranchia [true sea-slugs]

Opisthobranchs which range in size from minute forms living between sand grains to large Pacific species, some of which may weigh as much as 1.5 kg. There is no shell present, but calcareous spicules are often embedded in the skin. The body, which incorporates the visceral mass, mantle and foot, is externally bilaterally symmetrical with a slug-like or flattened form. Cerata are frequently present. There are usually two pairs of tentacles on the head, one pair of which may have external sheaths into which they can be withdrawn. A veil of skin may protect the mouth, and a radula of varying size and shape is usually present. The foot, which is often elongated, lacks parapodial lobes. The mantle cavity and true internal gill are absent, but in many species secondary respiratory structures have developed. These may be folds of tissue along the sides of the body or finger-like processes around the anus. The penis frequently bears a strong spine and an open sperm groove is lacking.

Nudibranchs are cosmopolitan, the majority living on the sea bottom with a few actively burrowing into soft sediments. One or two species are planktonic. They are all predators, feeding on a wide range of invertebrate animals. Some species are highly selective, feeding exclusively on one species of organism but others take a variety of prey. Some nudibranchs which feed on coelenterates use their stinging cells for defence, diverting them from the gut into the cerata where they are stored. Others use special skin glands which secrete distasteful fluids to repel possible predators such as fish. Several species have a conspicuous coloration, whilst others are extremely well camouflaged. The significance of this coloration is still not fully understood. Nudibranchs should be handled with care, as their secretions can often cause inflammation and blistering of the human skin.

## Suborder Dendronotacea

Nudibranchs which have special sheaths on the head into which one pair of tentacles can be withdrawn. A veil of skin, sometimes very elaborate, usually lies around the mouth. The radula, which is very variable in structure, is absent in a few species. A ridge is present on each side of the body, frequently bearing branched processes which probably function as gills.

The majority feed on coelenterates, but species with no radula eat small crustaceans, which they swallow whole.

Families: Tritoniidae, Marianinidae (=Aranucidae), Lomanotidae, Dendronotidae, Hancockiidae, Bornellidae, Dotoidae, Tethydidae (=Fimbriidae), Scyllaeidae, Phylliroidae.

### Family Tritoniidae

89/4   *Tritonia gracilis* (Risso) (=*Duvaucelia gracilis*). This species, which is rarely more than 1 cm long, has one pair of retractable branched tentacles and several pairs of branched cerata along the back. The veil in front of the mouth has six finger-shaped extensions. Found amongst undergrowth in shallow waters of the Mediterranean.

### Family Dendronotidae

89/2   *Dendronotus* sp. This beautiful Caribbean species, which is very similar to *D. frondosus*, has many tree-shaped cerata and many finger-like extensions of the head veil.

### Family Hancockiidae

89/5   *Doto fragilis* (Forbes). The body of this species reaches 3 cm in length. The concentric rings of tubercles around the cerata are characteristic. Occurs in rocky areas below tide level in the North-east Atlantic and Mediterranean, where it feeds on hydroids, particularly *Nemertesia*.

89/7   *Hancockia uncinata* (Hesse). A rare species, reaching 15 mm in length, with characteristic hand-like cerata containing nematocysts. Found in rocky areas in shallow waters of the Mediterranean and North-east Atlantic, where it feeds on hydroids.

### Family Tethydidae

89/9   *Tethys leporina* Linnaeus (=*Fimbria fimbria*). An exceptionally large and characteristic nudibranch, which may reach 20 cm in length. It envelopes small bottom-living crustaceans, bivalve molluscs and echinoderms with its large oral veil. It has, however, been seen gliding along the surface water film in calm conditions,

catching small planktonic animals. Found in the Mediterranean and adjacent Atlantic.

### Family Scyllaeidae

80/2   *Scyllaea pelagica* Linnaeus. A laterally flattened species up to 6 cm long which is normally found amongst floating *Sargassum* weed, where it preys upon attached hydroids. Found in the North Atlantic and Caribbean.

### Family Phylliroidae

89/10   *Phylliroe bucephala* Pèron and Lesueur. A planktonic transparent laterally flattened species up to 2 cm long without cerata but with a pair of tentacles. It is found frequently in the surface waters of the North Atlantic where it feeds on the medusa, *Zanclea costata*. Also reported from the East Pacific and Indo-Pacific.

## Suborder Doridacea

The largest group of nudibranchs, which contains dorsoventrally flattened forms, often with spicules in the skin. One pair of tentacles can be withdrawn into special cavities, but tentacular sheaths are rarely present. Defensive glands are frequently present on these tentacles. The radula, which is generally very broad with several rows of teeth, may be absent in a few species. External gills are present, typically arranged in a circle around the posteriorly situated anus. The gills are retracted on to the body surface when the animal is alarmed or can be withdrawn into a special pocket in more advanced forms.

They feed on a variety of animals including sponges, worms, barnacles, bryozoans and sea-squirts.

Superfamily: Gnathodoridoidea.
Families: Doridoxidae, Bathydorididae.

Superfamily: Anadoridoidea.
Families: Corambidae, Goniodorididae, Onchidorididae, Triophidae, Notodorididae, Polyceridae, Gymnodorididae, Vayssiereidae.

Superfamily: Eudoridoidea.
Families: Hexabranchidae, Cadlinidae, Chromodorididae, Actinocyclidae, Aldisidae, Rostangidae, Dorididae, Archidorididae, Homoiodorididae, Baptodorididae, Discodorididae, Kentrodorididae, Asteronotidae, Platydorididae.

Superfamily: Porodoridoidea.
Families: Phyllidiidae, Dendrodorididae.

### Family Goniodorididae

90/3   *Trapania maculata* Haefelfinger. A little-known species, up to 15 mm long, with characteristic bright yellow markings. The backward-pointing processes on either side of the lamellar tentacles and of the gills are also distinctive. It has been seen feeding on bryozoans in rocky subtidal areas of the Mediterranean and Northeast Atlantic.

### Family Notodorididae

90/4   *Notodoris megastima* Allan. A rigid-bodied species about 4 cm long, with a warty surface. The colour and markings are striking on this Indo-Pacific reef dweller.

### Family Polyceridae

90/1   *Polycera capensis* Quoy & Gaimard. Members of this genus have finger-like processes on the veil. This species, which may reach 5 cm in length, makes seasonal appearances in shallow reef areas of the Indo-Pacific.

90/2    *Polycera faeroensis* Lemche. This species, which reaches 4 cm in length, has eight frontal processes on the veil, which distinguishes it from other species in the genus occurring in the same locality. It is found in shallow subtidal rocky areas feeding on colonial bryozoans such as *Membranipora*. Occurs in the North-east Atlantic.

90/7    *Tambja affinis* Eliot. A distinctly coloured species, reaching 6 cm or more in length, that is found in subtidal reef areas of the Indo-Pacific, where it feeds on a variety of colonial animals.

90/5    *Tambja* sp. A strikingly beautiful specimen, photographed on coral rubble in the Red Sea.

### Family Gymnodorididae

90/6    *Nembrotha* sp. Members of this genus are often strikingly coloured, with prominent gills situated in the middle of the back. They reach a length of about 5 cm. This specimen was photographed on the Great Barrier Reef.

### Family Hexabranchidae

90/9    *Hexabranchus imperialis* Kent [Spanish dancer]. A distinctive species reaching 15 cm in length, which is often found on Indo-Pacific reefs.

90/8, 90/10,    *Hexabranchus sanguineus* Rüppell & Leuckart [Spanish
91/1    dancer]. This characteristic species, which reaches a length of 20 cm reveals the white rim of its mantle when swimming (Plate 91/1). It inhabits shallow reef areas in the Indo-Pacific where it lays strings of eggs in whorls (Plate 90/8).

### Family Chromodorididae

91/2    *Casella atromarginata* Bergh. A species reaching 4–5 cm in length, with a tough body surface. It occurs on Indo-Pacific reefs, where it feeds on sponges.

91/4    *Casella* sp. Similar to *C. atromarginata*, but with a different edge pattern on the mantle. Photographed in shallow waters of the Red Sea.

91/3    *Chromodoris coi* Risbec. A species up to 5 cm in length which has a characteristic pattern on its back. A common inhabitant of Indo-Pacific reefs.

91/5    *Chromodoris quadricolor* (Rüppell & Leuckart) [pyjama nudibranch]. Another unmistakable species reaching a length of 5 cm. It is found commonly in shallow reef areas of the Indo-Pacific, where it feeds on colonial invertebrates.

91/6    *Chromodoris* sp. A beautifully coloured species found living on Indo-Pacific reefs.

91/7    *Chromodoris* sp. An unknown species, possibly *C. gracilis*, photographed in a cave in the Mediterranean.

91/10    *Hypselodoris tricolor* (Cantraine). An unmistakably marked species of chromodorid which is common in subtidal rocky areas of the Mediterranean, where it feeds on sponges.

### Family Dorididae

92/1    *Peltodoris atromaculata* Bergh. This strikingly spotted flattened species reaches a length of 6 cm. It feeds, apparently exclusively, on the sponge *Petrosia dura*, which occurs in subtidal rocky areas of the Mediterranean.

### Family Archidorididae

91/9    *Archidoris pseudoargus* (Rapp) [sea-lemon]. The mantle

of this species is covered by numerous small protruberances. Specimens ranging in size from 5 to 12 cm are found both intertidally and subtidally feeding on various sponges, particularly *Halichondria panicea*, in the North-east Atlantic.

91/8    *Archidoris tuberculata* Cuvier [warty doris]. The large warty protruberances on the thick mantle of this species are characteristic, although the colour is unusually dark. Found on a variety of bottoms in the Mediterranean.

### Family Asteronotidae

92/4    *Asteronotus brassica* Allan. The mantle pattern of this 3–4 cm long species is distinctive. It is found in shallow reef areas of the Indo-Pacific.

### Family Phyllidiidae

92/2    *Phyllidia bourguini* Risbec. The warty protruberances on the upper surface of this specimen are characteristic of the genus. This species reaches a length of 10 cm and is found on and around reefs in the Indo-Pacific.

92/5    *Phyllidia ocellata* Cuvier. The 'eye spots' along the back are characteristic of this species, which may reach 10 cm in length. Found amongst undergrowth on Indo-Pacific reefs.

92/3    *Phyllidia varicosa* Lamarck. The many small warts around the edge of the body in this 10 cm long species distinguish it from other closely related forms. It exudes a toxic mucus from the body surface which repels predators. Found on Indo-Pacific reefs, where it is thought to feed on sponges.

### Family Dendrodorididae

92/6    *Dendrodoris limbata* (Cuvier). A dark-coloured species which may reach 20 cm in length. It is found subtidally, frequently feeding on the sponge *Suberites*, in the Mediterranean.

## Suborder Arminacea

A group of nudibranchs whose members possess several features of the other suborders. However, there is usually only one pair of tentacles present and these do not have external sheaths. A veil of skin may be situated around the mouth, whilst the radula is often narrow with several rows of teeth. Cerata are frequently present and the anus is often situated away from the posterior end on the right hand side of the body.

A variety of organisms including hydroids, shelled molluscs and bryozoans form the diet of arminaceans.

Superfamily: Euarminoidea.
Families: Heterodorididae, Doridomorphidae, Arminidae.

Superfamily: Metarminoidea.
Families: Madrellidae, Dironidae, Antiopellidae, Gonieolididae, Charcotiidae, Heroidae.

### Family Dironidae

92/7    *Dirona albolineata* Cockerell & Eliot. The light coloration and white-edged pointed cerata are characteristic of this species which may reach 5–6 cm in length. It is found in shallow rocky areas of the North-east Pacific, feeding on bryozoans and sea-squirts as well as on small gastropods.

## Suborder Aeolidacea

Nudibranchs whose most conspicuous feature is rows

or clusters of dorsal ceratal processes. One of the two pairs of tentacles are often very long, but there are no tentacular sheaths. A mouth-veil of skin is rarely present and the narrow radula frequently has only one row of teeth. The cerata contain branches of the mid-gut gland and pockets at their tips for holding coelenterate stinging cells.

These nudibranchs are often brightly coloured, with characteristic cerata in each species. The majority feed on coelenterates, but a few species attack other opisthobranchs and crustaceans or eat mollusc and fish eggs. The few planktonic nudibranchs belong to this group.

Superfamily: Protoaeolidoidea.
Family: Notaeolidiidae.

Superfamily: Euaeolidoidea.
Families: Coryphellidae, Nossidae, Protoaeolidiellidae, Caloriidae, Phidianidae, Facelinidae, Favorinidae, Myrrhinidae, Glaucidae, Pteraeolidiidae, Herviellidae, Aeolidiidae, Spurillidae, Eubranchidae, Cumanotidae, Flabellinidae, Pseudovermidae, Fionidae, Calmidae.

### Family Coryphellidae

92/8　*Coryphella pedata* (Montagu). A delicate species reaching 4 cm in length, which feeds on colonial hydroids in rocky subtidal areas. Found in the North-east Atlantic and Mediterranean.

### Family Caloriidae

92/9　*Caloria maculata* Trinchese. This species, which rarely exceeds 3 cm in length, is characterised by the white spots over its body. Like its fellow aeolidians, it is protected from the unwanted attentions of larger animals by the hydroid stinging cells which are held in the tips of the cerata. These have been derived from the colonial hydroids on which it feeds. Found in rocky subtidal areas of the Mediterranean.

### Family Facelinidae

92/10　*Hermissenda crassicornis* (Eschscholtz). The length of this species occasionally exceeds 5 cm. It feeds on a variety of invertebrates including hydroids, bryozoans, molluscs and sea-squirts. This is perhaps the commonest aeolid in the North-east Pacific, being found on both intertidal and subtidal rocky bottoms as well as in sea-grass beds.

### Family Glaucidae

92/11　*Glaucus atlanticus* Forster. An oceanic species, up to 5 cm in length, which cannot be readily confused with any other. It glides along the surface water film seeking out floating coelenterates such as *Velella* and *Porpita* which it devours voraciously, being immune to their powerful stinging cells. Cosmopolitan in warm oceanic waters.

### Family Aeolidiidae

93/1　*Aeolidia papillosa* (Linnaeus). Many rows of unbranched cerata occur on the back of this species, which may reach 10 cm or more in length. It is common in intertidal and subtidal rocky areas, where it feeds on several species of anemone. Found in the North Atlantic and North Pacific.

### Family Spurillidae

93/4　*Spurilla* sp. This specimen, with the inflated curving cerata characteristic of most members of the family, is feeding on the polyps of a soft coral, *Lobophytum*, in the Red Sea.

### Family Flabellinidae

*Flabellina affinis* (Gmelin). A species reaching 2 cm in　93/6 length that is sometimes confused with *Coryphella pedata*, but which can be distinguished by the lamellate tentacles above the eyes, and by the branching cerata. Feeds on colonial hydroids in rocky subtidal areas of the Mediterranean.

*Hervia costai* Haefelfinger. Bears a superficial resem-　93/5 blance to *Flabellina* but can be readily distinguished by the lack of rings on the tentacles above the eyes and by the orange patches of pigment on the head. A common Mediterranean species occurring in shallow rocky areas where it feeds particularly on the colonial hydroid, *Eudendrium*.

### Family Fionidae

*Fiona pinnata* (Eschscholtz). An oceanic species up to　93/2 6 cm in length, with many finger-like cerata situated along the lateral borders of the body. The cerata have a characteristic undulating membrane along their inner border. It is known to feed on *Velella* and the goose-barnacle, *Lepas*. Cosmopolitan in warm oceanic waters.

## Order Onchidacea

Oval-shaped opisthobranchs which resemble the chitons in their habits. The head has one pair of tentacles with eyes at their tips and a pair of skin folds on either side of the mouth. A radula is present. The extensive mantle projecting beyond the large broad foot frequently has warts on its dorsal surface; it may bear eyes or bunches of gills. A large cavity, which functions as a lung, is located in the posterior part of the body, opening to the exterior near the anus. A penis is usually situated near the right tentacle.

Onchidaceans live either in mangrove flats, or in crevices in rocky intertidal areas. Rock-dwellers feed by scaping organisms from rocks with the radula, emerging from their crevices when the tide is out. They appear to breathe equally well in air or in water.

Family: Onchidiidae.

### Family Onchidiidae

*Onchidium daemelli* Semper. This Indo-Pacific species,　93/3 which reaches 4 cm in length, lives near the high-tide mark in hollows in the mud. It is thought to browse on algal debris left by the receding tide.

## Subclass Pulmonata

Mainly freshwater and terrestrial gastropods with only a few marine species. Detorsion or partial detorsion has occurred in many pulmonates. The shell, which is either spirally coiled or uncoiled and conical, seldom has an operculum. Reduction of the shell may occur and it is absent altogether in some species. The head usually bears one or two pairs of tentacles and the radula is well developed. The foot generally occupies the entire ventral surface forming a flat creeping sole. Gills are absent and the anterior part of the mantle cavity has become adapted for respiration in air or water.

Pulmonates are hermaphrodite but copulation with exchange of sperm usually takes place. A miniature adult develops directly from the egg.

The few marine pulmonates live in rocky coastal waters of both temperate and tropical seas.

## Order Basommatophora

Shelled aquatic pulmonates with a single pair of tentacles, having an eye near their base.

Some of the marine forms are limpet-like, living attached to rocks, whilst others have spirally coiled shells and live mainly in rock crevices in the intertidal zone.

Superfamily: Siphonariacea.
Families: Trimusculidae, Siphonariidae.

Superfamily: Amphibolacea.
Family: Amphibolidae.

Superfamily: Ellobiacea.
Families: Ellobiidae, Otinidae.

# Class Bivalvia [bivalves]

approx. 7500 species

Marine and freshwater molluscs which include the well known mussels, clams, oysters, cockles and scallops. The body is laterally compressed and enclosed within two calcareous valves which are hinged at their dorsal margins and form the typical bivalved shell. Each valve bears a protuberance on the dorsal surface (the beak or umbo) which is the oldest part of the shell. The shell valves are united by an elastic ligament which controls their opening and are held together by two large muscles (adductors) which act antagonistically to the ligament (Figure 31). The mantle, which is in the form of two hanging folds, is joined along the upper line and encloses a large mantle cavity containing a pair of gills. In most bivalves, the posterior edges of the mantle folds are fused and modified to form a pair of tubes (siphons) through which water currents flow in and out of the mantle cavity. The incoming water currents carry oxygen and often also food, which is collected by the cilia of greatly enlarged and modified gills.

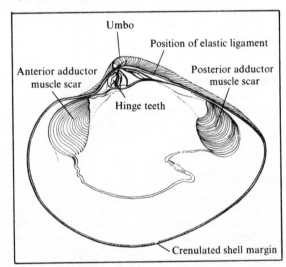

**Figure 31** *View of the inside of the right shell valve of* Venus striatula.

These is no obvious head or radula, but the lips of the mouth are often extended to form sensory 'palps'. In burrowing bivalves, the foot is well developed but is often reduced in the non-burrowing forms such as the oysters. In several species, e.g. mussels, the foot produces long threads (byssus threads) by which they attach to rocks, pier pilings etc.

The sexes are separate in the majority of bivalves, although a few hermaphrodite forms exist. Fertilisation of the egg normally occurs in the seawater, but in some species, e.g. the European oyster, it occurs in the mantle cavity, the sperm being brought in with the water currents. The young hatch as free-swimming larvae which then develop into a more advanced form, the active dispersal phase of the life cycle.

These bottom-living animals are found in all seas from the intertidal zone to great depths. The majority are associated with sand or mud, but several are adapted to life on hard bottoms. Most bivalves feed by filtering food from the water with the gills, but a few scavenging and carnivorous forms occur. They are usually fairly immobile, but some species, e.g. scallops, can swim by flapping the shell valves together.

Bivalves are an important source of food for many animals and several species are farmed commercially in coastal regions. Some species, e.g. the ship worm *Teredo*, which burrows into wood, cause extensive damage in many harbours of the world. Valuable pearls are obtained from a number of bivalves. These are formed by the laying down of mother-of-pearl around a foreign body, such as a large sand grain, which has become trapped between the shell and mantle.

## Subclass Palaeotaxodonta

A group which is considered to contain the most primitive living bivalves. The shell valves are similar to one another, often triangular or oval in shape, reaching a length of a few centimetres. Numerous short teeth occur along the hinge between the two shells and the two adductor muscles are equal in size. The relatively large foot has a flattened ventral surface. The simple gills, which consist of a series of broad leaf-like filaments on either side of a central axis (Figure 32a), may have adjacent filaments connected by tufts of cilia. The gills are used solely for respiration and not for feeding as in many of the other bivalves.

Representatives are found in all seas, sometimes at considerable depths. They burrow into soft sediments, feeding on organic debris which is collected by two pairs of large palps which are pushed out between the valves of the shell. The single order contains living as well as fossil forms.

### Order Nuculoida

Characters identical to those of the subclass.

Superfamily: Nuculacea.
Family: Nuculidae.

Superfamily: Nuculanacea.
Families: Malletiidae, Nuculanidae.

## Subclass Cryptodonta

A group of primitive bivalves with thin shells which can reach a length of several centimetres. The hinge may possess a row of short teeth, but these are often absent. Two adductor muscles are present. The gills consist of a row of leaf-like filaments on either side of a central axis (Figure 32a). They collect organic debris from the water current passing through the animal as well as performing a normal respiratory function. The palps are small and unpaired.

Only one order possesses living forms.

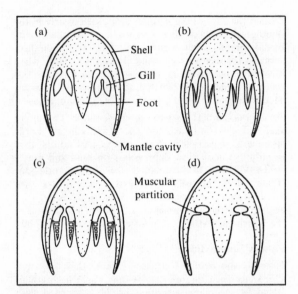

**Figure 32** *The range of gill form in the bivalves.* (a) *Palaeotaxodonta, Cryptodonta;* (b) *or* (c) *Pteriomorphia, Palaeoheterodonta;* (c) *Heterodonta;* (d) *Anomalodesmata (Poromyacea only).*

## Order Solemyoida

Bivalves with oblong or oval shell valves, with the beaks situated towards the shorter posterior end. There are no teeth along the hinge and the anterior adductor muscle is larger than the posterior one. An opening is present between the edges of the shell when the valves are drawn together by the adductor muscles. The outer edges of the shell valves are uncalcified and they can be tucked in between the valves by special muscles when the shell is closed. The large foot, which resembles a flattened sole, is used for burrowing rapidly into the sediment.

Representatives of this order live buried in muddy or sandy bottoms.

Superfamily: Solemyacea.
Family: Solemyidae.

## Subclass Pteriomorphia

Fixed or sedentary bivalves including the economically important mussels, scallops and oysters. The shell valves may have a wing-like process extending from the hinge area. The anteriorly situated beaks of each valve are often well separated from one another by a triangular area on the dorsal margin of one or both valves. There may be a row of teeth in the hinge, but this is frequently reduced or absent. In some forms the two adductor muscles are of equal size, but in others the anterior one is reduced or absent altogether. The foot is small and many of these bivalves are cemented to the substratum by one shell valve or attached by byssal threads. The edges of the mantle are not fused together to form siphons; even so the water currents which enter and leave the mantle cavity are often localised. The gills, which are used for feeding and respiration, are long and narrow and each is bent back upon itself, thus making the complete gill W-shaped in cross-section (Figure 32*b*). The ascending and descending arms of each filament are held together by sheets of tissue. In addition, adjacent filaments are joined by interlocking cilia or by tissue

connections (Figure 32*c*). Food particles are strained from the incoming water current by this structure and carried along food grooves to the mouth.

## Order Arcoida

Bivalves with circular or angular heavy shells which are often finely sculptured. They can reach a length of 15 cm. The shell valves, which often appear velvety, are of similar shape and have a row of teeth on the hinge. Two adductor muscles of equal size are present.

They are found from the lower shore to moderate depths, attached to rocks by byssus threads or buried in the sand.

Superfamily: Arcacea.
Families: Arcidae, Cucullaeidae, Noetiidae.

Superfamily: Limopsacea.
Families: Limopsidae, Glycymerididae, Manzanellidae, Philobryidae.

### Family Arcidae [ark shells]

*Arca foliata* Forskål. The shell of members of this family has a characteristic long straight hinge. The larger of the two specimens, seen here with the green ormer *Haliotis tuberculata*, is about 3 cm long. Found attached to subtidal rock by byssal threads in the Indo-Pacific.  93/9

## Order Mytiloida

An order of bivalves which contains the mussels and pen shells. The similar shell valves are either wedge- or fan-shaped and can attain lengths of 90 cm in the pen shells. The beaks are situated at, or close to, the anterior end of each valve. The hinge teeth and the anterior adductor muscle are greatly reduced or absent altogether.

They occur in all seas, commonly attached to rocks and solid objects by the byssal threads. A few species are found embedded in soft rocks and corals.

Superfamily: Mytilacea.
Family: Mytilidae.

Superfamily: Pinnacea.
Family: Pinnidae.

### Family Mytilidae

*Lithophaga teres* Philippi [chocolate date mussel]. A species up to 7 cm in length, found boring in coral both intertidally and subtidally. Occurs in the Indo-Pacific.  93/7

*Lithophaga* sp. The coral head of *Psammocora* has been broken open to reveal the 2 cm-long shell of this boring bivalve from the Indo-Pacific.  93/8

*Modiolus barbatus* (Linnaeus) [bearded horse-mussel]. The shell may reach 5 cm or more in length and is characterised by many whiskers over its surface. It lives both on the shore and subtidally under rocks and in the holdfasts of *Laminaria*. Found in the North-east Atlantic and Mediterranean.  93/10

*Musculus discors* (Linnaeus). The shell of this North Atlantic species, with a distinctive array of radiating posterior ribs, rarely exceeds 7 cm in length. Commonly found intertidally under rocks and amongst the calcareous alga *Corallina*.  93/11

*Mytilus edulis* Linnaeus [common mussel]. Th shell may reach 15 cm or more in length, but rarely exceeds  94/1

10 cm. A common species, often found in dense clusters on the shore attached to rocks by byssal threads. Individuals on exposed coasts rarely attain sizes comparable to those in estuarine beds. Found in the North Atlantic and North Pacific. Collected extensively for food.

### Family Pinnidae

94/2    *Atrina vexillum* Born. The triangular shell of this species may reach 30 cm in length and is normally found embedded in coral growths. The wide upper part of the shell seen here is gaping to reveal the striped mantle edge. Found in the Indo-Pacific.

94/8, 94/9    *Pinna nobilis* Linnaeus [fan mussel]. The unmistakable fan-shaped shell reaches 40 cm in length. In young specimens the external ornamentation of scales is better developed (Plate 94/8). It is often found upright in sand and mud, attached to an object below the sediment surface by its byssal threads. The adductor muscle of this species is prized as food. Found in the Mediterranean and adjacent Atlantic.

## Order Pterioidea

Representatives of this order, which contains the familiar scallops and oysters, often have thickened dissimilar shell valves which may be strongly sculptured, ribbed or have spiny projections. Wing-like processes of variable size and shape commonly occur on the shell valves, although they are not present in the oysters. There are few or no teeth along the hinge and the anterior adductor muscle is very small or absent altogether.

They are found in both shallow and deep waters, where many are attached to solid objects by byssal threads or by one shell valve. Some lie freely on the bottom and are capable of swimming by rapid opening and closing of the shell valves.

Superfamily: Pteriacea
Families: Pteriidae, Isognomonidae, Pulvinitidae, Malleidae.

Superfamily: Pectinacea.
Families: Pectinidae, Plicatulidae, Spondylidae, Dimyidae.

Superfamily: Anomiacea.
Family: Anomiidae.

Superfamily: Limacea.
Family: Limidae.

Superfamily: Ostreacea.
Families: Gryphaeidae, Ostreidae.

### Family Pteriidae [wing oysters]

94/5    *Pteria aegyptiaca* ('Chemnitz'). The winged shell valves of this 5 cm long specimen nestling amongst the fire coral *Millepora*, are covered externally by algal growth. Internally the shell has a pearly appearance. Found in the Indo-Pacific.

94/7    *Pteria* sp. An unnamed species, similar in size and appearance to *P. aegyptiaca*, attached to coral off Mombasa.

### Family Pectinidae

94/3    *Chlamys opercularis* (Linnaeus) [Queen scallop]. The upper shell valve of this species, which may reach 9 cm across, is more convex than the lower one. The pink gonads and the large number of well developed eyes

around the edge of the mantle can be seen clearly in the photographed specimens. This species is attached by its byssal threads when young, but lives freely on firm subtidal sediments as an adult. It swims actively when disturbed by predators such as the starfish *Asterias rubens*. Found in the North-east Atlantic and adjacent Mediterranean. Fished commercially in certain areas.

*Pecten maximus* (Linnaeus) [great scallop]. The upper    94/4, 94/ shell valve is flattened and the lower convex. Large specimens may reach 15 cm across. As adults they usually live in shallow depressions on sand and gravel below tide level. In the detailed view of the mantle edge (Plate 94/4) the well developed eyes can be seen amongst short sensory tentacles. This species is not such an accomplished swimmer as *C. opercularis*. Found in the North-east Atlantic. Collected extensively for food.

### Family Spondylidae

*Spondylus americanus* Hermann [Atlantic thorny oyster].    95/1 The spiny shell of this species may reach a length of 10 cm or more, and is often heavily encrusted with hydroids and sponges. Occurs attached to rocks and corals in the Caribbean.

*Spondylus aurantius* Lamarck. A close-up photograph    95/2 of this large Indo-Pacific species, reaching 20 cm across, reveals the frilled mantle and the large number of small eyes. Specimens are often found attached beneath rocky overhangs and on cave walls. One valve of this specimen is almost completely covered by the red sponge, *Microciona*, whilst the other bears hydroids, bryozoans and sponges.

### Family Limidae [file shells]

*Lima hians* (Gmelin) [gaping file shell]. The dirty white    95/3 shell of this species has a pronounced gape anteriorly. The mantle fringe of non-retractable reddish-orange annulated tentacles is a feature of this species. Specimens are able to swim with the aid of these sticky easily detachable tentacles, which are also used to ward off predators. Found on sand and shell-gravel bottoms below tide level in the North-east Atlantic and Mediterranean.

*Lima scabra* (Born) [rough file shell]. This species,    95/4 which reaches 5 cm or more in length, exists in two forms, one having a more delicate and more finely ribbed shell than the other. Like most members of the genus, it is able to weave a protective nest of byssal threads and bottom debris around itself. It is found between coral rocks in shallow Caribbean waters.

*Promantellum vigens* Iredale. This 3–4 cm long Indo-    95/6 Pacific species normally lives on sand beneath coral boulders. When disturbed it is able to swim vigorously by flapping its long mantle tentacles.

### Family Ostreidae

*Crassostrea gigas* (Thunberg) [Japanese oyster]. A large    95/7 oyster whose shell may reach a length of 20 cm or more. Specimens are usually cemented to hard bottoms by the left valve. The shell may have a very distorted growth form, although the fluted nature of the shell is usually apparent. This species, which is grown commercially in estuaries throughout the North Pacific, is not considered to be the gastronomic equal of Atlantic oysters by connoisseurs.

*Lopha cristagalli* (Linnaeus) [cock's-comb or rooster-    95/5, 95 comb oyster]. A distinctive species with a zig-zag edge to the shell, which may reach 10 cm across. One or both shell valves are often encrusted by other organisms, particularly by the red sponge *Microciona*. Found subtidally in the Indo-Pacific attached to rocks and coral, often beneath overhangs.

95/9   *Ostrea edulis* Linnaeus [common European oyster]. The shell valves of this species often have a rounded outline and are up to 10 cm across. The two valves are dissimilar; the upper (right) one being flattened, the lower convex. It lives naturally from low water downwards on rocks and firm muddy gravel and sand. It is also grown commercially on man-made hard bottoms. Found in the North-east Atlantic and Mediterranean.

95/10  *Ostrea hyotis* Linnaeus (= *Pycnodonta hyotis*) [fluted oyster]. The wavy edge of the shell of this specimen is gaping to reveal the beautifully patterned mantle. Lives subtidally attached to coral or rock in the Indo-Pacific.

96/1   *Saccostrea cucullata* Born. This species, with a fluted shell up to 8 cm long, has been eaten by man since prehistoric times. The upper valve of the photographed specimen shows evidence of boring by the sponge *Cliona* and the bristleworm *Polydora*. It is found in a variety of habitats in the Indo-Pacific, particularly in estuaries, attached to rocks, mangrove roots, and even on open mud.

## Subclass Palaeoheterodonta

A group of mainly freshwater bivalves with only a few marine species. The shell valves are similar, often circular to oval in shape. Two adductor muscles are commonly present, but the number of teeth in the hinge is often small. The edges of the mantle are free or incompletely fused together and the gills, which are used for feeding, have their filaments joined by cilia or sheets of tissue.

## Order Trigonioida

Small bivalves with three-cornered oval shells which bear ribs radiating from the beak region. The anterior adductor muscle is small and there are only two or three teeth in the hinge of each valve. The edges of the mantle are not fused and the gill filaments have ciliary connections.

The only living genus, *Neotrigonia*, occurs in southern latitudes.

Superfamily: Trigoniacea.
Family: Trigoniidae.

## Subclass Heterodonta

A large subclass of bivalves which contains the cockles, venus and razor shells, the giant clam and the boring shipworms and piddocks. The shells, which are of varying shape, range in length from a few millimetres to well over one metre in the giant Indo-Pacific clam. This clam may weigh over 250 kg.

There is often a flat or concave area between the beak and hinge margin of each shell valve which is partly or wholly occupied by the ligament. The hinge may be strong, containing teeth of two kinds – large teeth situated just below the beaks and smaller subsidiary teeth along the dorsal margins of the valves, or be weak with few or no teeth. In the latter case the ligament is usually also reduced and the shell gapes open.

Two adductor muscles of similar size are usually present although the anterior one may be reduced or absent in some species. The mantle edges show varying degrees of fusion and siphons are commonly present. The highly developed gills are used for feeding as well as for respiration and have tissue connections between the filaments (Figure 32c).

Representatives of this subclass are found in all seas, where many burrow into soft sediments (Figure 33). Several occur on the surface of sand and mud whilst others are attached to or bore into rocks.

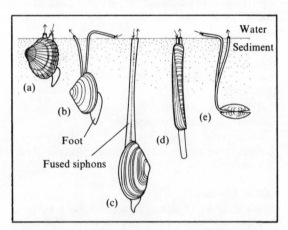

**Figure 33** *Section through sediment revealing the attitude most commonly adopted by* (a) Cerastoderma, (b) Macoma, (c) Mya, (d) Ensis, (e) Tellina. *Arrows mark the direction of water flow.*

## Order Veneroida

Bivalves with similarly shaped shell valves, whose beaks are often towards the anterior end. The hinge usually has two types of teeth and there are often two equally sized adductor muscles. The ventral mantle edges are fused in one or more places and siphons are usually present. The foot may be elongated in the burrowing forms or be smaller with byssal threads in attached bivalves.

This order contains bivalves living on the surface of the sediment as well as many burrowing forms. Several others attach to rocks or corals.

Superfamily: Lucinacea.
Families: Lucinidae, Thyasiridae, Mactromyidae, Fimbriidae, Ungulinidae, Cyrenoididae.

Superfamily: Chamacea.
Family: Chamidae.

Superfamily: Leptonacea.
Families: Erycinidae, Kelliidae, Leptonidae, Montacutidae, Galaeommatidae.

Superfamily: Chlamydoconchacea.
Family: Chlamydoconchidae.

Superfamily: Cyamiacea.
Families: Cyamiidae, Turtoniidae, Sportellidae, Neoleptonidae.

Superfamily: Carditacea.
Families: Carditidae, Condylocardiidae.

Superfamily: Crassatellacea.
Families: Astartidae, Crassatellidae, Cardiniidae.

Superfamily: Cardiacea.
Families: Cardiidae, Lymnocardiidae.

Superfamily: Tridacnacea.
Family: Tridacnidae.

Superfamily: Mactracea.
Families: Mactridae, Anatinellidae, Cardiliidae, Mesodesmatidae.

Superfamily: Solenacea.
Families: Solenidae, Cultellidae.

Superfamily: Tellinacea.
Families: Tellinidae, Donacidae, Psammobiidae, Scrobiculariidae, Semelidae, Solecurtidae.

Superfamily: Dreissenacea.
Family: Dreissenidae.

Superfamily: Gaimardiacea.
Family: Gaimardiidae.

Superfamily: Arcticacea.
Families: Arcticidae, Bernardinidae, Kelliellidae, Trapeziidae.

Superfamily: Glossacea.
Families: Glossidae, Vesicomyidae.

Superfamily: Veneracea.
Families: Veneridae, Petricolidae, Cooperellidae, Glauconomidae.

## Family Chamidae

7/1    *Chama gryphoides* Linnaeus [Mediterranean jewel box]. This species is attached to rocks below tide level by its deeply convex left valve, whilst the upper, more flattened valve acts as a lid on the 'box'. The shell is often encrusted by other organisms such as the red sponge *Microciona*. Found in the Mediterranean.

## Family Cardiidae

96/5    *Acanthocardia echinata* (Linnaeus) [prickly cockle]. A cockle up to 6 cm long with radiating ribs on the shell valves bearing short spines. The large muscular pink foot of the specimen in the foreground of the photograph can be clearly seen. It normally lives buried or partly buried in subtidal muddy sand or muddy gravel in the North-east Atlantic and Mediterranean.

96/10   *Cerastoderma edule* (Linnaeus) (=*Cardium edule*) [common edible cockle]. Large specimens may reach a length of 5 cm. The solid shell valves have radiating ridges without spines and fine concentric lines. It commonly inhabits intertidal clean and muddy sand and gravel. Specimens may be exposed or, more usually, are buried beneath the sediment surface. The species is collected extensively by man as food, and is also preyed upon heavily by natural predators such as oyster-catchers and plaice. Found in the North-east Atlantic and Mediterranean.

## Family Tridacnidae

96/2, 96/4   *Tridacna crocea* Lamarck. One of the smaller species, rarely more than 15 cm long, which is found buried in coral rock, with only the fluted edge of the shell exposed (Plate 96/4). Although filter-feeders, they derive some if not most of their nutriment, as do most *Tridacna* species, from zooxanthellae embedded in the fleshy mantle (Plate 96/2). Found in shallow waters of the Indo-Pacific.

96/3    *Tridacna gigas* Linnaeus [giant clam]. This massive species may reach a width of over 1 m and weigh in excess of 200 kg. This clam, which has an undeserved reputation as a man-trap as a result of Hollywood

movies, lives in shallow reef areas with some, or all, of its shell exposed. Found in the Indo-Pacific.

*Tridacna maxima* Roeding. A species which lives in   96/7
shallow reef areas of the Indo-Pacific and rarely reaches more than 20–30 cm across.

## Family Mactridae

*Mactra corallina* (Linnaeus) [rayed trough shell]. A   96/6
thin-shelled species reaching 5 cm in length, with a smooth shiny surface. The specimen seen here has been killed by the necklace shell *Natica alderi*, which has bored a round hole through the oldest part of the shell into the underlying adductor muscle. Found in clean intertidal and subtidal sand in the North-east Atlantic and Mediterranean.

## Family Cultellidae

*Ensis siliqua* (Linnaeus) [razor shell]. The long   96/8
rectangular-shaped valves of members of this family are distinctive. This species, which may reach 20 cm in length, is normally found buried, with its long axis vertical, in clean sand on the lower shore and in shallow water. A North-east Atlantic and Mediterranean species.

## Family Solecurtidae

*Pharus legumen* (Linnaeus). Somewhat similar in form   96/9
to *Ensis*, but with the ends of the valves rounded and with the long edges curved. This species, which rarely exceeds 12 cm in length, burrows in sand on the lower shore and in shallow water in the North-east Atlantic and Mediterranean.

## Family Veneridae

*Venerupis decussata* (Linnaeus) [carpet shell]. A solidly   97/1
built shell up to 7 cm in length and of varying colour. The concentric rings and radiating lines make a characteristic squared pattern which is particularly noticeable at the anterior and posterior ends of the shell. Lives buried in muddy sand and clay on the shore and in shallow water in the North-east Atlantic and Mediterranean.

*Venus striatula* (da Costa) [striped Venus]. A thick-   97/3
shelled species, with specimens rarely exceeding 4 cm in length. The shell has distinctive concentric ridges and usually three dark brown radiating rays. The specimen seen here, which has been dug from the clean sand in which it normally lives, has its two siphons united. Found in the North-east Atlantic and Mediterranean.

# Order Myoida

Thin-shelled burrowing forms whose shell valves are often dissimilar to one another. There are few or no teeth in the hinge and the ligament is often reduced, causing the shell to gape open. The anterior adductor muscle may be reduced, but it is well developed in the rock-boring piddocks, where it aids the drilling process. The ventral edges of the mantle are completely fused except where the foot emerges, and long, well developed siphons are present.

Members of this order burrow into soft sediments, rocks, corals or wood.

Superfamily: Myacea.
Families: Myidae, Corbulidae, Spheniopsidae.

Plate 97

Superfamily: Gastrochaenacea.
Family: Gastrochaenidae.

Superfamily: Hiatellacea.
Family: Hiatellidae.

Superfamily: Pholadacea.
Families: Pholadidae, Teredinidae.

### Family Gastrochaenidae

97/2  *Gastrochaena cuneiformis* Spengler [flask shell]. A characteristically shaped Indo-Pacific species, up to 3 cm long, that lives in burrows in coral or soft rock with only the tips of its siphons exposed.

### Family Hiatellidae

97/4  *Hiatella arctica* (Linnaeus). There is a great deal of variation in shell outline in this species, but the majority tend to be rectangular and not more than 3–4 cm long. Lives in kelp holdfasts, rock crevices or in rock borings attached by its byssal threads. Found from the lower shore downwards in the Arctic, North Atlantic and Mediterranean, North Pacific and Panamanian regions.

### Family Pholadidae

97/8  *Pholas dactylus* Linnaeus [common piddock]. The thin elongated shell valves of this species may reach 15 cm in length. Anteriorly, concentric and radiating ribs on the valve surfaces assist in the boring activities of this bivalve. It bores into a variety of soft rocks, wood and clay to provide the animal with a protective chamber from which the siphons protrude. The specimen in the photograph was revealed when the rock was broken open. The animal has a blue-green phosphorescence when viewed in the dark. Found in the North-east Atlantic and Mediterranean.

### Family Teredinidae

97/6  *Bankia setacea* (Tyron). This species bores deeply into wood, using two small anteriorly placed roughened shell valves which are small in relation to the rest of the body. The valves are rotated with respect to the foot, carving out a circular burrow which is then lined by a calcareous secretion of the mantle. Although a certain amount of wood enters the digestive tract it is not clear how much this aids the nutrition of the animal, as it still possesses the filter-feeding mechanism found in other bivalves. A piece of timber has been broken open to reveal the limy burrows within. Found in the North-east Pacific.

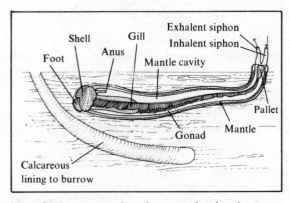

Figure 34 *Cross-section through a piece of timber showing an old burrow, and a new one being excavated by the shipworm,* Teredo.

*Teredo navalis* Linnaeus [shipworm]. Similar in habits to the above species, with a limy tube which may reach 20 cm in length. As in *Bankia*, the entrance to the burrow can be closed by limy structures (pallets) when the siphons are withdrawn (Figure 34). Widely distributed in the northern hemisphere.  97/7

## Subclass Anomalodesmata

The shell valves, which are often elongated, are dissimilar to one another and the hinge teeth are reduced or absent. Two equally sized adductor muscles are commonly present, although the anterior one is reduced in a few species. The mantle edges are fused and the siphons are often long and joined together. In the bizarre watering-pot shells (Superfamily Clavagellacea) the long siphons are enclosed within a calcareous tube. In several species, the gills, which have their filaments joined by sheets of tissue, are used both for feeding and respiration, but in the Superfamily Poromyacea the gills are modified into a pair of muscular partitions which drive water through the mantle cavity (Figure 32*d*). As a result the mantle has taken over the respiratory function. These bivalves are carnivores and scavengers, feeding on small animals and organic debris swept into the mantle cavity in the water current. Some hermaphrodite species occur within this subclass.

They are found in many seas either burrowing into soft sediments or lying on the surface of sand and mud.

### Order Pholadomyoida

Characters similar to those of the subclass.

Superfamily: Pholadomyacea.
Family: Pholadomyidae.

Superfamily: Pandoracea.
Families: Pandoridae, Cleithothaeridae, Laternulidae, Lyonsiidae, Myochamidae, Periplomatidae, Thraciidae.

Superfamily: Poromyacea.
Families: Poromyidae, Cuspidariidae, Verticordiidae.

Superfamily: Clavagellacea.
Family: Clavagellidae.

## Class Scaphopoda [elephant's tusk shells]
approx. 350 species

An exclusively marine group of molluscs whose shell length ranges from a few millimetres to several centimetres. The shell, which resembles an elephant's tusk, is a curved tapered tube, open at both ends. The elongated body of the mollusc is surrounded by the mantle and a large mantle cavity extends along the complete ventral surface. There are no gills present. The head and foot project from the larger anterior end of the shell. The head which has no eyes, carries two clusters of club-shaped tentacles which are used as sensory structures as well as for food capture. The cylindrical foot is used for burrowing and in some species its tip can form an anchoring disc (Figure 35).

The sexes are separate and the eggs and sperm are shed into the water, where fertilisation occurs. The fertilised egg develops into a planktonic larva which passes through a more advanced larval stage before developing into an adult.

Scaphopods are found in all seas from shallow to abyssal depths. They burrow into soft sediments where

Plate 97

Phylum Mollusca

they lie with the smaller posterior end of the shell just projecting above the bottom. They feed on microscopic organisms (particularly foraminiferans) in the sand or surrounding water; food is captured and transferred to the mouth by the club-shaped tentacles.

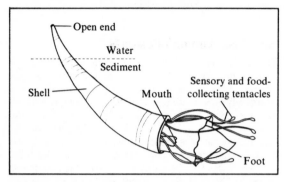

**Figure 35** *The scaphopod,* Dentalium, *in its natural position in the sediment.*

## Order Dentalioida

Animals with smooth or ribbed shells up to several centimetres in length. The short cone-shaped foot is thrust into the sand as the animal burrows.

Families: Dentaliidae, Laevidentaliidae.

### Family Dentaliidae

97/5   *Dentalium dentale* Linnaeus. The gently curving ribbed shell of this species, which reaches 3 cm or more in length, normally lies buried in soft sediments with only the narrow posterior tip protruding. One of these two specimens removed from Mediterranean sand has its muscular digging foot extended.

## Order Siphonodentalioida

Animals with small, usually smooth, shells often only a few millimetres in length. The elongated worm-like foot has an expanded frilled disc at its end which acts as an anchoring device.

Families: Siphonodentaliidae, Cadulidae.

## Class Cephalopoda
approx. 600 species

An entirely marine group of highly developed molluscs, which includes the octopus, squid and cuttlefish. Members range in size from 15 mm to the reported 18 m length of the giant squid, the largest living invertebrate. The body is bilaterally symmetrical and has no obvious foot. An external chambered shell is present in the primitive genus *Nautilus*, but in other cephalopods the shell is reduced and internal, or absent altogether. The well developed head bears tentacles which often have one or more rows of suckers. The mouth, which lies at the centre of the tentacle ring, is armed with a strong parrot-like beak. The prominent pair of eyes frequently resemble those of vertebrates in their complexity. A muscular mantle surrounds the visceral mass and the ventrally positioned mantle cavity with its gills and anus. At the head end of the body, the mantle has a free border (collar). A funnel, a modification of the foot, leads from the mantle cavity and opens under the head. The water which enters the mantle cavity

around the edge of the collar can be ejected rapidly through the funnel when the mantle contracts and seals the collar against the body. The animal thus has a jet-propulsive mechanism for rapid movement when required. Many cephalopods have an ink-sac opening into the mantle cavity which produces a 'smoke-screen' of ink in the water when the animal is disturbed. Pigment cells in the skin make rapid colour changes possible in many species, and light-producing organs are present in deep-sea forms.

Sexes are separate in cephalopods, and prior to copulation there is often an elaborate mating ritual involving colour changes and touching of the tentacles. One (occasionally more) of the tentacles of the male has become modified (hectocotylus) and is used to transfer sperm to the mantle cavity of the female. The large eggs, which are either laid singly or in jelly-like clusters, develop directly into a tiny adult form. Some species, e.g. the octopus, may brood the eggs.

Representatives of this class range from shallow to abyssal depths in all seas. Some are solitary bottom-living animals, e.g. octopuses, whilst others live in groups freely swimming in the open water, e.g. squids. All are active predators, feeding mainly on crustaceans and fish. Cephalopods form an important part of the diet of many fish and toothed whales. They are also eaten by man in many parts of the world.

## Subclass Nautiloidea

Cephalopods with an external many-chambered shell which is coiled in the only living genus, *Nautilus*. The body occupies only the last chamber of the shell, gas being present in the other chambers (Figure 36*a*). The gas gives buoyancy to the animal. The numerous tentacles on the head have no suckers and the large eyes, which are situated on short stalks, are simple, consisting of an open pit with no lens. The funnel is composed of two separate lobes. Ejection of water through the funnel when the animal swims is caused by contraction of the funnel muscles and withdrawal of the body into the shell rather than by contractions of the mantle as in other cephalopods. Two pairs of gills are present in the mantle cavity. It has no ink-sac or pigment cells in the skin.

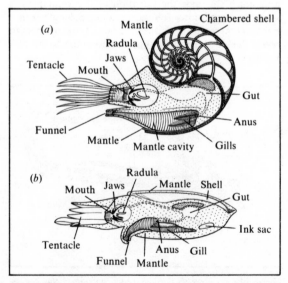

**Figure 36** *Longitudinal section through the body of* (a) *the primitive cephalopod,* Nautilus, *and* (b) *the cuttlefish,* Sepia.

Four tentacles of the male are modified for copulation, and the fertilised eggs are attached singly to the substratum by the female.

*Nautilus* is the only survivor of a large group of cephalopods with external shells which flourished over 100 million years ago. It lives in the open water of tropical regions ranging from near the surface to depths greater than 500 m. It is commonly found in groups and is active at night when it uses its many tentacles to capture prey.

Family: Nautilidae.

### Family Nautilidae

98/1  *Nautilus macromphalus* Sowerby. The shell of this species reaches 20 cm across. This specimen, photographed swimming at the surface, normally lives in deeper water feeding on various types of crustaceans and fish. Restricted to the region of New Caledonia in the Indo-Pacific.

98/2  *Nautilus pompilius* Linnaeus (= *N. repurtus*). Similar in size and habits to the species listed above. This specimen was photographed at night after it had swum up from deeper water using its powerful jet-propulsive mechanism. Restricted to the region of Palau in the Indo-Pacific.

## Subclass Coleoidea

Cephalopods whose shells are internal or absent altogether (Figure 36b). The head has eight or ten suckered tentacles and a pair of highly developed eyes. The funnel is always a closed tube and there is a single pair of gills in the mantle cavity. An ink-sac is present in most species and there are usually pigment cells in the skin.

## Order Sepioidea [cuttlefishes and bottle-tailed squids]

Cephalopods with an internal calcareous shell which is either coiled and chambered, or straight, or with a vestigial horny shell, or with no shell at all. The body is usually fairly short and broad with lateral fins. The head has ten tentacles, two of which are longer and more mobile than the rest. These longer tentacles, which bear suckers at their spoon-shaped ends only, can be retracted into special pits, and are used for capturing prey. The eyes are covered by a thin layer of skin and in addition are usually equipped with eyelids.

Sepioids live mainly in coastal waters of tropical and temperate regions, but there are a few species living in the open ocean.

Families: Spirulidae, Sepiidae, Sepiadariidae, Sepiolidae, Idiosepiidae.

### Family Spirulidae

98/4  *Spirula spirula* Linnaeus. An oceanic species, rarely longer than 7 cm, whose spirally coiled internal shell can be seen protruding near the rear end. They have a light-producing organ at the apex of the body near the posterior end. It lives in shoals at depths from 100–2000 m, feeding on a variety of planktonic animals. Occurs in tropical and subtropical regions of the Atlantic, Indian and Pacific Oceans.

### Family Sepiidae

97/9  *Sepia officinalis* Linnaeus [common cuttlefish]. The oval-shaped species, which may reach 30 cm in length, is capable of rapid colour change. It is found over sand and sea-grass, sometimes in large shoals, where it seeks out prey such as prawns. The internal 'shell' of the dead animal is frequently washed up on the beach. Found in the North-east Atlantic and Mediterranean.

*Sepia plangon* Gray. The large number of pigment cells  98/5 in the skin and the beautiful eye coloration can be seen in this photograph of the front end of a specimen 20 cm long. Sometimes common over sand in shallow waters of the Indo-Pacific.

*Sepia* sp. The outline of this 10 cm-long well  98/3 camouflaged specimen is broken up by ragged protruberances from the body surface. Found in the Indo-Pacific.

### Family Sepiolidae

*Heteroteuthis* sp. Like other members of the family, this  98/6 oceanic form is short in relation to its width (about 5 cm long) and has large, posteriorly situated, lateral fins. Clouds of luminescent particles may be emitted from the body, possibly as a decoy, when the animal is disturbed. Ranges in depth from 50–750 m in the Atlantic.

*Sepiola atlantica* d'Orbigny. The large eyes characteristic  98/7 of a predatory species can be clearly seen in this photograph, as can the internal shell. This species, which may reach a length of 5 cm, occurs over sand and in sea-grass beds in shallow inshore waters of the North-east Atlantic.

## Order Teuthoidea [squids]

Cephalopods with an internal horny shell which is in the form of a flattened pen. The body is elongated, often torpedo-shaped, with lateral fins. The head has ten tentacles, two of which may be elongated, but these cannot be retracted into pits. The eyes may or may not be covered by skin.

Squids are found in coastal waters and in the open ocean.

### Suborder Myopsida

Squids in which the lateral fins are united at the posterior end of the body. Two of the tentacles on the head are elongated and the large eyes are covered by thin skin.

Members of this suborder, which contains the common squid, are restricted to coastal waters.

Families: Pickfordiateuthidae, Loliginidae.

### Family Loliginidae

*Loligo vulgaris* Lamarck [common squid]. This species  98/8 reaches 50 cm in length and has paired triangular fins, joined posteriorly, on the rear half of the body. The internal horny shell is long and thin. They are found in shoals over sand and attach their egg strands to hard objects on the sea bed.

*Loligo* sp. A beautifully coloured individual, close to the  98/9 sandy bottom in the Red Sea.

### Suborder Oegopsida

Squids in which the tentacles are often all the same length and their suckers modified to form hooks. The eyes are not covered by skin.

This suborder, which contains the giant squids, has representatives in abyssal depths as well as in surface waters.

Families: Lycoteuthidae, Enopleuthidae, Octopo-teuthidae, Onychoteuthidae, Cycloteuthidae, Gona-tidae, Psychroteuthidae, Lepidoteuthidae, Archi-teuthidae, Histioteuthidae, Alluroteuthidae, Bathy-teuthidae, Ctenopterygidae, Brachioteuthidae, Bato-teuthidae, Ommastrephidae, Thysanoteuthidae, Chiro-teuthidae, Mastigoteuthidae, Promachoteuthidae, Grimalditeuthidae, Joubiniteuthidae, Cranchiidae.

### Family Enopleuthidae

99/1   *Abraliopsis morrisi* (Vérany). This oceanic species, which rarely has a mantle length greater than 3 cm, is characterised by its fins, which form a broad arrow-head. A red prawn has been caught by this specimen. These common North Atlantic squids normally hunt in shoals in the top 100 m or so.

99/2   *Ancistrocheirus lesueuri* (d'Orbigny). An oceanic species with a mantle length of up to 35 cm and with rounded fins. It has been taken from the stomachs of sperm whales. Recorded from the Atlantic and Indo-Pacific.

99/3, 99/6   *Pyroteuthis margaritifera* (Rüppell). The mantle of this oceanic species rarely reaches more than 4 cm in length. Light-organs are present over the body. It has been recorded from the surface to depths of 1000 m in the Atlantic, Mediterranean and Indo-Pacific.

### Family Onychoteuthidae

99/4   *Onychia carribaea* Lesueur. A small species with a mantle length rarely exceeding 3 cm. It is one of the few species of oceanic squid that is normally caught at the surface during the day. Numerous records from the Atlantic and Caribbean.

99/5   *Onychoteuthis banksi* (Leach). Adult females may have a mantle length approaching 30 cm. The bite of this species is toxic to man. It ranges in depth from the surface (particularly at night) to 500 m or more. It occurs in sufficient abundance to be sold for human consumption in Japan. It has a world-wide distribution.

### Family Histioteuthidae

99/7   *Calliteuthis meleagroteuthis* Chun. The largest members of this species have a mantle length of 6–7 cm. This specimen has the left eye 4–5 times larger than the other. Caught at depths of about 500 m. From the few catch records available it appears to have a cosmopolitan distribution.

99/8   *Histioteuthis bonellii* Férussac. A ventral view of this specimen shows the densely distributed light-producing organs characteristic of this family. When viewed from below in the depths at which it normally occurs (400–500 m) the amount of light produced by these organs is similar to that filtering through from the surface. Hence the squid becomes invisible from below. Never-theless, it still falls victim to sperm whales in large numbers in the North Atlantic. Recorded from the Atlantic, Mediterranean and Indian Ocean.

### Family Brachioteuthidae

100/1   *Brachioteuthis riisei* (Steenstrup). During growth, a juvenile stage occurs with an extraordinary long neck that shortens during later development. The specimen here is one such juvenile stage. An oceanic species with a world-wide distribution.

### Family Cranchiidae

100/3   *Bathothauma lyromma* Chun. A rare translucent deep-sea squid with eyes situated on stalks. Members of this family seem to retain many of the features associated with larval forms, even when they have apparently reached adulthood. Specimens with a mantle length of over 10 cm have been recorded. It has a world-wide distribution usually below 500 m.

*Cranchia scabra* Leach. A bulbous species of oceanic squid whose mantle may reach 11 cm in length. It has red pigment cells scattered over the mantle surface. One of the commonest species in this family, which has been recorded from the surface to depths in excess of 1000 m. It has a cosmopolitan distribution.   100/2

*Phasmatopsis* sp. The silvery reflecting layer coating the opaque eyes and liver of this otherwise transparent squid help it to elude predators by making it less visible at the depths at which it normally occurs. Found in the Atlantic.   100/4

## Order Octopoda [octopuses]

Cephalopods with short rounded bodies which, except in some deep-sea forms, have no fins. The internal shell is usually absent, but an external shell is found in the female *Argonauta*. The head has eight tentacles, often of equal length, which may be joined by a web of skin at their bases. Long mobile filaments (cirri) occur on the tentacles of some species.

The majority are found on sandy and rocky bottoms in shallow waters, although a few are oceanic. They are usually night-active predators, some species exhibiting a high degree of intelligence.

### Suborder Cirrata

Deep-sea forms which frequently have fins. The tentacles, which are often of unequal length, possess 'sensory' cirri between the muscular suckers.

Cirrate octopods occur at great depths (over 2000 m) and are rarely captured. They have been observed from a deep-sea vehicle, hovering a few metres above the bottom with the aid of fins, and swimming by jet-propulsion with the tentacles and web of skin closed together to form a tapering tail.

Families: Cirroteuthidae, Stauroteuthidae, Opistho-teuthidae.

### Suborder Incirrata

This suborder contains the familiar octopuses of shallow waters. Fins and tentacular cirri are absent.

Families: Bolitaenidae, Amphitretidae, Idioctopodidae, Vitreledonellidae, Octopodidae, Tremoctopodidae, Ocythoidae, Argonautidae, Allopsidae.

### Family Bolitaenidae

*Bolitaenia* sp. Members of this octopod family are unusual in that they are oceanic with a gelatinous body. The eyes are widely separated and the jaws are relatively soft. Little is known of their habits, but it is possible that they float with the tentacles uppermost, catching small crustaceans. They have been recorded from depths of 100–1500 m in the Pacific and Atlantic.   100/6

### Family Octopodidae

*Eledone cirrhosa* (Lamarck) [lesser octopus]. A night-active species with a body length rarely exceeding 50 cm. The arms have a single row of suckers. It feeds   100/5 on crabs and molluscs which are cracked open with its strong beak. During the day it hides amongst rocks in shallow waters. Found in the North-east Atlantic.

100/7 *Hapalochlaena maculosa* Hoyle [blue-ringed octopus]. This distinctive species reaches only 10 cm in length, but it has a deadly bite. The toxin, which acts rapidly on the neuromuscular system, has been known to cause death of incautious humans within 15 minutes. It is common on reefs and under stones from low-tide level to 20 m in the Indo-Pacific.

100/8 *Octopus cyaneus* Gray. This common species may reach a length of 30 cm or more. Like other members of this genus it has a double row of suckers on the arms. Found both intertidally and subtidally around reefs and rocky areas in the Indo-Pacific.

100/9, *Octopus vulgaris* Lamarck [common octopus]. A relatively
100/11 large species of octopus which may reach a length of 1 m. The upper side of the body is warty. Like other members of this family it is capable of changing its colour to match that of its background (Plate 100/9). It is most active at night, seeking out and enveloping its prey (particularly crabs) with the web uniting the base of the arms (Plate 100/11), before breaking the carapace open with its horny beak. Usually hides in its lair amongst rocks in the subtidal zone during the day. Found in the Mediterranean and North Atlantic.

100/10 *Octopus horridus* d'Orbigny. A delicately coloured species with an almost translucent web of tissue between the arms. Found in the Indo-Pacific.

**Family Tremoctopodidae**

101/1 *Tremoctopus* sp. An oceanic octopod which spreads its arms to form a large catchment web. This species is found in the Atlantic.

**Family Argonautidae**

*Argonauta argo* Linnaeus [paper nautilus]. The female   101/3 of this oceanic species secretes a thin protective shell (not homologous to the normal mollusc shell) by means of a pair of specially modified glandular arms. After fertilisation, her eggs are brooded within this shell, which also serves as protection for the adult. The smaller male does not secrete a shell. A cosmopolitan species in warm waters.

## Order Vampyromorpha [vampire squids]

Small cephalopods with eight long arms and two small thread-like arms which can be withdrawn into pockets. The eight long arms are united by a web of skin and each bears a single row of suckers alternating with paired cirri. The body is short and stout and has relatively large fins. The eyes are large and light-producing organs are present in the skin. There is no ink-sac.

Vampire squids are deep-water forms occurring at depths of 300–3000 m in all tropical and subtropical oceans. They are thought to be plankton-feeders, entangling small prey in the web of skin uniting the arms.

Family: Vampyroteuthidae.

**Family Vampyroteuthidae**

*Vampyroteuthis infernalis* Chun. A rare oceanic species   101/2 of deep-living cephalopod rarely more than 3–4 cm in length. All but the tips of the tentacles are united by a web of skin.

# Phylum Bryozoa

(= *Ectoprocta or Polyzoa*) approx. 4000–5000 species

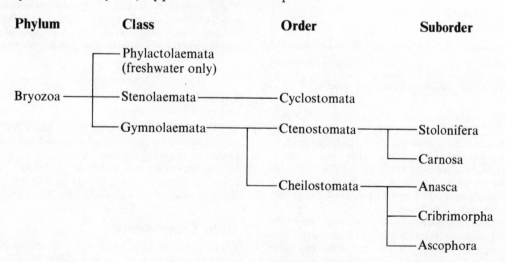

Classification follows Ryland (1970) supplemented by Ryland & Hayward (1977) and P. Cook (personal communication) for the Cheilostomata.

**Plate 101**                                                                 **Phylum Bryozoa**

A group of colonial animals almost all of which are attached to a substratum and live in the sea, although one Class and a few other species occur in freshwater or waters of variable salinity. The colonies vary in height or diameter from less than one millimetre to over one metre, and occur as flat sheets, plant-like tufts, fleshy lobes or coral-like growths. The small members of a colony (zooids) have body walls which are horny (cuticular), calcareous or gelatinous, and colonies may include from a few to many millions of zooids. Part of the body wall is a circular or horseshoe-shaped structure (lophophore) which bears ciliated tentacles and can be pushed out through an orifice to gather food. The gut is U-shaped and the anus opens just outside the lophophore which includes a central mouth. Body cavities of zooids are connected through gaps or pores in the walls. Although many of the zooids are capable of feeding, some are modified for other functions. They may not have any internal structure and form joints, creeping branches or rootlets for attachment; others have modified lophophores or other organs and are supportive, protective or have a cleaning function. In most species modified zooids or groups of zooids form a brood chamber for the young (Figure 37).

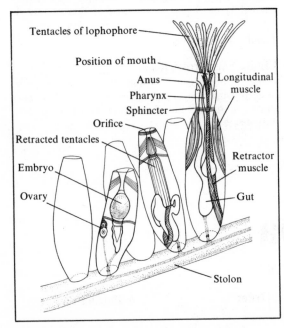

**Figure 37** *Part of a bryozoan colony showing a zooid modified as a brood chamber, and two feeding zooids – one with the tentacles retracted, the other with them expanded.*

Colonies are hermaphrodite, though some zooids may be male or female. In a few species the fertilised egg develops into a free-swimming larva which may feed. In most species the fertilised eggs pass into a brood chamber. A brooded larva has a shorter free-living life and does not feed. Most larvae attach themselves to a surface and change into a zooid from which the colony develops by asexual budding.

Bryozoans are cosmopolitan, occurring from the intertidal zone to great depths. Colonies are often found on rock, dead shells and seaweeds, but some species can live on muddy or sandy bottoms, or encrust living crustacean and mollusc shells or shells inhabited by hermit crabs. They feed on small plankton and organic particles which are collected from the water

currents produced by the ciliated tentacles. They are important fouling organisms, and some species are very resistant to anti-fouling paints. They are found on ships' hulls, pier pilings, legs of oil rigs and in the water intake-pipes of power stations.

## Class Stenolaemata
approx. 1000 species

Marine bryozoans with cylindrical zooids with calcified walls which do not deform to push out the lophophore. The orifices are terminal and circular.

## Order Cyclostomata

Zooids with circular lophophores, and orifices closed by an unspecialised terminal membrane and muscles. The lophophore is pushed out by a complex internal muscle and body cavity system. The zooids are connected through open pores. Modified zooids include joints, rootlets, cleaning zooids, and brooding zooids which protect a large number of embryos.

Families: Tubuliporidae, Diastoporidae, Idmoneidae, Terviidae, Hastingsiidae, Entalophoridae, Corymboporidae, Fascigeridae, Crisiidae, Crisinidae, Stegohorneridae, Pseudidmoneidae, Calvetiidae, Heteroporidae, Lichenoporidae, Disporellidae.

### Family Tubuliporidae
*Tubulipora flabellaris* Fabricius. The calcified colonies, which are up to 3 cm across, have radiating rows of tubular zooids. Found attached to seaweeds, shells and stones in the Mediterranean and North-east Atlantic.   101/4

### Family Idmoneidae
*Hornera frondiculata* Lamouroux. Forms small branching gorgonian-shaped colonies 5 cm or more across. The zooids are arranged in longitudinal rows and are free at their anterior ends, giving the colony a prickly appearance. Found attached to rocks, usually below 20 m, in the Mediterranean.   101/7

### Family Crisiidae
*Crisia denticulata* (Lamarck). A common undergrowth species on hard objects below the kelp zone in the North Atlantic. The photograph also shows a colony of *Bugula turbinata* in the bottom right hand corner (see page 111).   101/6

## Class Gymnolaemata
approx. 3000–4000 species

Mostly marine bryozoans with zooids whose body walls are uncalcified or partially calcified and are deformed in part to push out the lophophore. The orifice has a closing apparatus and zooids are connected through pores in the walls plugged by special cells. Many different kinds of zooids usually form the colony.

## Order Ctenostomata

Cylindrical or box-like zooids with cuticular walls and a terminal or subterminal orifice closed by a flap or a folded collar. Modified zooids include joints, rootlets and sometimes enlarged brooding zooids which protect several embryos.

## Suborder Stolonifera

Colonies with erect or creeping modified zooids from which feeding zooids arise singly or in groups. Some species bore into shells.

Families: Valkeriidae, Mimosellidae, Triticellidae, Hypophorellidae, Vesiculariidae, Buskiidae, Terebriporidae, Penetrantiidae, Monobryozoontidae.

## Suborder Carnosa

Colonies forming a continuous or discrete crust, or erect fleshy lobes. Modified zooids include joints, rootlets, spines and brooding zooids.

Families: Flustrellidriidae, Alcyonidiidae, Arachnidiidae, Immergentiidae, Nolellidae, Victorellidae, Paludicellidae, Benedeniporidae, Lobiancoporidae.

### Family Alcyonidiidae

101/5    *Alcyonidium hirsutum* (Fleming). Forms encrusting growths, particularly over intertidal brown seaweeds. The surface of the colony has characteristic protruberances. The colony in the photograph is seen with several of the food-catching lophophores protruded. Found in the North Atlantic.

## Order Cheilostomata

Colonies composed of cylindrical or box-like zooids with partially calcified walls. The orifices are usually subterminal and closed by a hinged lid (operculum). Modified zooids include joints, rootlets, and brood chambers formed by one or several zooids. The chambers usually protect only one embryo. Some zooids have modified opercula which may be protective beaks, or bristles which are supportive or which clean silt from the colony surface.

## Suborder Anasca

Zooids in which the body wall bearing the operculum is uncalcified, or only partially calcified, and is deformed by muscles to push out the lophophore. This cuticular frontal wall may be ringed by spines or unprotected, but may also have an internal calcareous shield below it.

Families: Aetidae, Scrupariidae, Eucrateidae, Membraniporidae, Electridae, Flustridae, Calloporidae, Chaperiidae, Onychocellidae, Setosellidae, Cupuladriidae, Microporidae, Thalamoporellidae, Steganoporellidae, Chlidoniidae, Cellariidae, Epistomiidae, Scrupocellariidae, Bicellariellidae, Beaniidae, Bugulidae.

### Family Membraniporidae

101/8    *Membranipora membranacea* (Linnaeus) [sea-mat]. A species commonly found encrusting kelp plants. The photograph shows the rectangular box shape of the individuals making up the colony. It has a world-wide distribution in temperate waters.

### Family Electridae

102/1,   *Electra pilosa* (Linnaeus) [hairy sea-mat]. The colonies
102/3   have a hirsute appearance due to the spines and bristles surrounding the individual zooids, which have a more rounded outline than those of *M. membranacea*. It is found encrusting seaweeds, shells and stones in the intertidal zone and in shallow waters of the North Atlantic, Mediterranean and Zealandic region.

### Family Flustridae

*Flustra foliacea* (Linnaeus) [hornwrack]. Forms leaf-
102/2
shaped branching flexible colonies up to 15 cm high with rectangular zooids covering both sides of the branches. The colonies are firmly attached to rocks and stones in shallow coastal waters of the North Atlantic and Mediterranean.

### Family Bugulidae

*Bugula turbinata* Alder. Forms flexible colonies up to
101/6,
5 cm long with branches arranged in a spiral. Each
102/6
zooid has a protective snapping beak. Colonies are usually found under rocky overhangs in shallow coastal waters of the Mediterranean and North-east Atlantic.

## Suborder Cribrimorpha

Zooids with a calcified shield of fused hollow spines above an uncalcified orifice and with a body wall which deforms to push out the lophophore.

Family: Cribrilinidae.

## Suborder Ascophora

Zooids with an extensive calcified shield forming a distinct orifice above an uncalcified part of the frontal body wall which deforms to push out the lophophore.

### (*a*) Umbonuloid Ascophora

The shield is a partially calcified fold, formed above a primary orifice and uncalcified body wall, which deforms by muscle action to push out the lophophore.

Families: Umbonulidae, Exochellidae, Exechonellidae, Metrarabdotosidae, Adeonidae.

### (*b*) Gymnocystidean Ascophora

The shield is formed by calcification of the cuticular body wall followed by development of a pouch opening beneath the orifice, which is developed from body wall cells. Muscles attached to the wall of the pouch and to the side walls of the zooid, enlarge the pouch which fills with sea-water and pushes out the lophophore.

Families: Hippothoidae, Chorizoporidae, Eurystomellidae, possibly also Savignyellidae, Pasytheidae.

### (*c*) Cryptocystidean Ascophora

The shield is formed by an internal calcified layer below the uncalcified body wall, followed by development of a pouch beneath the shield, which functions in the same way as that in the gymnocystidean Ascophora.

Families: Gigantoporidae, Stomachetosellidae, Schizoporellidae, Watersiporidae, Hippoporinidae, Smittinidae, Microporellidae, Escharellidae, Sertellidae, Adeonellidae, Cheiloporinidae, Crepidacanthidae, Cleidochasmatidae, Petraliidae, Petraliellidae, Orbituliporidae, Conescharellinidae.

### Family Hippoporinidae

*Pentapora fascialis* (Pallas) (=*Hippodiplosia fascialis*).
102/5,
The flattened branching colonies of this species are
102/7
somewhat reminiscent of those of *Flustra foliacea* but are heavily calcified and brittle (Plate 102/7). Found on

rocks and open ground forming colonies up to 20 cm high (Plate 102/5). A Mediterranean species.

102/9 *Pentapora foliacea* (Ellis & Solander) [ross or rose coral]. Similar to the above species, but the flattened branches of the colonies fuse into a network of plates between which numerous small crabs find shelter. Domed colonies up to 50 cm across may be found attached to rocks below 10 m, particularly in areas where currents are fast. Found in the North-east Atlantic.

### Family Smittinidae

102/4 *Porella cervicornis* (Pallas). The stiff branching colonies of this species, which reach a height of 5–10 cm, are found attached to rocks beneath overhangs and in caves in the Mediterranean.

### Family Escharellidae

102/8 *Schizobrachiella sanguinea* (Norman). Forms irregular encrusting colonies over hard objects in fast-flowing water or has a more erect growth form with trumpet-shaped extensions in still water. A fast-growing species which sometimes fouls ships' hulls. Found in the Mediterranean.

### Family Sertellidae

*Sertella beaniana* (King) (= *Retepora beaniana*) [mermaid's veil]. Colonies form delicate net-like growths, up to 10 cm across, beneath rocky overhangs and in caves in the Mediterranean and North-east Atlantic.  103/4

### Family Adeonellidae

*Adeonella polystomella* (Reuss). The colonies, which have calcified upright flattened branches, are found beneath rocky overhangs and in caves in the Mediterranean.  103/1, 103/2

## (d) Unplaced families

Vittaticellidae, Margarettidae, Phylactoliporidae, Phylactellidae, Parmulariidae, Celleporariidae, Myriaporidae, Euthyrisellidae, Sclerodomidae, Onchoporidae, Bifaxariidae, Mamilloporidae, Lekythoporidae.

### Family Myriaporidae

*Myriapora truncata* (Pallas) [false coral]. Forms clumps of cylindrical branches with characteristically truncated tips. The colonies, which may reach 10 cm in height, are found attached to rock in shaded areas in the Mediterranean.  103/3

# Phylum Phoronida

[horseshoe worms] approx. 15 species

A group of marine worms which live in chitinous tubes. The cylindrical body, which can reach a length of 20 cm, has a horseshoe-shaped row of ciliated tentacles (lophophore) encircling the mouth (Figure 38). The gut is U-shaped and the anus opens to the exterior on the dorsal surface above the lophophore. The posterior end of the body is thickened and serves to anchor the animal permanently in its tube. The anterior end with the tentacles is thrust out for feeding, but is quickly withdrawn when the animal is disturbed.

Most species are hermaphrodite, but a few forms with separate sexes are known. Fertilisation of the eggs occurs either in the seawater or in the body cavity. Some species brood their eggs amongst the tentacles. The free-swimming larva is common in the plankton at certain times of the year. After several weeks of planktonic life, it sinks to the bottom and rapidly changes into a small adult which immediately secretes the tube.

Phoronids occur intertidally and in shallow waters of both tropical and temperate regions. Their tubes may be buried vertically in soft sediments, or may occur in tangled masses attached to rocks and other objects. Some species burrow into rocks and shells. They feed by filtering small organisms and organic debris from the water using the cilia of the tentacles and mucus secretions. They bear a superficial resemblance to the tube-dwelling polychaete worms but their bodies lack segments and bristles.

Two genera only: *Phoronis* and *Phoronopsis*.

103/5 *Phoronis australis* Haswell. The lophophore of specimens of this colonial form may be up to 2 cm across when fully expanded. This species can often be located in

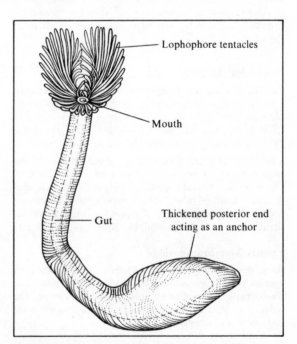

**Figure 38** *The general body form of the horseshoe worm,* Phoronis.

clumps associated with the tubes of *Cerianthus*. It has been found in shallow waters of the Mediterranean and Indo-Pacific.

*Phoronis* sp. Specimens with black lophophores protruding from coral sand on the Great Barrier Reef.  103/6

1 *Clathrina coriacea*
2 *Leucosolenia botryoides*
3 *Clathrina coriacea*
4 *Leucosolenia* sp.
5 *Scypha ciliata*
6 *Scypha compressa*
7 *Oscarella lobularis*
8 *Leucettusa imperfecta*
9 *Scypha gelatinosa*

1 *Myxilla incrustans*
2 *Aurora rowi*
3 *Chondrilla sacciformis*

4 *Myxilla incrustans*
5 *Melanchora elliptica*
6 *Tethya diploderma*
7 *Pachymatisma johnstoni*
8 *Polymastia boletiformis*
9 *Tethya aurantium*
10 *Suberites domuncula*
11 *Tethya* sp.
12 *Suberites domuncula*

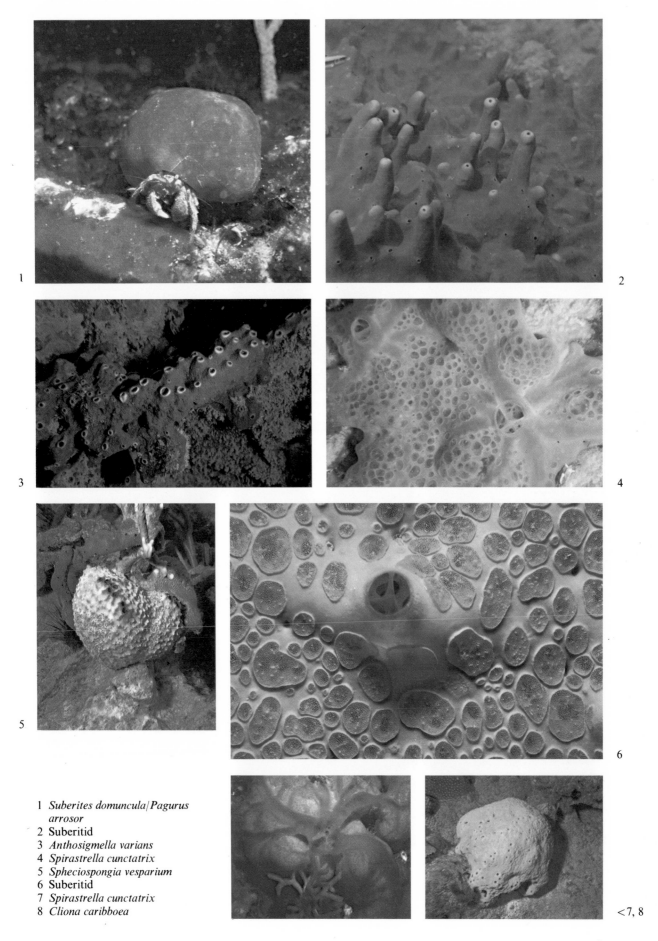

1 *Suberites domuncula/Pagurus arrosor*
2 Suberitid
3 *Anthosigmella varians*
4 *Spirastrella cunctatrix*
5 *Spheciospongia vesparium*
6 Suberitid
7 *Spirastrella cunctatrix*
8 *Cliona caribboea*

<7, 8

1

2

3, 4 >

5

6, 7 >

8

9, 10 >

11

| 1 | *Cliona celata* | 5 | *Phakellia* sp. | 9 | *Axinella polypoides* |
|---|---|---|---|---|---|
| 2 | *Cliona lampa* | 6 | *Axinella verrucosa* | 10 | *Agelas clathrodes* |
| 3 | *Cliona lampa* | 7 | Axinellid | 11 | *Agelas clathrodes* |
| 4 | *Axinella damicornis* | 8 | *Raspailia hispida* | | |

**4**

1 *Agelas clathrodes*
2 *Hymeniacidon perleve*
3 *Hemimycale columella*
4 *Crambe crambe*
5 *Mycale macilenta*
6 *Agelas clathrodes*
7 *Halichondria panicea*

8 *Mycale* sp.
9 *Amphilectus fucorum*
10 *Crambe crambe*

1

2

3

4

5

6, 7 >

10 >

9

8

1  *Clathraria rubrinodis*
2  *Echinoclathria gigantea*
3  Poecilosclerid
4  *Myxilla incrustans*
5  Poecilosclerid
6  *Adocia* sp.
7  *Myxilla* sp.
8  *Haliclona (Amphimedon)*
    *compressa*
9  *Cribrochalina* sp.
10  *Adocia* sp.

6

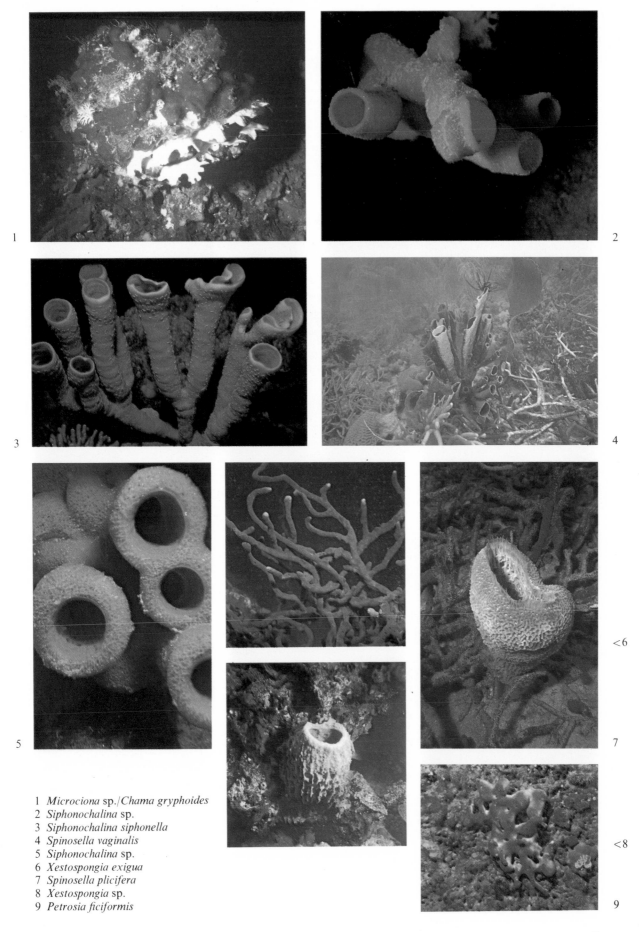

1  *Microciona* sp./*Chama gryphoides*
2  *Siphonochalina* sp.
3  *Siphonochalina siphonella*
4  *Spinosella vaginalis*
5  *Siphonochalina* sp.
6  *Xestospongia exigua*
7  *Spinosella plicifera*
8  *Xestospongia* sp.
9  *Petrosia ficiformis*

**Phylum Porifera**

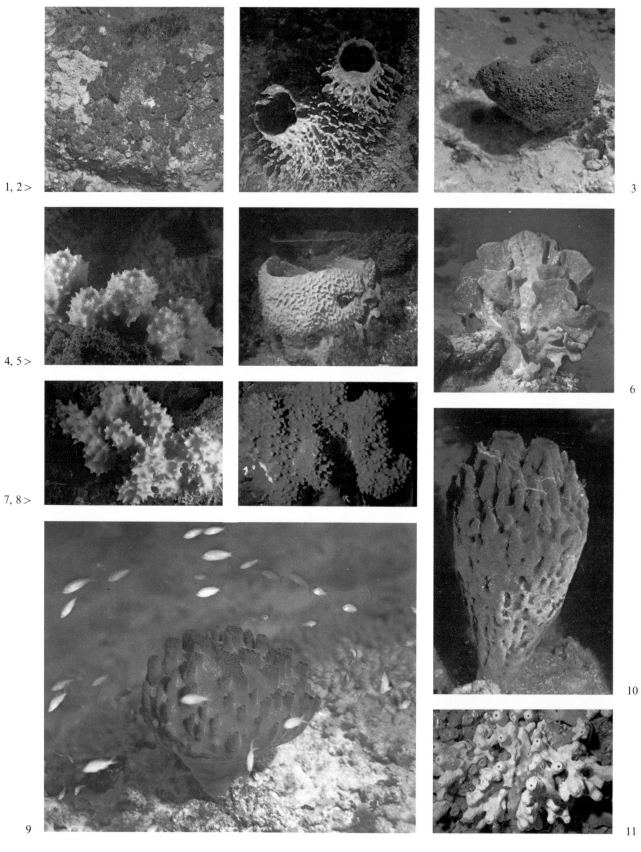

1, 2 >

4, 5 >

7, 8 >

3

6

9

10

11

1  *Petrosia ficiformis*
2  *Petrosia testudinaria*
3  *Hippospongia communis*
4  *Dysidea* sp.

5  *Ircinia campana*
6  *Phyllospongia* sp.
7  *Dysidea avera*
8  *Dysidea etheria*

9  *Rhizochalina ramsayi*
10  Spongiid
11  *Aplysina aerophoba*

**8**

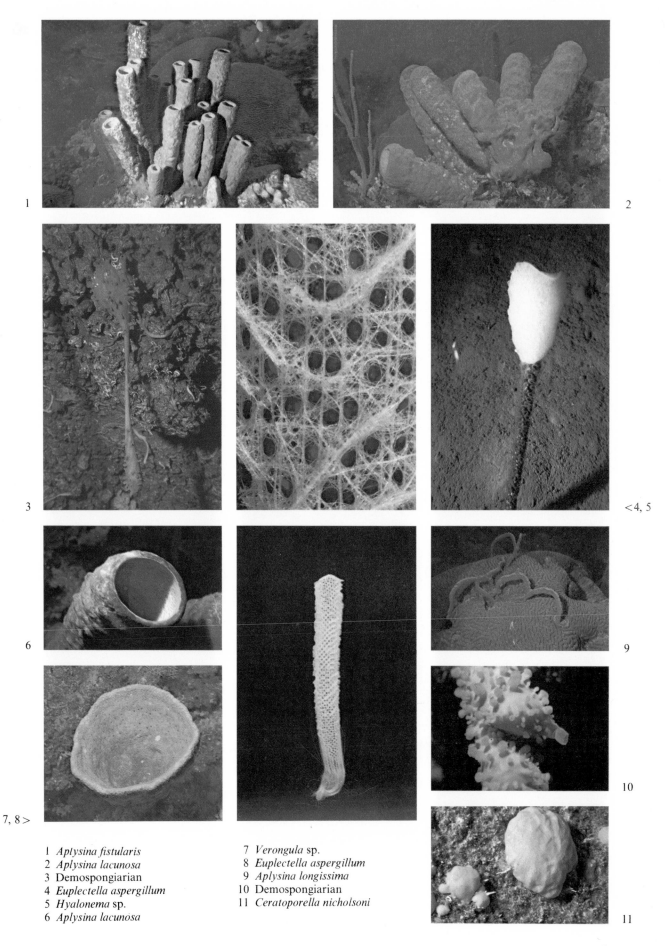

1
2
3
< 4, 5
6
9
7, 8 >
10
11

1  *Aplysina fistularis*
2  *Aplysina lacunosa*
3  Demospongiarian
4  *Euplectella aspergillum*
5  *Hyalonema* sp.
6  *Aplysina lacunosa*

7  *Verongula* sp.
8  *Euplectella aspergillum*
9  *Aplysina longissima*
10  Demospongiarian
11  *Ceratoporella nicholsoni*

1 *Corymorpha nutans*
2 *Tubularia indivisa*
3 *Corymorpha nutans*
4 *Tubularia mesembryanthemum*
5 Tubulariid
6 *Halocordyle disticha*
7 *Pennaria* sp.
8 *Pennaria* sp.
9 *Porpita porpita*

1

2

< 4, 5

3

< 6, 7

8, 9 >

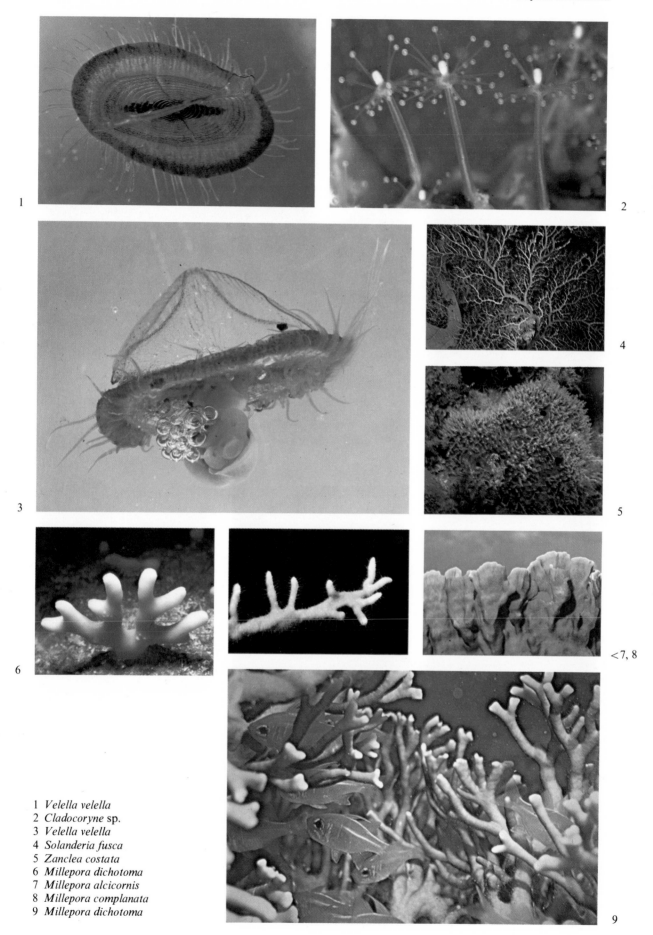

1 *Velella velella*
2 *Cladocoryne* sp.
3 *Velella velella*
4 *Solanderia fusca*
5 *Zanclea costata*
6 *Millepora dichotoma*
7 *Millepora alcicornis*
8 *Millepora complanata*
9 *Millepora dichotoma*

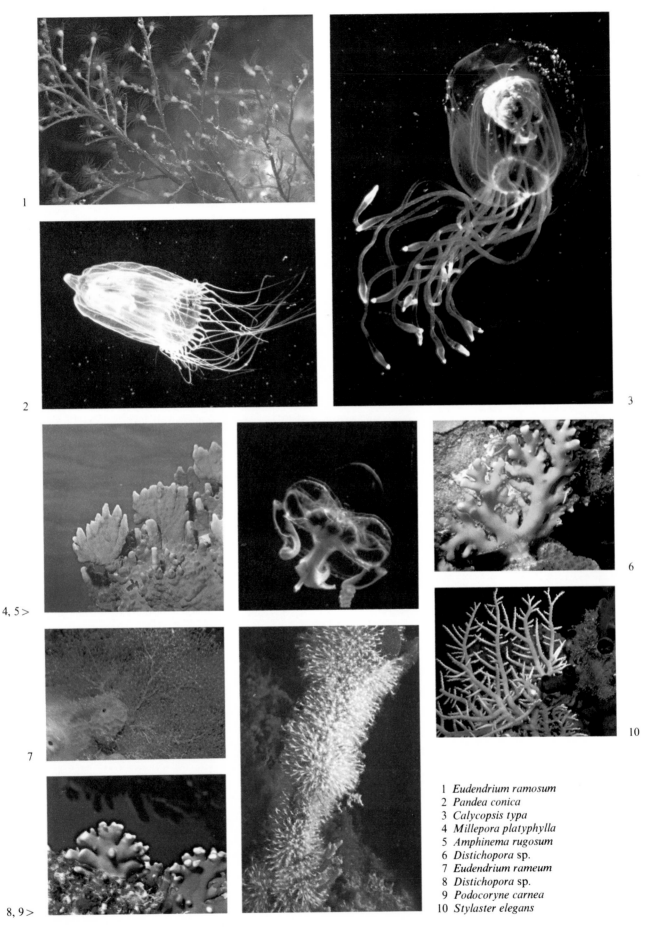

1

2

3

4, 5 >

6

7

10

8, 9 >

1 *Eudendrium ramosum*
2 *Pandea conica*
3 *Calycopsis typa*
4 *Millepora platyphylla*
5 *Amphinema rugosum*
6 *Distichopora* sp.
7 *Eudendrium rameum*
8 *Distichopora* sp.
9 *Podocoryne carnea*
10 *Stylaster elegans*

1   *Clava squamata*
2   *Obelia geniculata*
3   *Stylaster* sp.
4   *Obelia geniculata*
5   *Abietinaria abietina*

6   *Dynamena pumila*
7   *Hydrallmania falcata*
8   *Sertularella polyzonias*
9   *Synthecium evansi*
10   *Halecium halecinum*

1, 2 >

3

5

6

4

7

8

9

<10, 11

1 *Sertularia cupressina*
2 *Aglaophenia cupressina*
3 *Aglaophenia kirchenpaueri*
4 *Aglaophenia elongata*
5 *Aglaophenia octodonta*
6 *Aglaophenia pluma*

7 *Antenella* sp.
8 *Gymnangium montagui*
9 *Lytocarpus philippinus*
10 *Lytocarpus myriophyllum*
11 *Lytocarpus* sp.

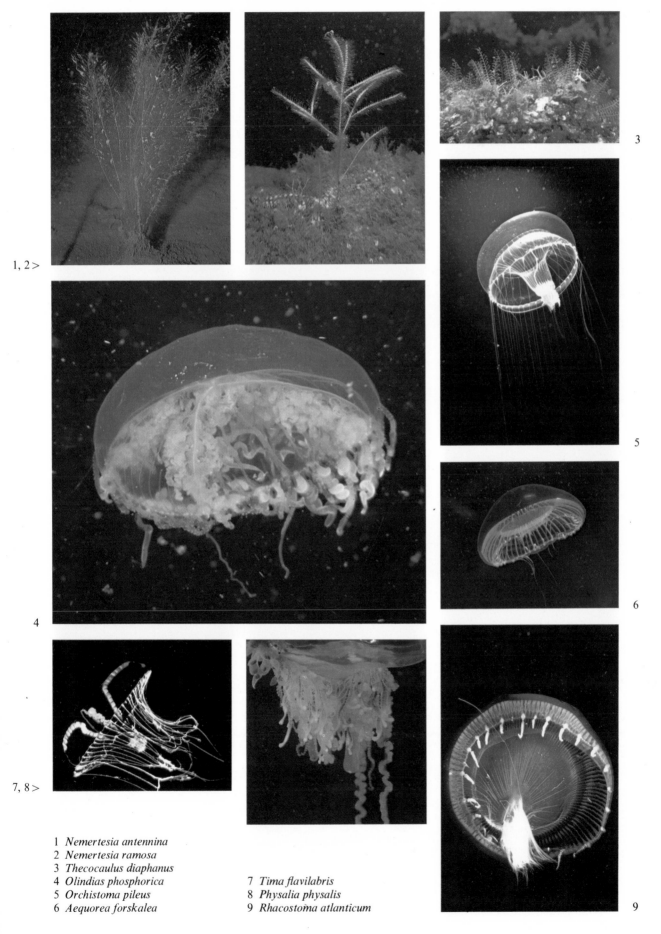

1, 2 >

3

4

5

6

7, 8 >

9

1  *Nemertesia antennina*
2  *Nemertesia ramosa*
3  *Thecocaulus diaphanus*
4  *Olindias phosphorica*
5  *Orchistoma pileus*
6  *Aequorea forskalea*

7  *Tima flavilabris*
8  *Physalia physalis*
9  *Rhacostoma atlanticum*

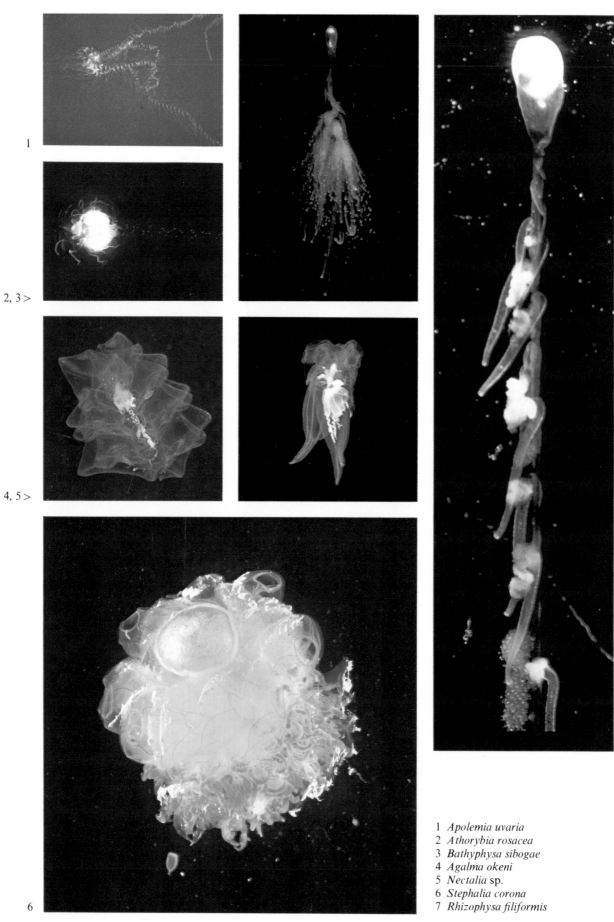

1

2, 3 >

4, 5 >

6

7

1  *Apolemia uvaria*
2  *Athorybia rosacea*
3  *Bathyphysa sibogae*
4  *Agalma okeni*
5  *Nectalia* sp.
6  *Stephalia corona*
7  *Rhizophysa filiformis*

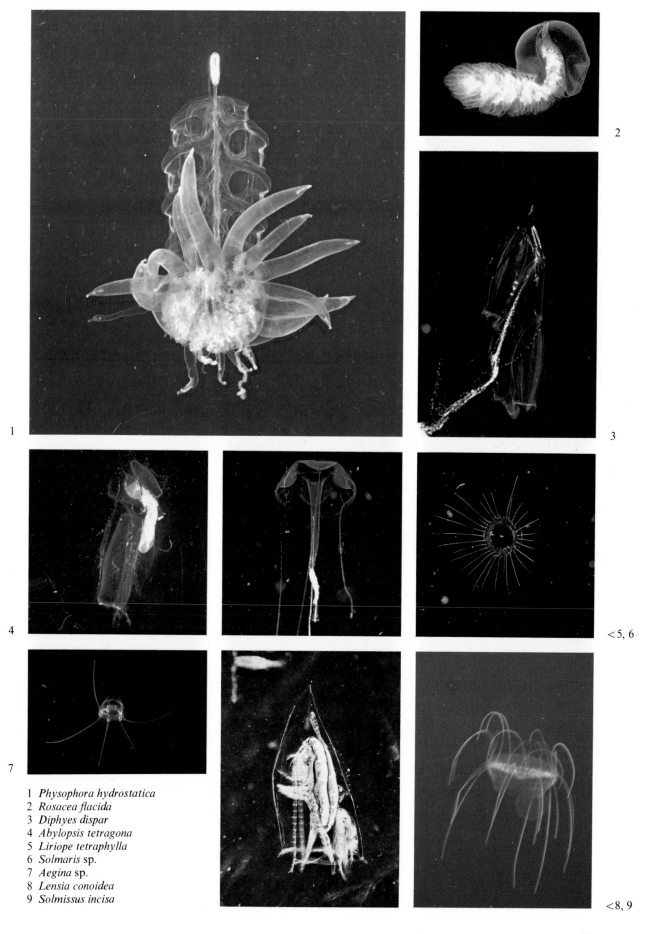

1 *Physophora hydrostatica*
2 *Rosacea flacida*
3 *Diphyes dispar*
4 *Abylopsis tetragona*
5 *Liriope tetraphylla*
6 *Solmaris* sp.
7 *Aegina* sp.
8 *Lensia conoidea*
9 *Solmissus incisa*

**17**

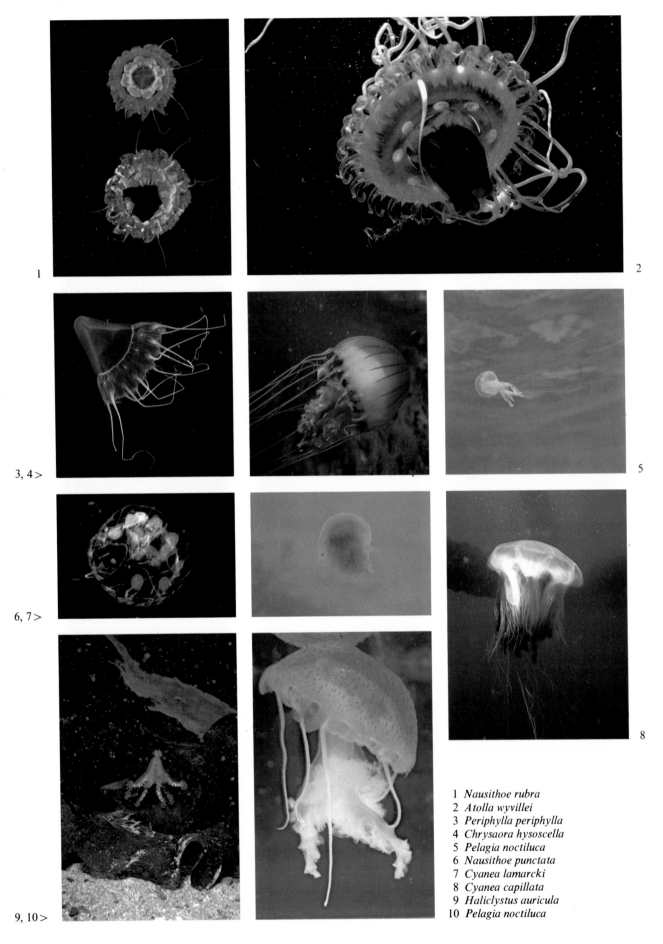

1

2

3, 4 >

5

6, 7 >

9, 10 >

8

1 *Nausithoe rubra*
2 *Atolla wyvillei*
3 *Periphylla periphylla*
4 *Chrysaora hysoscella*
5 *Pelagia noctiluca*
6 *Nausithoe punctata*
7 *Cyanea lamarcki*
8 *Cyanea capillata*
9 *Haliclystus auricula*
10 *Pelagia noctiluca*

1 *Aurelia aurita*
2 *Rhizostoma pulmo*
3 *Cassiopeia andromeda*

4 *Cassiopeia andromeda*
5 *Carybdea rastoni*
6 *Cotylorhiza tuberculata*
7 *Phyllorhiza punctata*
8 Rhizostomean
9 *Cotylorhiza tuberculata*

**19**

1, 2 >

3

4

5

6

7

8

<9, 10

1  *Chironex fleckeri*
2  *Antipathes* sp.
3  *Cirrhipathes anguina*
4  *Cirrhipathes anguina*

5  *Cerianthus membranaceus*
6  *Pachycerianthus maua*
7  *Cirrhipathes* sp.
8  *Stichopathes luetkeni*

9  *Antipathes dichotoma*
10  Antipatharian

1

<2, 3

4

<5, 6

7

8

9, 10 >

1  *Cerianthus membranaceus*
2  *Isarachnanthus* sp.
3  *Coelogorgia palmosa*
4  Cerianthid
5  *Clavularia* sp.
6  *Tubipora musica*
7  *Cornularia cornucopiae*
8  *Tubipora musica*

9   *Alcyonium digitatum*
10  *Telesto multiflora*
11  *Alcyonium digitatum*

11

**21**

1, 2>

5, 6>

3

4

7

8

9

10

1  *Alcyonium palmatum*
2  *Parerythropodium coralloides/
   Eunicella cavolinii*
3  *Sarcophyton trocheliophorum*

4  *Sinularia* sp.
5  Alcyoniid
6  *Dendronephthya klunzingeri*
7  *Dendronephthya rubeola*

8  *Sarcophyton trocheliophorum*
9  *Sarcophyton trocheliophorum*
10  *Dendronephthya klunzingeri*

1

<2, 3

4

5

6

<7, 8

9

1   *Dendronephthya klunzingeri*
2   *Cactogorgia simpsoni*
3   *Subergorgia hicksoni*
4   *Anthelia glauca*
5   *Xenia elongata*
6   *Dendronephthya rubeola*
7   *Heliopora coerulea*
8   *Briareum asbestinum*
9   *Scleronephthya* sp.
10  *Melithaea squamata*

10

1 *Melithaea squamata*
2 *Subergorgia* sp.
3 *Corallium rubrum*
4 *Paramuricea clavata*
5 *Mopsella ellisi*
6 *Echinomuricea klavereni*
7 Plexaurids
8 *Paramuricea clavata*
9 *Paramuricea clavata*
10 *Paramuricea macrospina*

1

2

5

6

<3, 4

<7, 8

1 *Paramuricea* sp.
2 *Eunicella verrucosa*
3 *Eunicea clavigera*
4 *Eunicella cavolinii*
5 *Muricea muricata*
6 *Plexaurella dichotoma*
7 *Eunicella cavolinii*
8 *Eunicella singularis*
9 *Eunicella singularis*
10 *Eunicella verrucosa*

<9, 10

**25**

1, 2 >

3

4

5

6

9

7

< 8, 10

11, 12 >

1  *Rumphella* sp.
2  *Gorgonia flabellum*
3  *Gorgonia ventalina*
4  *Ellisella erythrea*
5  Ellisellid
6  *Pseudopterogorgia americana*
7  *Gorgonia ventalina*
8  *Pseudopterogorgia acerosa*
9  *Lophogorgia sarmentosa*
10  *Juncella fragilis*
11  *Mopsea australis*
12  *Primnoella australasiae*

1 Gorgonian
2 *Cavernularia* sp.
3 *Cavernularia obesa*
4 *Veretillum cynomorium*
5 *Scytaliopsis ghardagensis*
6 *Funiculina quadrangularis*
7 *Pennatula phosphorea*
8 *Cavernularia obesa*
9 *Virgularia mirabilis*

1   *Pennatula phosphorea*
2   *Ptilosarcus gurneyi*
3   *Sarcoptilus* sp.
4   *Pteroeides spinosum*
5   Pteroeidid
6   *Boloceroides mcmurrichi*
7   *Bunodeopsis globulifera*
8   *Pteroeides* sp./*Virgularia* sp.
9   *Ilyanthus mitchelli*

28

1  *Actinia equina* var. *mesembryanthemum*
2  *Actinia equina* var. *mesembryanthemum*
3  *Actinia equina* var. *mesembryanthemum*
4  *Actinia equina* var. *fragacea*

5  *Anemonia sargassiensis*
6  *Anthopleura thallia*
7  *Andresia parthenopea*
8  *Anemonia sulcata*

9  *Anemonia sulcata*
10  *Anthopleura xanthogrammica*
11  *Bunodactis verrucosa*

**29**

1 *Bunodosoma cavernata*
2 *Condylactis gigantea*
3 *Tealia felina* var. *coriacea*
4 *Condylactis aurantiaca*
5 *Tealia felina* var. *coriacea*

6 *Physobrachia douglasi*
7 *Tealia felina* var. *lofotensis*
8 *Triactis producta*
9 Actiniid

1

&lt;2, 3

4

5                                                                                 6

7                                                                                 &lt;8, 9

| 1 *Alicia mirabilis* | 4 *Stoichactis kenti* | 7 *Lebrunia danae* |
|---|---|---|
| 2 *Alicia* sp. | 5 *Stoichactis kenti* | 8 *Radianthus koseirensis* |
| 3 *Gyrostoma helianthus* | 6 *Radianthus ritteri* | 9 *Gyrostoma quadricolor* |

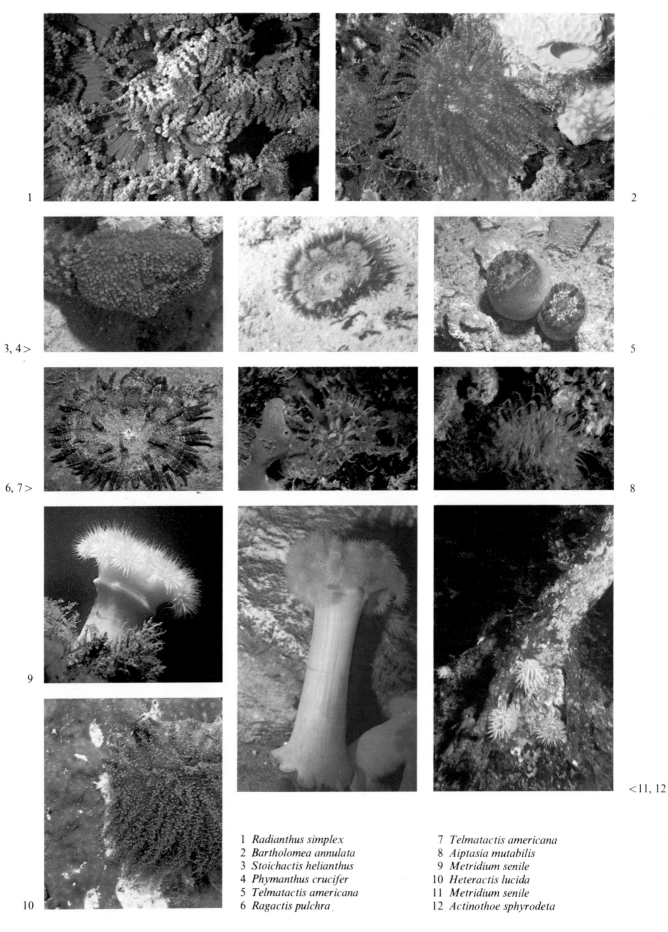

1  *Radianthus simplex*
2  *Bartholomea annulata*
3  *Stoichactis helianthus*
4  *Phymanthus crucifer*
5  *Telmatactis americana*
6  *Ragactis pulchra*

7  *Telmatactis americana*
8  *Aiptasia mutabilis*
9  *Metridium senile*
10  *Heteractis lucida*
11  *Metridium senile*
12  *Actinothoe sphyrodeta*

1 *Cereus* sp.
2 *Sagartia elegans* var. *nivea*
3 *Sagartiogeton undata*
4 *Sagartia elegans* var. *venusta*
5 Hormathiids
6 *Corynactis viridis*

7 *Corynactis viridis*
8 *Adamsia palliata/Calliactis parasitica*
9 *Corynactis australis*
10 *Anthothoe* sp.

**Phylum Coelenterata**

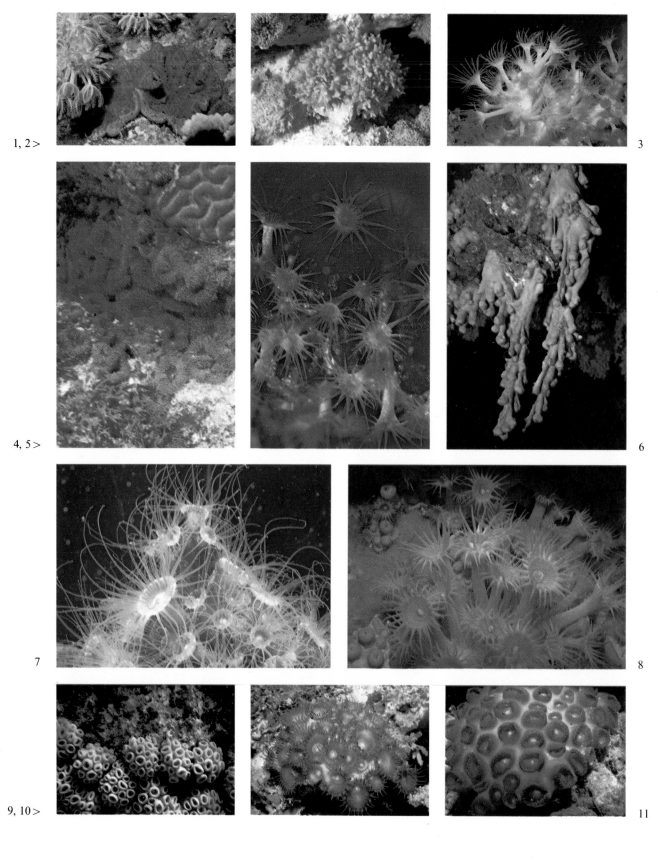

1, 2 >

4, 5 >

3

6

7

8

9, 10 >

11

1  *Actinodiscus nummiformis*
2  *Rhodactis sanctithomae*
3  *Parazoanthus axinellae*
4  *Ricordia florida*

5  *Epizoanthus* sp.
6  *Parazoanthus axinellae*
7  *Epizoanthus paxii*
8  *Parazoanthus* sp.

9   *Palythoa mammillosa*
10  *Palythoa tuberculosa*
11  *Palythoa tuberculosa*

1  *Palythoa* sp.
2  *Zoanthus sociatus*
3  Zoanthid
4  Scleractinans
5  *Pocillopora damicornis*

6  *Stylophora pistillata*
7  *Stephanocoenia michelini*
8  *Acropora cervicornis*
9  *Seriatopora hystrix*
10 *Acropora cervicornis*

**35**

1 *Acropora humilis*
2 *Acropora palmata*
3 *Acropora cervicornis*
4 *Acropora variabilis*

5 *Acropora palmata*
6 *Acropora* sp.
7 *Acropora pulchra*
8 *Montipora* sp.

9 *Agaricia lamarcki*
10 *Montipora foliosa*
11 *Montipora* sp.
12 *Agaricia lamarcki*

1 *Leptoseris* sp.
2 *Pachyseris* sp.
3 *Pachyseris rugosa*
4 *Fungia fungites*
5 *Cycloseris cyclolites*
6 *Cycloseris cyclolites*

7 *Pavona varians*
8 *Fungia fungites*
9 *Fungia scutaria*
10 *Fungia actiniformis*
11 *Fungia echinata*

37

1   *Porites astreoides*        6   *Goniopora lobata*
2   *Porites lutea*            7   *Goniopora lobata*
3   *Herpolitha limax*       8   *Goniopora* sp.
4   *Parahalomitra robusta*    9   *Porites porites*
5   *Goniopora* sp.

1 *Porites* sp.
2 *Cladocora cespitosa*
3 *Colpophyllia natans*
4 *Diploria labyrinthiformis*
5 *Cyphastrea japonica*
6 *Diploria labyrinthiformis*

7 *Diploria clivosa*
8 *Colpophyllia natans*
9 *Diploria strigosa*
10 *Echinopora horrida*

7, 8 >

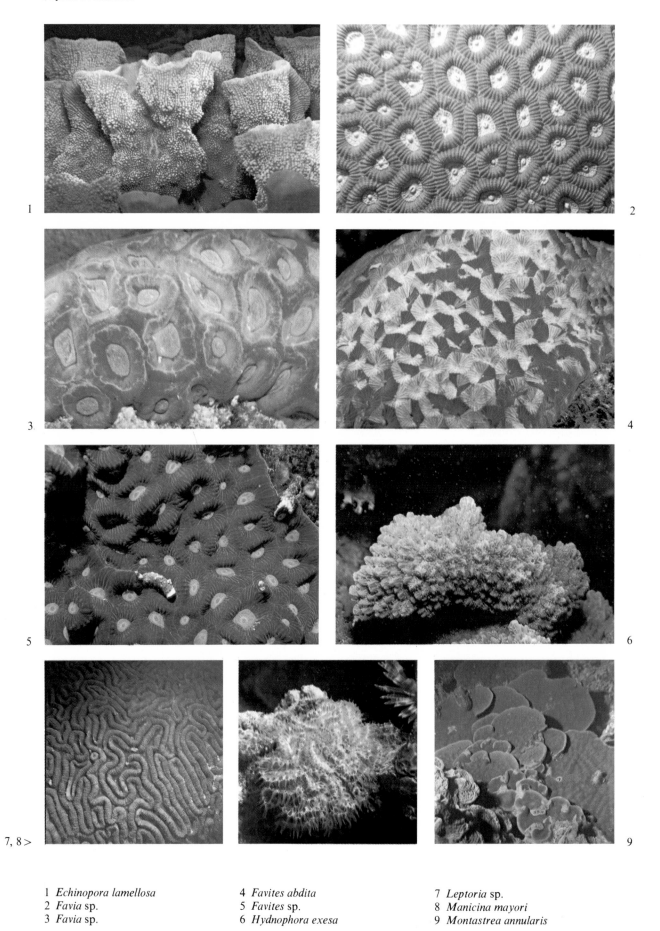

1  *Echinopora lamellosa*       4  *Favites abdita*        7  *Leptoria* sp.
2  *Favia* sp.                  5  *Favites* sp.           8  *Manicina mayori*
3  *Favia* sp.                  6  *Hydnophora exesa*      9  *Montastrea annularis*

**40**

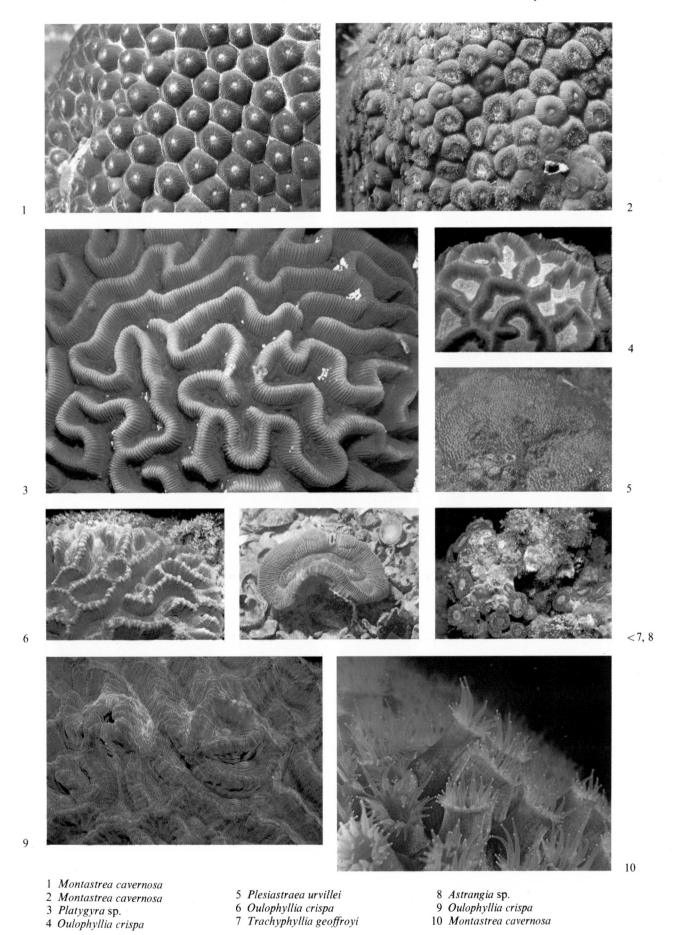

1 *Montastrea cavernosa*
2 *Montastrea cavernosa*
3 *Platygyra* sp.
4 *Oulophyllia crispa*

5 *Plesiastraea urvillei*
6 *Oulophyllia crispa*
7 *Trachyphyllia geoffroyi*

8 *Astrangia* sp.
9 *Oulophyllia crispa*
10 *Montastrea cavernosa*

1, 2>

3

5

4

6

7

8

9, 10>

1 *Phyllangia mouchezii*
2 *Acrohelia horrescens*
3 *Galaxea clavus*
4 *Dendrogyra cylindrus*
5 *Galaxea fascicularis*
6 *Meandrina brasiliensis*
7 *Galaxea fascicularis*
8 *Meandrina brasiliensis*
9 *Oculina patagonica*
10 *Dichocoenia stokesi*

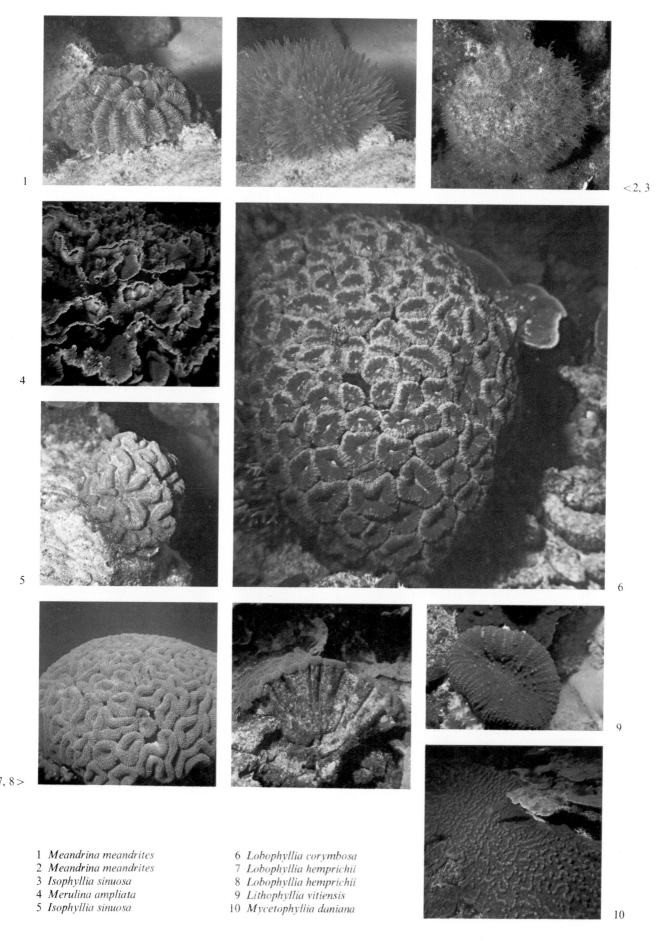

1

< 2, 3

4

5

6

7, 8 >

9

10

1 *Meandrina meandrites*
2 *Meandrina meandrites*
3 *Isophyllia sinuosa*
4 *Merulina ampliata*
5 *Isophyllia sinuosa*

6 *Lobophyllia corymbosa*
7 *Lobophyllia hemprichii*
8 *Lobophyllia hemprichii*
9 *Lithophyllia vitiensis*
10 *Mycetophyllia daniana*

**43**

1, 2 >

3

4

5

6

7

8

9

< 10, 11

| | |
|---|---|
| 1 *Mussa angulosa* | 7 *Pectinia lactuca* |
| 2 *Protolobōphyllia* sp. | 8 *Symphyllia* sp. |
| 3 *Scolymia cubensis* | 9 *Scolymia lacera* |
| 4 *Echinophyllia* sp. | 10 *Scolymia lacera* |
| 5 *Mycedium elephantotum* | 11 *Echinophyllia* sp. |
| 6 *Mycedium elephantotum* | |

1 *Pectinia* sp.
2 *Caryophyllia smithi*
3 *Caryophyllia smithi*
4 *Physogyra liechtensteina*
5 *Euphyllia* sp.

6 *Pterogyra sinuosa*
7 Bubble coral
8 *Eusmilia fastigiata*
9 *Euphyllia* sp.
10 *Eusmilia fastigiata*

1 *Astroides calycularis*
2 *Balanophyllia gemmifera*
3 *Balanophyllia bairdiana*
4 *Dendrophyllia gracilis*

5 *Balanophyllia gemmifera*
6 *Balanophyllia italica*
7 *Balanophyllia regia*
8 *Dendrophyllia gracilis*

9 *Heteropsammia michelini*
10 *Leptopsammia pruvoti*
11 *Dendrophyllia* sp.

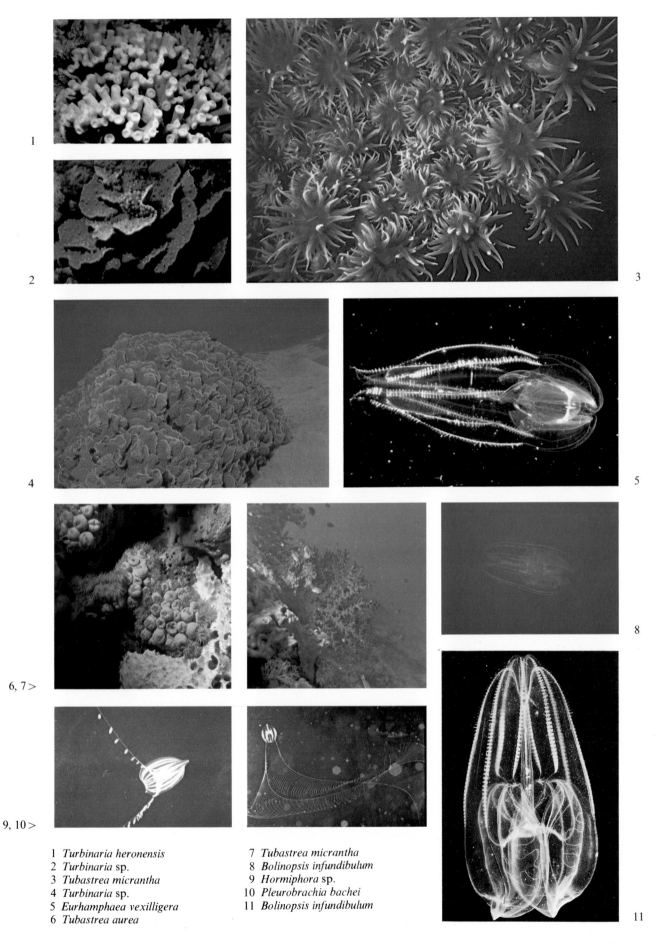

1 *Turbinaria heronensis*
2 *Turbinaria* sp.
3 *Tubastrea micrantha*
4 *Turbinaria* sp.
5 *Eurhamphaea vexilligera*
6 *Tubastrea aurea*

7 *Tubastrea micrantha*
8 *Bolinopsis infundibulum*
9 *Hormiphora* sp.
10 *Pleurobrachia bachei*
11 *Bolinopsis infundibulum*

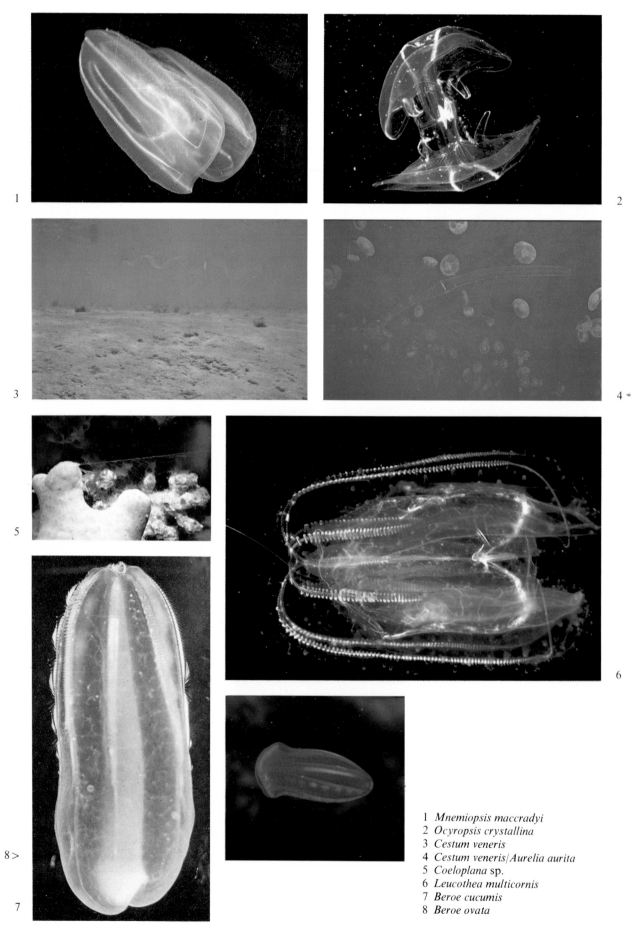

1  *Mnemiopsis maccradyi*
2  *Ocyropsis crystallina*
3  *Cestum veneris*
4  *Cestum veneris/Aurelia aurita*
5  *Coeloplana* sp.
6  *Leucothea multicornis*
7  *Beroe cucumis*
8  *Beroe ovata*

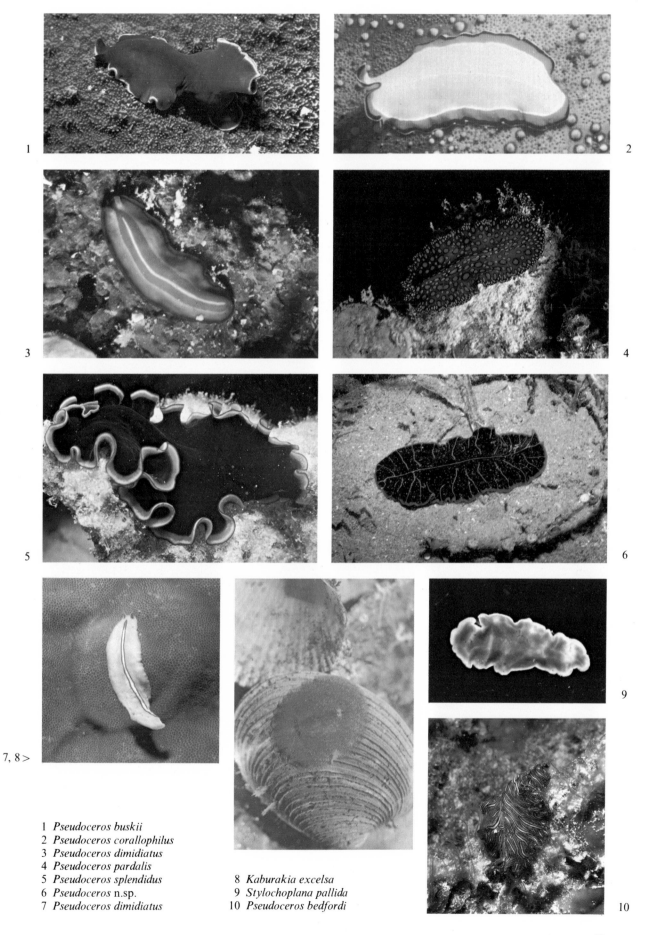

1   *Pseudoceros buskii*
2   *Pseudoceros corallophilus*
3   *Pseudoceros dimidiatus*
4   *Pseudoceros pardalis*
5   *Pseudoceros splendidus*
6   *Pseudoceros* n.sp.
7   *Pseudoceros dimidiatus*

8   *Kaburakia excelsa*
9   *Stylochoplana pallida*
10   *Pseudoceros bedfordi*

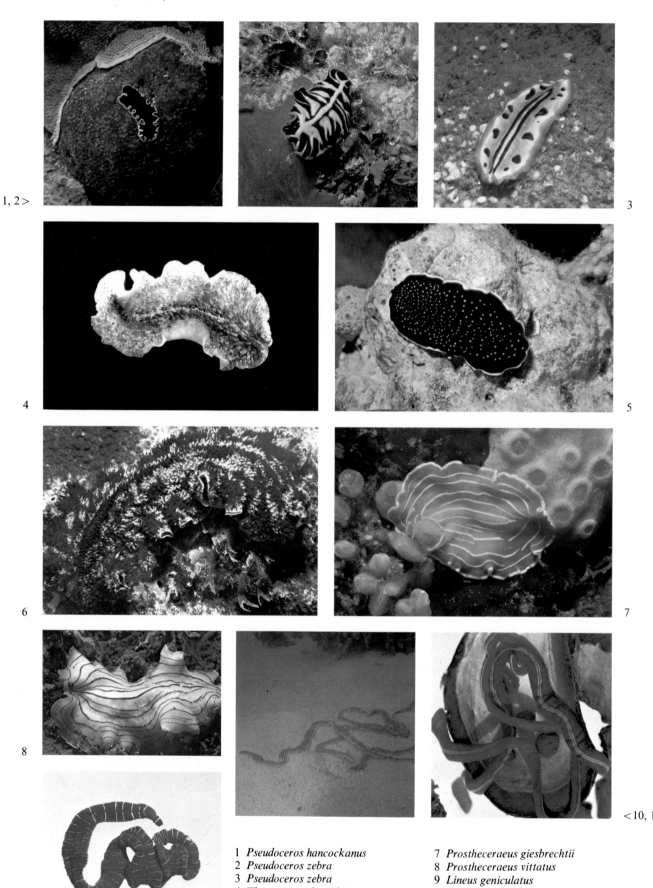

1,2 >

3

4

5

6

7

8

< 10, 11

9

1  *Pseudoceros hancockanus*
2  *Pseudoceros zebra*
3  *Pseudoceros zebra*
4  *Thysanozoon brocchii*
5  *Thysanozoon flavomaculatum*
6  *Thysanozoon* n.sp.

7  *Prostheceraeus giesbrechtii*
8  *Prostheceraeus vittatus*
9  *Lineus geniculatus*
10  *Baseodiscus quinquelineatus*
11  *Lineus bilineatus*

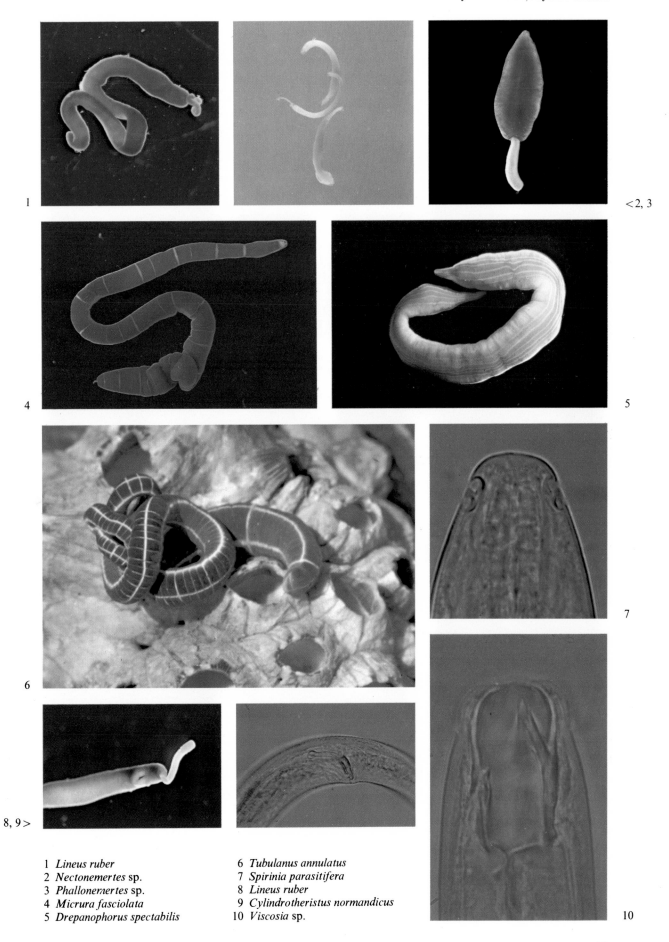

1
<2, 3
4
5
6
7
8, 9 >
10

1  *Lineus ruber*
2  *Nectonemertes* sp.
3  *Phallonemertes* sp.
4  *Micrura fasciolata*
5  *Drepanophorus spectabilis*

6  *Tubulanus annulatus*
7  *Spirinia parasitifera*
8  *Lineus ruber*
9  *Cylindrotheristus normandicus*
10  *Viscosia* sp.

**51**

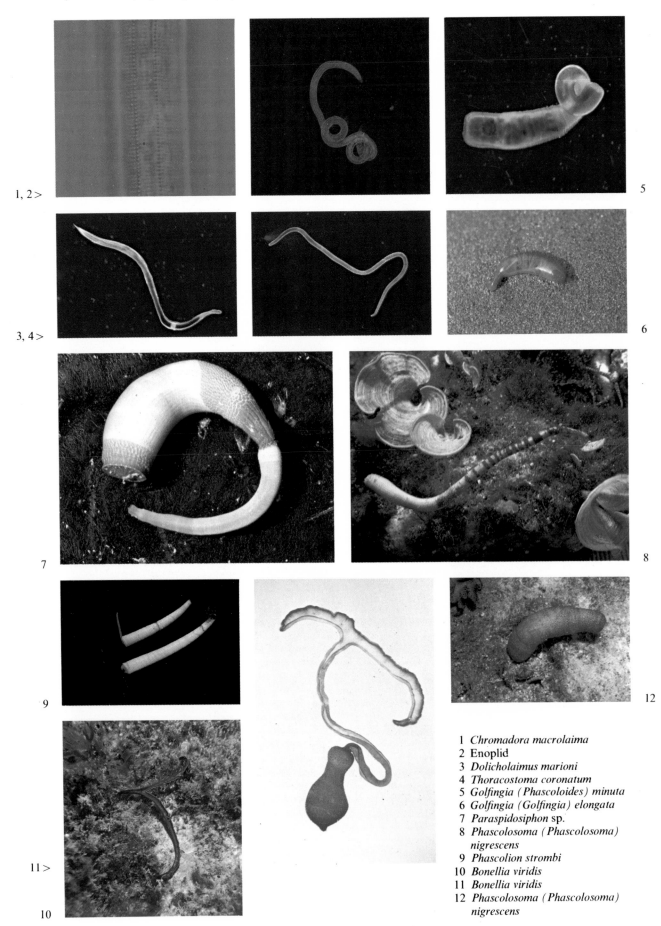

1, 2 >

3, 4 >

5

6

7

8

9

11 >

10

12

1 *Chromadora macrolaima*
2 Enoplid
3 *Dolicholaimus marioni*
4 *Thoracostoma coronatum*
5 *Golfingia ( Phascoloides) minuta*
6 *Golfingia (Golfingia) elongata*
7 *Paraspidosiphon* sp.
8 *Phascolosoma (Phascolosoma) nigrescens*
9 *Phascolion strombi*
10 *Bonellia viridis*
11 *Bonellia viridis*
12 *Phascolosoma ( Phascolosoma) nigrescens*

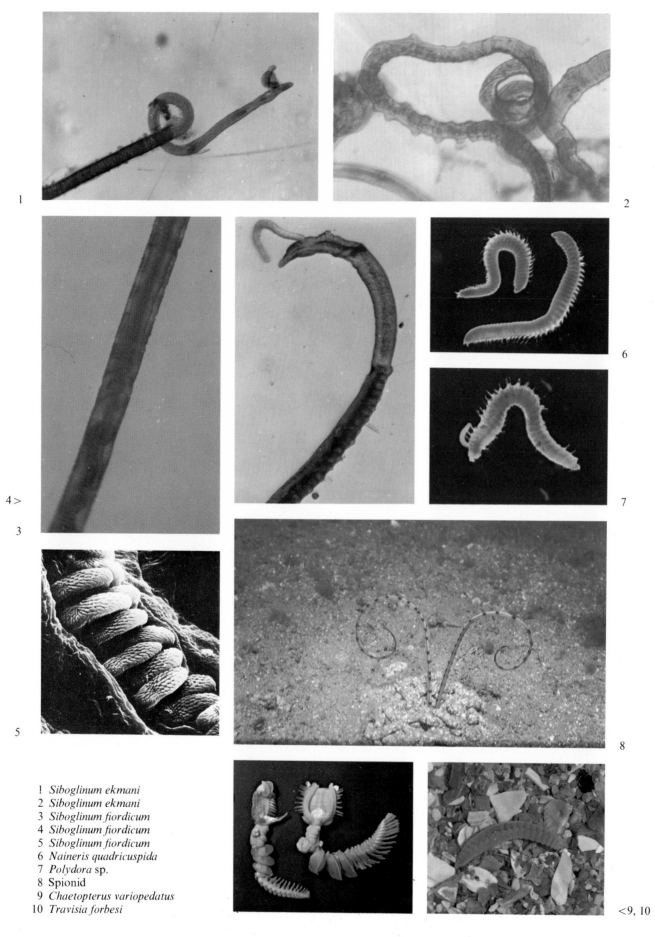

1 *Siboglinum ekmani*
2 *Siboglinum ekmani*
3 *Siboglinum fiordicum*
4 *Siboglinum fiordicum*
5 *Siboglinum fiordicum*
6 *Naineris quadricuspida*
7 *Polydora* sp.
8 Spionid
9 *Chaetopterus variopedatus*
10 *Travisia forbesi*

**53**

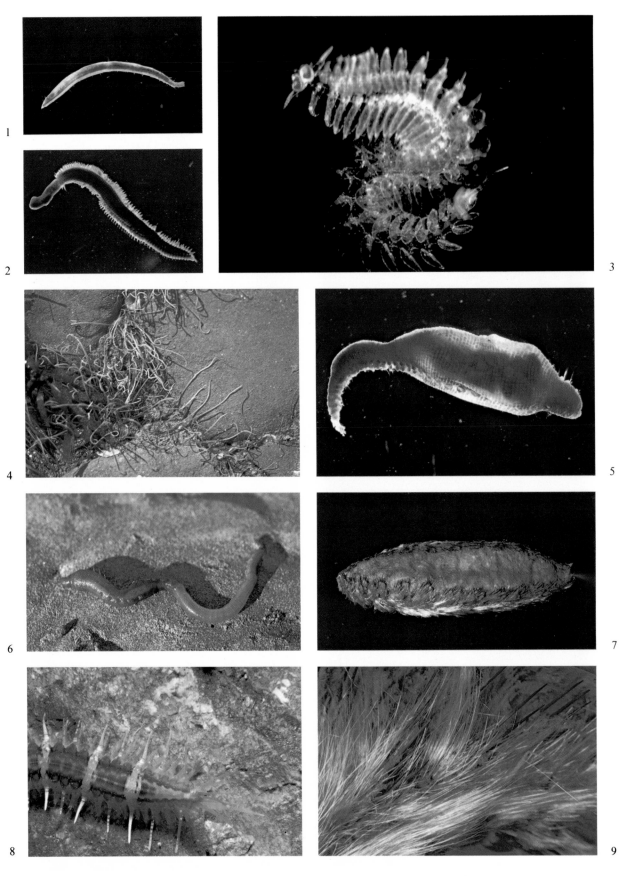

1  *Ophelina acuminata*
2  *Eulalia viridis*
3  Alciopid
4  *Cirriformia tentaculata*
5  *Scalibregma inflatum*
6  *Arenicola marina*
7  *Aphrodita aculeata*
8  *Alentia gelatinosa*
9  *Aphrodita aculeata*

1  Lepidonotus clava
2  Hermodice carunculata
3  Hermodice carunculata
4  Hermodice carunculata
5  Nereis pelagica
6  Kefersteinia cirrata
7  Neanthes virens
8  Glycera alba
9  Autolytus prolifera
10  Glycera alba

<6, 7

<9, 10

**55**

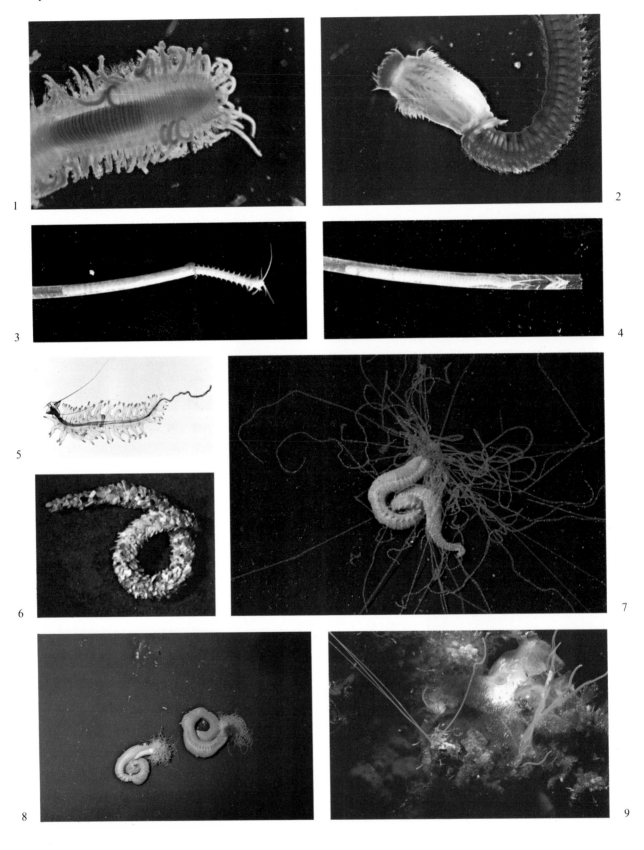

1 *Trypanosyllis zebra*
2 *Nephtys cirrosa*
3 *Hyalinoecia tubicola*

4 *Hyalinoecia tubicola*
5 *Tomopteris helgolandica*
6 *Owenia fusiformis*

7 *Eupolymnia nebulosa*
8 *Amphitrite johnstoni*
9 *Eupolymnia nebulosa*

1                                                    < 2, 3

4                                                     5

6                                                   < 7, 8

                                                            < 10, 11

4  *Sabellastarte magnifica*
5  *Sabellastarte magnifica*
6  *Lanice conchilega*
7  *Terebellides stroemi*
8  *Fabricia sabella*
9  *Lanice conchilega*
10  *Myxicola infundibulum*
11  *Myxicola infundibulum*
12  *Sabella penicillus*

9

1  *Sabellaria alveolata*
2  *Loimia medusa*
3  *Bispira volutacornis*

12

1 > 
2 > 
4, 5 > 
6, 7 > 
9, 10 > 
3 
8 

1 *Sabella* sp.
2 *Filograna implexa*
3 *Spirographis spallanzani*
4 *Sabellastarte sanctijosephi*
5 *Filograna implexa*
6 *Sabellastarte sanctijosephi*
7 *Spirographis spallanzani*
8 *Hydroides norvegica*
9 *Pomatoceros triqueter*
10 *Pomatoceros triqueter*

**58**

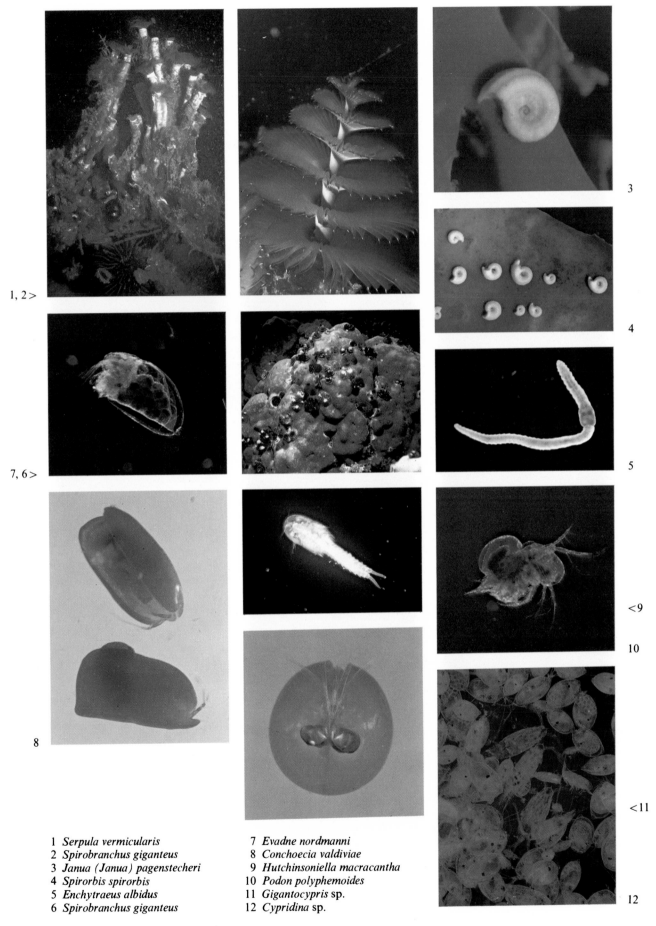

1,2>

7,6>

8

3

4

5

<9

10

<11

12

1  *Serpula vermicularis*
2  *Spirobranchus giganteus*
3  *Janua (Janua) pagenstecheri*
4  *Spirorbis spirorbis*
5  *Enchytraeus albidus*
6  *Spirobranchus giganteus*

7  *Evadne nordmanni*
8  *Conchoecia valdiviae*
9  *Hutchinsoniella macracantha*
10  *Podon polyphemoides*
11  *Gigantocypris* sp.
12  *Cypridina* sp.

1, 2 >

3

4

5

6

9, 10 >

< 7, 8

1 *Bathycalanus* sp.
2 *Megacalanus princeps*
3 *Hemirhabdus* sp.
4 *Valdiviella brevicornis*
5 *Scalpellum scalpellum*
6 *Microlaophonte trisetosa*
7 *Polticipes polymerus*
8 *Lepas anatifera*
9 *Sapphirina* sp.
10 *Microlaophonte trisetosa*

1 *Lepas anserifera*
2 *Balanus balanoides*
3 *Lepas fascicularis*
4 *Chthamalus stellatus*
5 *Tetraclita squamosa*
6 *Acasta spongites*
7 *Balanus crenatus*
8 *Balanus perforatus*
9 *Balanus balanoides*
10 *Nebaliopsis typica*

1   *Gnathophausia* sp.
2   *Limnoria quadripunctata*
3   Mysidae
4   *Squilla mantis*
5   *Limnoria quadripunctata*
6   *Eurydice pulchra*
7   *Dynamena* sp.
8   *Limnoria quadripunctata*
9   *Idotea baltica*

1 *Idotea baltica*
2 *Elasmopus pocillimanus*
3 *Idotea emarginata*/*Idotea granulosa*
4 *Ligia oceanica*
5 *Ampithoe ramondi*
6 *Meteusiroides* sp.

7 *Ampithoe rubricata*
8 *Microdeutopus gryllotalpa*
9 *Microdeutopus gryllotalpa*
10 *Apherusa jurinei*
11 *Corophium volutator*
12 *Gammarus locusta*
13 *Marinogammarus marinus*

1 *Melita palmata*
2 *Hyale nilssoni*
3 *Jassa falcata*
4 *Parajassa pelagica*
5 *Leucothoe spinicarpa*
6 *Orchestia gammarella*

7 *Orchestia gammarella*
8 *Scina* sp.
9 *Megalanceola* sp.
10 *Hyale schmidti*
11 *Podocerus variegatus*

1 *Scypholanceola* sp.
2 *Caprella linearis*
3 *Lycaea bovallioides*
4 *Phronima* sp.
5 *Primno macropa*
6 *Cystisoma* sp.
7 *Vibilia stebbingi*
8 *Pegohyperia princeps*
9 *Parathemisto gaudichaudii*
10 *Lycaea nasuta*

<2, 3

6

8

<9, 10

1 = Euphausia sanzoi  
2 = Thysanopoda sp.  
3 = Streetsia sp.  
4 = Parapenaeus longirostris  
5 = Funchalia villosa  

6 = Sergestes armatus  
7 = Plesiopenaeus edwardsianus  
8 = Acanthephyra purpurea  
9 = Notostomus gibbosus  

<8, 9

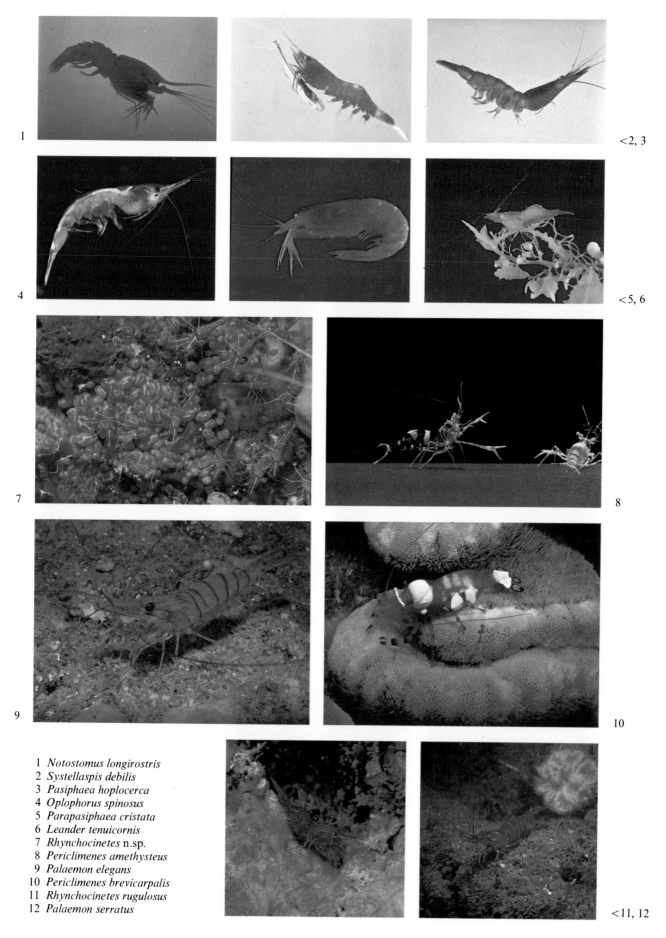

1                                             <2, 3

4                                             <5, 6

7                                             8

9                                             10

1  *Notostomus longirostris*
2  *Systellaspis debilis*
3  *Pasiphaea hoplocerca*
4  *Oplophorus spinosus*
5  *Parapasiphaea cristata*
6  *Leander tenuicornis*
7  *Rhynchocinetes* n.sp.
8  *Periclimenes amethysteus*
9  *Palaemon elegans*
10  *Periclimenes brevicarpalis*
11  *Rhynchocinetes rugulosus*
12  *Palaemon serratus*

<11, 12

**67**

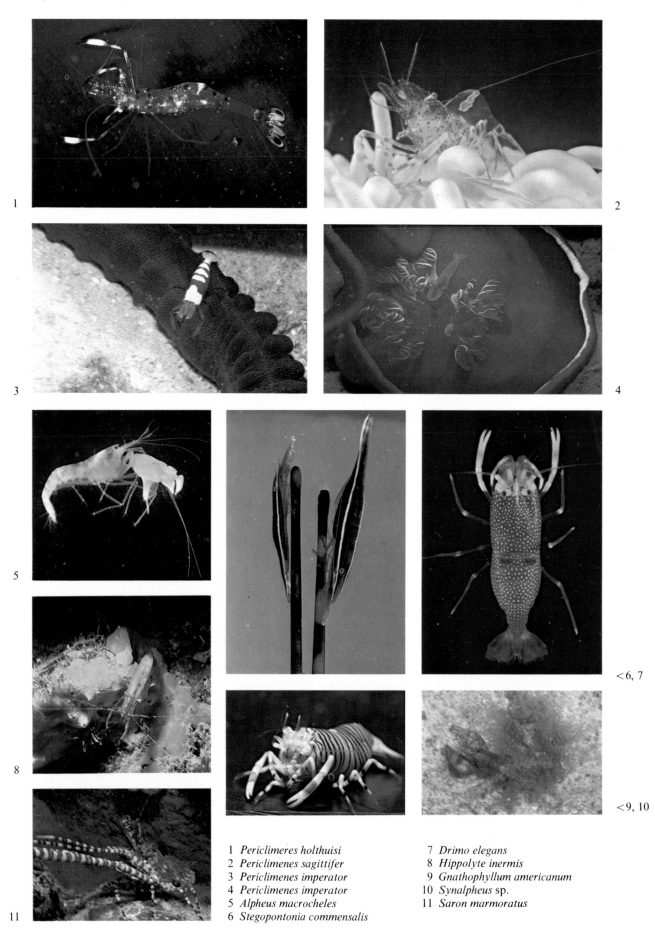

1  *Periclimeres holthuisi*
2  *Periclimenes sagittifer*
3  *Periclimenes imperator*
4  *Periclimenes imperator*
5  *Alpheus macrocheles*
6  *Stegopontonia commensalis*

7  *Drimo elegans*
8  *Hippolyte inermis*
9  *Gnathophyllum americanum*
10  *Synalpheus* sp.
11  *Saron marmoratus*

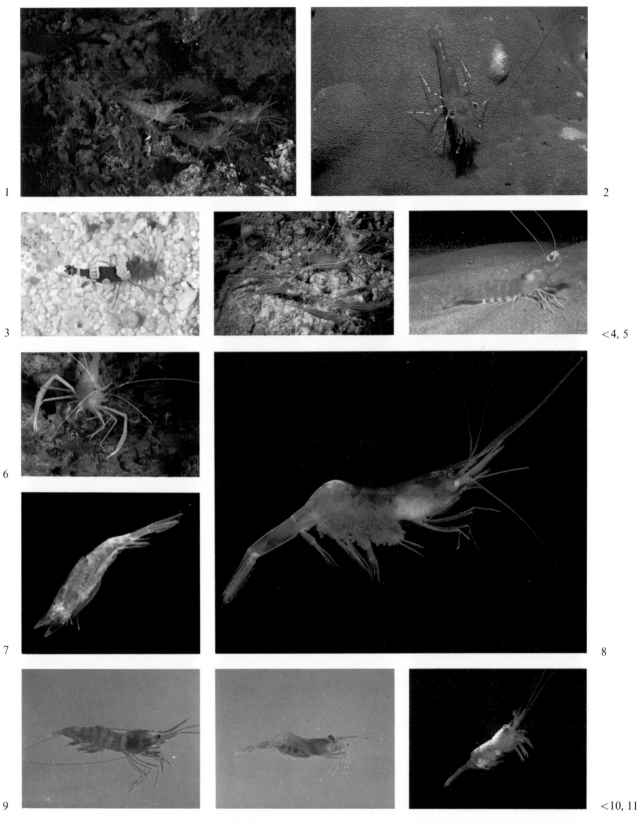

1 *Lysmata seticaudata*
2 *Saron* sp.
3 *Thor amboinensis*
4 *Parapandalus narval*
5 *Saron* sp.
6 *Stenopus spinosus*

7 *Physetocaris* sp.
8 *Parapandalus richardi*
9 *Heterocarpus grimaldii*
10 *Plesionika acanthonotus*
11 Velella *prawn*

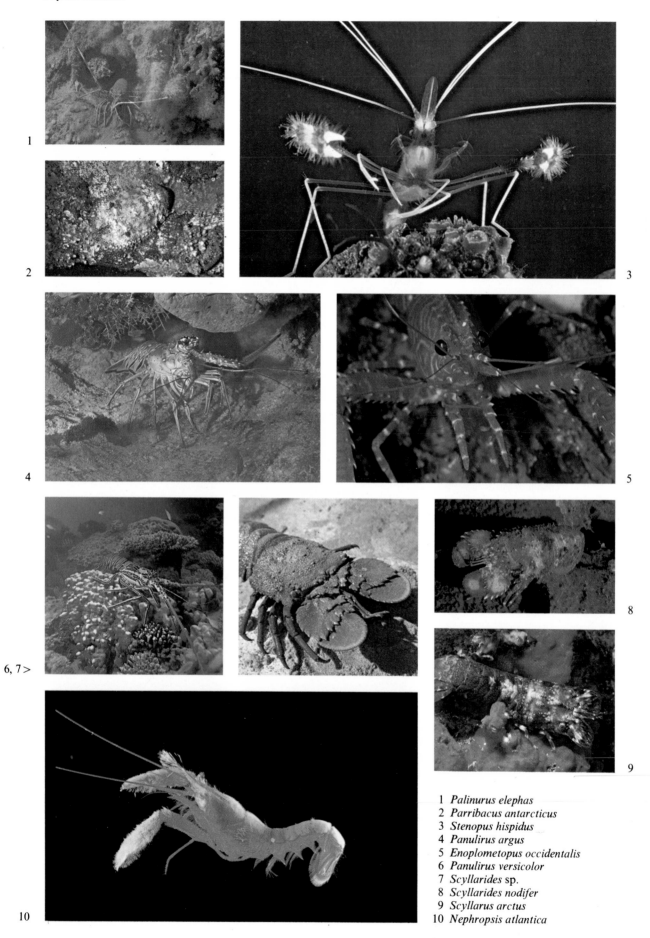

1 *Palinurus elephas*
2 *Parribacus antarcticus*
3 *Stenopus hispidus*
4 *Panulirus argus*
5 *Enoplometopus occidentalis*
6 *Panulirus versicolor*
7 *Scyllarides* sp.
8 *Scyllarides nodifer*
9 *Scyllarus arctus*
10 *Nephropsis atlantica*

1  *Dardanus lagopodes*
2  *Coenobita clypeatus*
3  *Homarus gammarus*

4  *Coenobita jousseaumei*
5  *Aniculus maximus*
6  *Dardanus megistos*

7  *Paguritta harmsi*
8  *Pagurus bernhardus*

1 *Pagurus calidus*
2 *Munida bamffica*
3 *Galathea strigosa*
4 *Galathea* sp.
5 *Petrochirus diogenes*
6 *Hippa* sp.
7 *Dromidia antillensis*
8 *Pisidia longicornis*
9 *Ranina ranina*
10 *Neopetrolisthes ohshimai*

1 *Dromia personata*
2 *Calappa flammea*
3 *Calappa* sp.
4 *Ilia nucleus*

5 *Achaeus cranchi*
6 *Macropodia longirostris*
7 *Maja squinado*
8 *Macropodia* sp.

9 *Maja squinado*
10 *Mithrax spinosissimus*

1  *Pisa tetraodon*
2  *Cancer pagurus*
3  *Stenorhynchus seticornis*

4  *Cancer pagurus*
5  *Lissocarcinus orbicularis*
6  *Cancer productus*

7  *Callinectes sapidus*
8  *Carcinus maenas*
9  *Eriphia sebana*

1 *Carpilius corallinus*
2 *Lophozozymus pictor*
3 *Pilumnus hirtellus*
4 *Xantho poressa*
5 *Macropipus corrugatus*
6 *Macropipus puber*
7 *Macropipus depurator*
8 *Etisus splendidus*

<7, 8

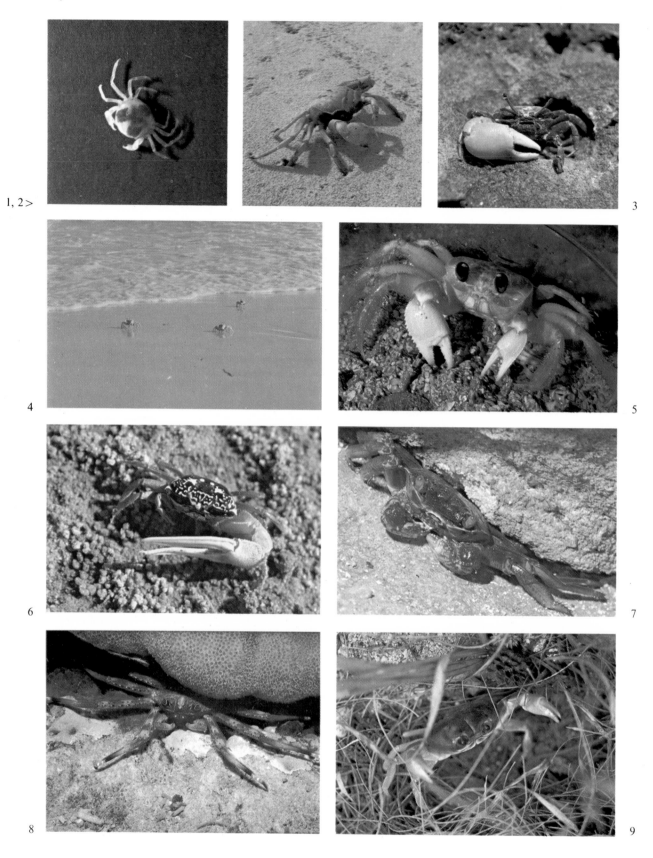

1, 2 >

3

4

5

6

7

8

9

1  *Pinnotheres pinnotheres*
2  *Ocypode ceratopthalma*
3  *Ocypode saratan*

4  *Ocypode cordimana*
5  *Ocypode quadrata*
6  *Uca* sp.

7  *Grapsus albolineatus*
8  *Percnon planissimum*
9  *Gecarcinus ruricola*

<2, 3

6

<7, 8

10

1  *Limulus polyphemus*
2  *Achelia* sp.
3  *Colossendeis colossea*
4  *Endeis pauciporosa*

5  *Colossendeis* sp.
6  *Lepidopleurus cajetanus*
7  *Acanthopleura gemmata*
8  *Acanthopleura haddoni*

9  *Tonicella lineata*
10 *Chiton olivaceus*

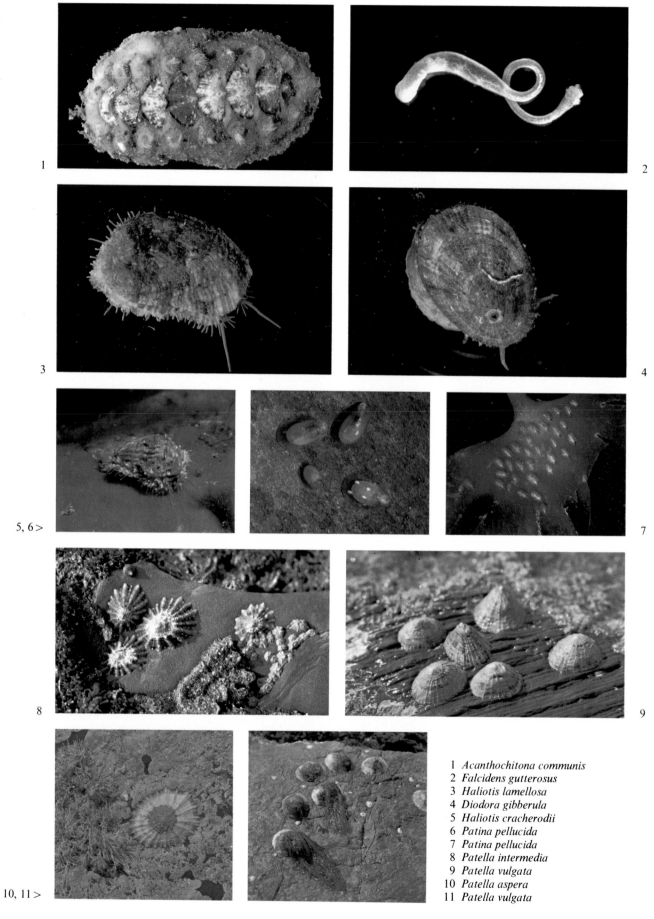

1 *Acanthochitona communis*
2 *Falcidens gutterosus*
3 *Haliotis lamellosa*
4 *Diodora gibberula*
5 *Haliotis cracherodii*
6 *Patina pellucida*
7 *Patina pellucida*
8 *Patella intermedia*
9 *Patella vulgata*
10 *Patella aspera*
11 *Patella vulgata*

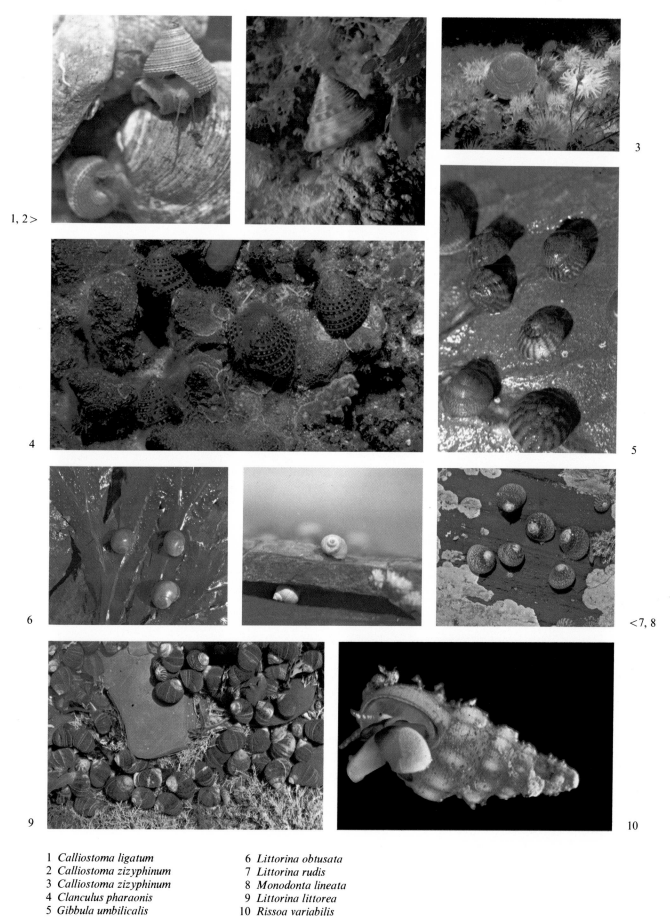

1, 2>

3

4

5

6

<7, 8

9

10

1 *Calliostoma ligatum*
2 *Calliostoma zizyphinum*
3 *Calliostoma zizyphinum*
4 *Clanculus pharaonis*
5 *Gibbula umbilicalis*

6 *Littorina obtusata*
7 *Littorina rudis*
8 *Monodonta lineata*
9 *Littorina littorea*
10 *Rissoa variabilis*

1   *Triphora perversa*
2   *Janthina* sp./*Scyllaea pelagica*/
     *Anemonia sargassiensis*
3   *Aporrhais pespelicani*
4   *Lambis crocata*
5   *Vermetus adansonii*

6   *Vermetus* sp.
7   *Skeneopsis planorbis*
8   *Cerithium echinatum*
9   *Vermetus maximus*
10  *Velacumantis australis*

80

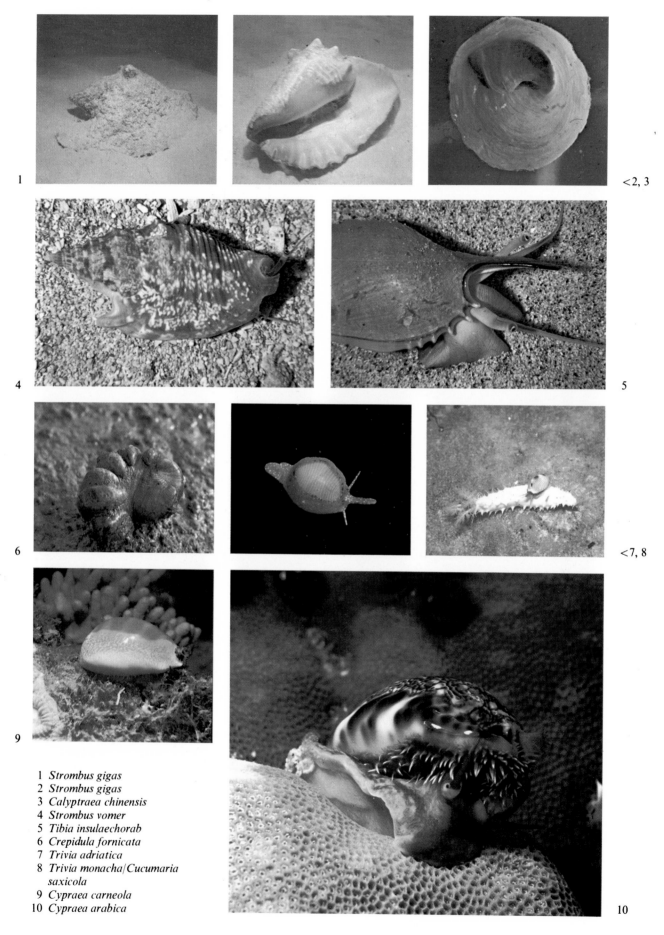

1
<2, 3
4
5
6
<7, 8
9

1 *Strombus gigas*
2 *Strombus gigas*
3 *Calyptraea chinensis*
4 *Strombus vomer*
5 *Tibia insulaechorab*
6 *Crepidula fornicata*
7 *Trivia adriatica*
8 *Trivia monacha/Cucumaria saxicola*
9 *Cypraea carneola*
10 *Cypraea arabica*

10

**81**

1, 2 >

3

4

5

6

< 8, 9

7

1 *Cypraea isabella*
2 *Cypraea lurida*
3 *Cypraea nucleus*
4 *Cypraea mappa*
5 *Cypraea pantherina*

6 *Cypraea talpa*
7 *Cypraea testudinaria*
8 *Cypraea tigris*
9 *Calpurnus verrucosus*

1  *Volva* n.sp.
2  *Cyphoma signatum*
3  *Jenneria pustulata*
4  *Cyphoma gibbosum*

5  *Natica alderi*
6  *Ovula ovum*
7  *Carinaria lamarcki*
8  *Natica hebraea*

9  *Casmaria ponderosa*
10  *Cassis madagascariensis*
11  *Cypraecassis rufa*

1, 2 >

3

4, 5 >

6

7

8

9

1  *Cabestana spengleri*
2  *Charonia rubicunda*
3  *Nucella lapillus*
4  *Murex trapa*
5  *Pyrene rustica*
6  *Euthria cornea*
7  *Tonna variegata*
8  *Charonia tritonis*
9  *Nucella* sp. radula
10 *Nassarius incrassatus*
11 *Fasciolaria tulipa*
12 *Buccinum undatum*

< 10, 11

12

1     *Fusinus polygonoides*       6     *Cymbiola magnifica*
2     *Pleuroploca filamentosa*    7     *Melo amphora*
3     *Ancillista velesiana*        8     *Persicula miliaris*
4     *Harpa amouretta*          9     *Mitra mitra*
5     *Oliva porphyria*          10     *Oliva flaviola*

1  *Clathurella purpurea*
2  *Conus betalinus*
3  *Conus geographus*
4  *Conus dalli/Pleuroploca filamentosa*
5  *Conus geographus*
6  *Conus geographus*
7  *Conus geographus*
8  *Conus striatus*
9  *Conus textile*
10  *Conus episcopus*

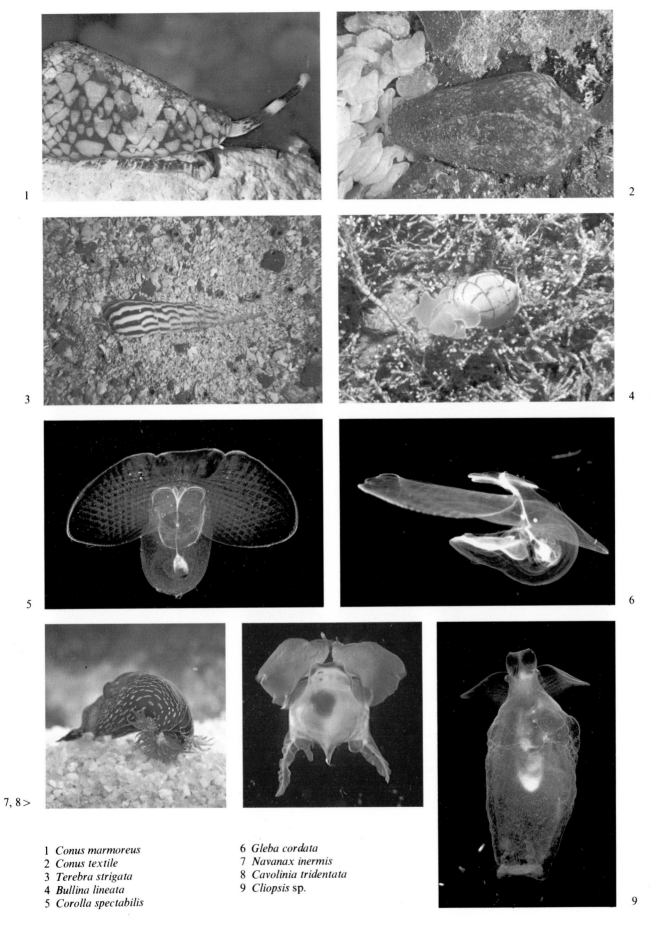

1   *Conus marmoreus*
2   *Conus textile*
3   *Terebra strigata*
4   *Bullina lineata*
5   *Corolla spectabilis*

6   *Gleba cordata*
7   *Navanax inermis*
8   *Cavolinia tridentata*
9   *Cliopsis* sp.

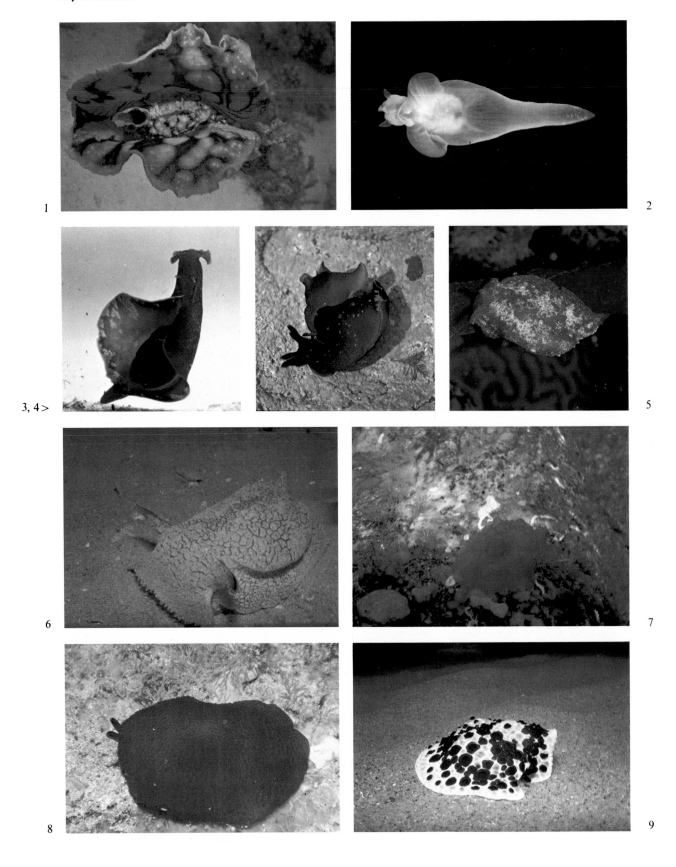

1 *Aplysia dactylomela*
2 *Clione limacina*
3 *Aplysia depilans*

4 *Aplysia punctata*
5 *Pleurobranchus areolatus*
6 *Pleurobranchaea meckeli*

7 *Pleurobranchus testudinarius*
8 *Pleurobranchus* sp.
9 *Pleurobranchus* sp.

1   *Thuridilla hopei*
2   *Dendronotus* sp.
3   *Tridachia crispata*
4   *Tritonia gracilis*
5   *Doto fragilis*
6   *Elysia viridis*
7   *Hancockia uncinata*
8   *Cyerce nigra*
9   *Tethys leporina*
10  *Phylliroe bucephala*

1  *Polycera capensis*
2  *Polycera faeroensis*
3  *Trapania maculata*
4  *Notodoris megastima*
5  *Tambja* sp.

6  *Nembrotha* sp
7  *Tambja affinis*
8  *Hexabranchus sanguineus* eggs
9  *Hexabranchus imperialis*
10  *Hexabranchus sanguineus*

1   *Hexabranchus sanguineus*      6   *Chromodoris* sp.
2   *Casella atromarginata*        7   *Chromodoris* sp.
3   *Chromodoris coi*              8   *Archidoris tuberculata*
4   *Casella* sp.                  9   *Archidoris pseudoargus*
5   *Chromodoris quadricolor*     10   *Hypselodoris tricolor*

1, 2>

3

4

5

6

<8, 9

7

10

11

1  *Peltodoris atromaculata*
2  *Phyllidia bourguini*
3  *Phyllidia varicosa*
4  *Asteronotus brassica*

5  *Phyllidia ocellata*
6  *Dendrodoris limbata*
7  *Dirona albolineata*
8  *Coryphella pedata*

9  *Caloria maculata*
10  *Hermissenda crassicornis*
11  *Glaucus atlanticus*

<2, 3

7

8

9

<10, 11

1 *Aeolidia papillosa*
2 *Fiona pinnata*
3 *Onchidium daemelli*
4 *Spurilla* sp.

5 *Hervia costai*
6 *Flabellina affinis*
7 *Lithophaga teres*
8 *Lithophaga* sp.

9 *Arca foliata/Haliotis tuberculata*
10 *Modiolus barbatus*
11 *Musculus discors*

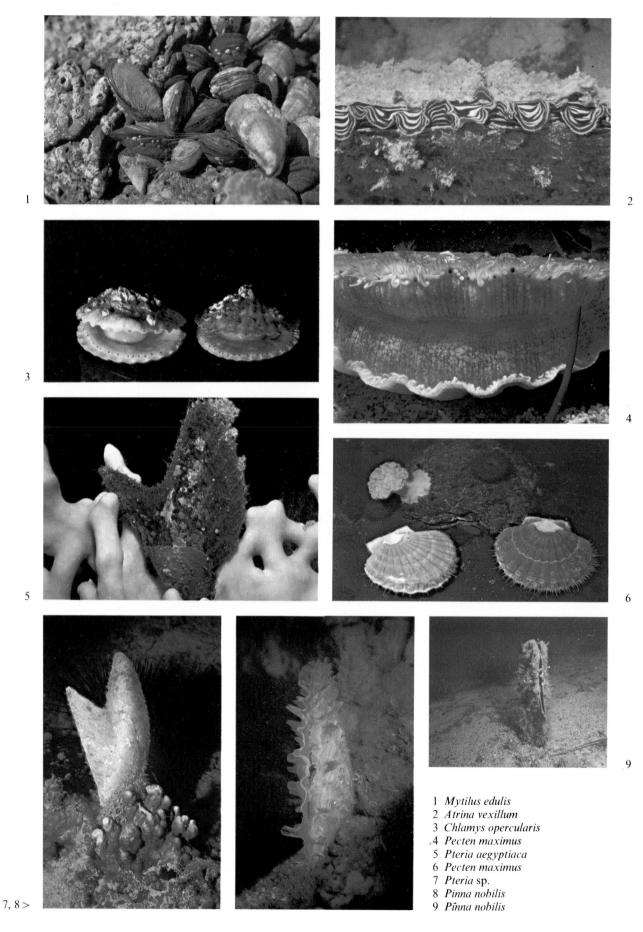

1   *Mytilus edulis*
2   *Atrina vexillum*
3   *Chlamys opercularis*
4   *Pecten maximus*
5   *Pteria aegyptiaca*
6   *Pecten maximus*
7   *Pteria* sp.
8   *Pinna nobilis*
9   *Pinna nobilis*

7, 8 >

**94**

1

2

3

<4, 5

6

<7, 8

9

10

1  *Spondylus americanus*
2  *Spondylus aurantius*
3  *Lima hians*
4  *Lima scabra*

5  *Lopha cristagalli*
6  *Promantellum vigens*
7  *Crassostrea gigas*
8  *Lopha cristagalli*

9  *Ostrea edulis*
10  *Ostrea hyotis*

1 *Saccostrea cucullata*
2 *Tridacna crocea*
3 *Tridacna gigas*
4 *Tridacna crocea*
5 *Acanthocardia echinata*

6 *Mactra corallina*
7 *Tridacna maxima*
8 *Ensis siliqua*
9 *Pharus legumen*
10 *Cerastoderma edule*

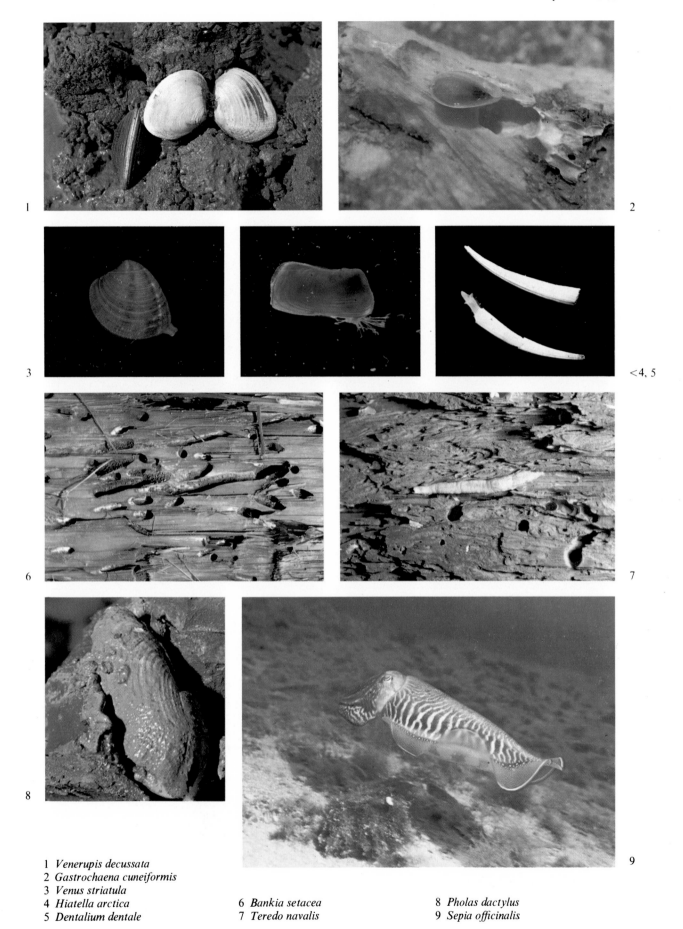

1 *Venerupis decussata*
2 *Gastrochaena cuneiformis*
3 *Venus striatula*
4 *Hiatella arctica*
5 *Dentalium dentale*
6 *Bankia setacea*
7 *Teredo navalis*
8 *Pholas dactylus*
9 *Sepia officinalis*

**97**

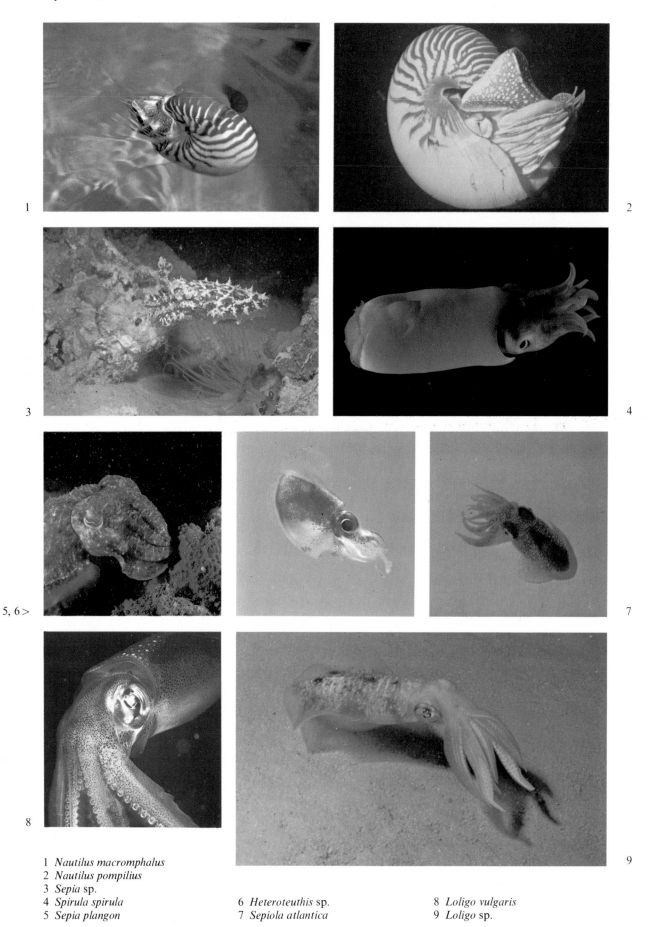

1 *Nautilus macromphalus*
2 *Nautilus pompilius*
3 *Sepia* sp.
4 *Spirula spirula*          6 *Heteroteuthis* sp.          8 *Loligo vulgaris*
5 *Sepia plangon*           7 *Sepiola atlantica*          9 *Loligo* sp.

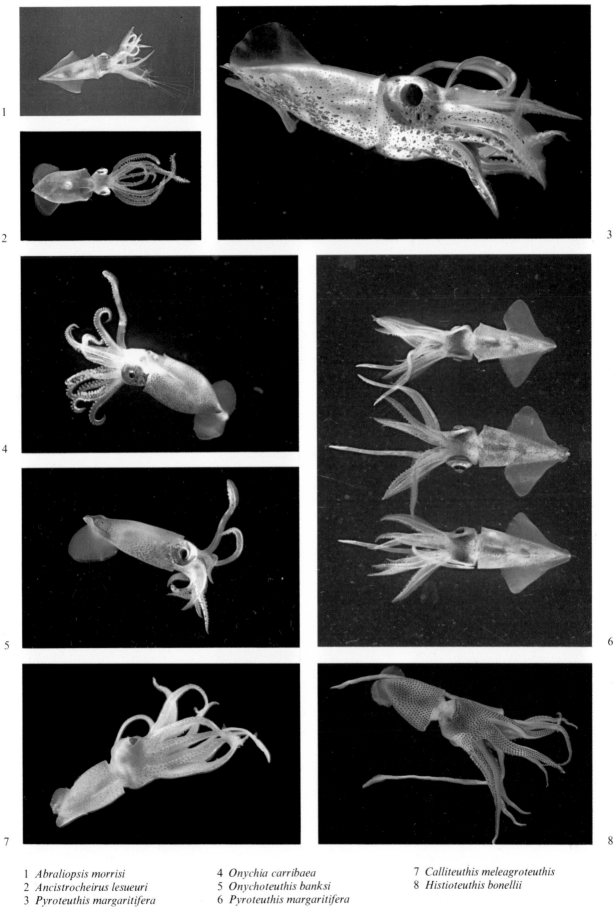

1 *Abraliopsis morrisi*
2 *Ancistrocheirus lesueuri*
3 *Pyroteuthis margaritifera*

4 *Onychia carribaea*
5 *Onychoteuthis banksi*
6 *Pyroteuthis margaritifera*

7 *Calliteuthis meleagroteuthis*
8 *Histioteuthis bonellii*

**99**

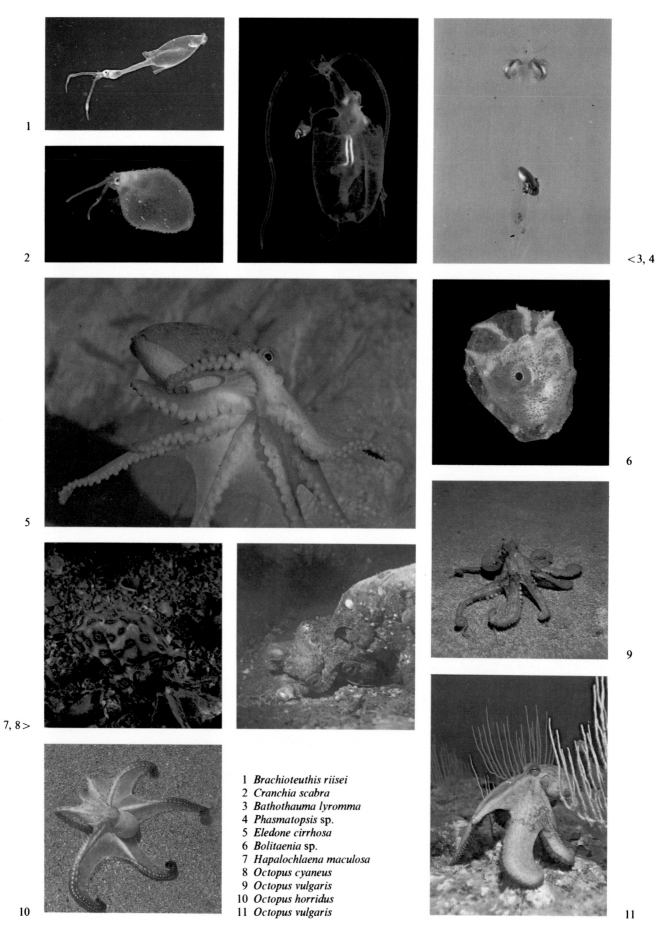

1 *Brachioteuthis riisei*
2 *Cranchia scabra*
3 *Bathothauma lyromma*
4 *Phasmatopsis* sp.
5 *Eledone cirrhosa*
6 *Bolitaenia* sp.
7 *Hapalochlaena maculosa*
8 *Octopus cyaneus*
9 *Octopus vulgaris*
10 *Octopus horridus*
11 *Octopus vulgaris*

1 *Tremoctopus* sp.
2 *Vampyroteuthis infernalis*
3 *Argonauta argo*

4 *Tubulipora flabellaris*
5 *Alcyonidium hirsutum*
6 *Crisia denticulata/Bugula turbinata*

7 *Hornera frondiculata*
8 *Membranipora membranacea*

1 *Electra pilosa*
2 *Flustra foliacea*
3 *Electra pilosa*

4 *Porella cervicornis*
5 *Pentapora fascialis*
6 *Bugula turbinata*

7 *Pentapora fascialis*
8 *Schizobrachiella sanguinea*
9 *Pentapora foliacea*

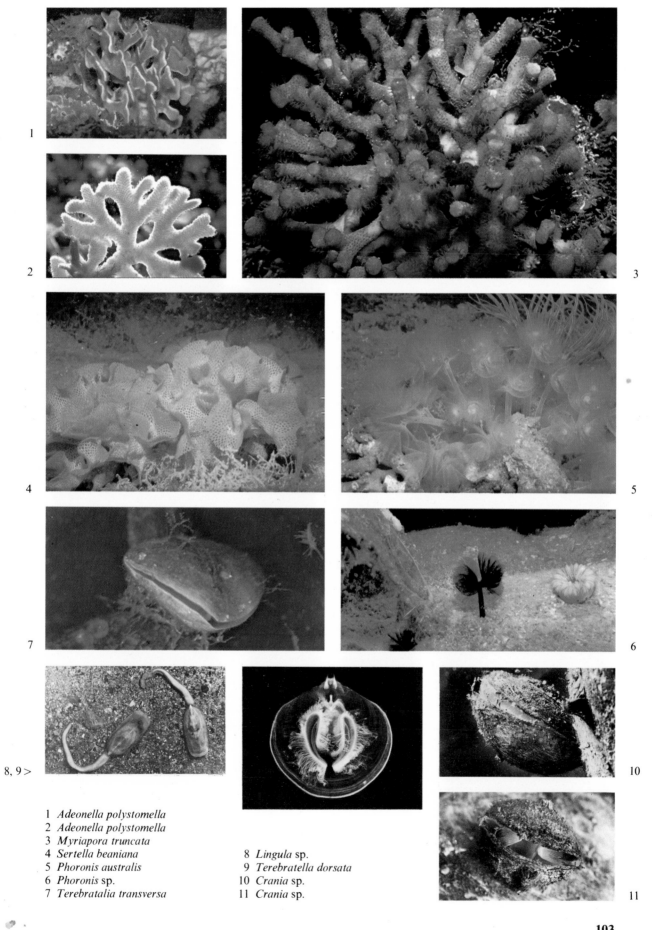

1  *Adeonella polystomella*
2  *Adeonella polystomella*
3  *Myriapora truncata*
4  *Sertella beaniana*
5  *Phoronis australis*
6  *Phoronis* sp.
7  *Terebratalia transversa*

8  *Lingula* sp.
9  *Terebratella dorsata*
10  *Crania* sp.
11  *Crania* sp.

**103**

1, 2>

3

4

5

6

7

9

8

1  *Eukrohnia fowleri*
2  *Sagitta zetesios*
3  *Sagitta* sp.
4  *Spadella cephaloptera*
5  *Comatula pectinata*

6  *Comatula* sp.
7  *Comatula pectinata*
8  *Comanthus* sp.
9  *Comanthina schlegeli*

1
2
3
4
5
6
7, 8 >
9

1 *Amphimetra* sp.
2 *Heterometra savignii*
3 *Nemaster rubiginosa*
4 *Himerometra robustipinna*
5 *Lamprometra klunzingeri*
6 *Himerometra* sp.
7 *Lamprometra klunzingeri/*
   *Heterometra savignii*
8 *Analcidometra armata*
9 *Lamprometra palmata*

1 *Oligometra carpenteri*
2 *Antedon bifida*
3 *Antedon bifida*
4 *Florometra seratissima*

5 *Antedon bifida*
6 *Ptilometra macronema*
7 *Luidia ciliaris*
8 Asteroid tube-feet

9 *Astropecten irregularis*
10 *Astropecten aranciacus*
11 *Astropecten polyacanthus*

1

2

3

<4, 5

6

7

8

<9, 10

1 *Astropecten spinulosus*
2 *Acodontaster* sp.
3 *Archaster laevis*
4 *Ceramaster placenta*

5 *Iconaster longimanus*
6 *Chaetaster longipes*
7 *Choriaster granulatus*
8 *Nectria ocellata*

9 *Pentagonaster duebeni*
10 *Choriaster granulatus*

**107**

1  *Culcita novaeguineae*
2  *Culcita novaeguineae*
3  *Fromia monilis*
4  *Culcita schmideliana*
5  *Culcita* sp.
6  *Fromia elegans*
7  *Protoreaster lincki*
8  *Protoreaster lincki*
9  *Fromia elegans*
10  *Fromia ghardaqana*

1  *Oreaster reticulatus*
2  *Pentaceraster mammillatus*
3  *Gomophia aegyptiaca*
4  *Hacelia attenuata*
5  *Nardoa novaecaledoniae*
6  *Nardoa pauciforis*
7  *Linckia laevigata*
8  *Ophidiaster confertus*
9  *Pharia pyramidata*

1 *Ophidiaster ophidianus*
2 *Tamaria megaloplax*
3 *Solaster endeca*
4 *Solaster stimpsoni*
5 *Euretaster cribrosus*
6 *Radiaster gracilis*
7 *Crossaster papposus*
8 *Anseropoda placenta*
9 *Asterina gibbosa*

8, 9 >

1  *Porania pulvillus*
2  *Dermasterias imbricata*
3  *Valvaster striatus*
4  *Echinaster sepositus*
5  *Patiriella exigua*

6  *Henricia oculata*
7  *Echinaster sepositus*
8  *Plectaster decanus*
9  *Henricia ornata*
10 *Acanthaster planci*

1, 2 >

3

4

5

6

7

8

< 9, 10

11

1  *Acanthaster planci*
2  *Metrodira subulata*
3  *Heliaster microbrachius*
4  *Acanthaster planci*
5  *Mithrodia clavigera*
6  *Asterias rubens*

7  *Evasterias troschelli*
8  *Asterias forbesi*
9  *Asterias rubens*
10  *Coscinasterias tenuispina*
11  *Marthasterias glacialis*

1  *Pycnopodia helianthoides*
2  *Pisaster ochraceus*
3  *Pycnopodia helianthoides*

4  *Uniophora granifera*
5  *Ophiocreas rhabdotum*
6  *Astrobrachion* sp.

7  *Astroboa nuda*
8  *Astrobrachion adhaerens*
9  *Astroboa nuda*

113

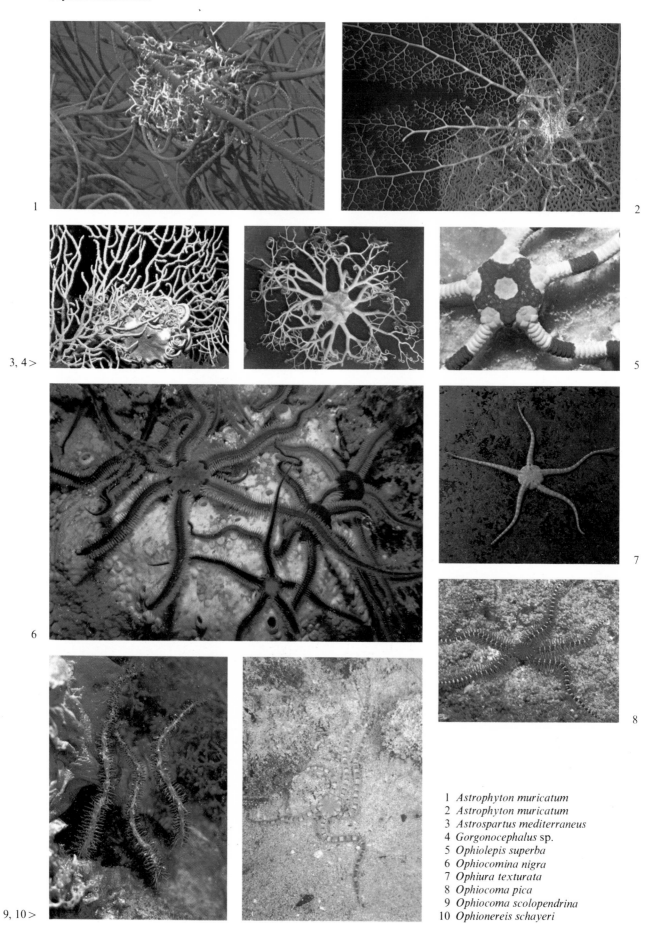

1 *Astrophyton muricatum*
2 *Astrophyton muricatum*
3 *Astrospartus mediterraneus*
4 *Gorgonocephalus* sp.
5 *Ophiolepis superba*
6 *Ophiocomina nigra*
7 *Ophiura texturata*
8 *Ophiocoma pica*
9 *Ophiocoma scolopendrina*
10 *Ophionereis schayeri*

**114**

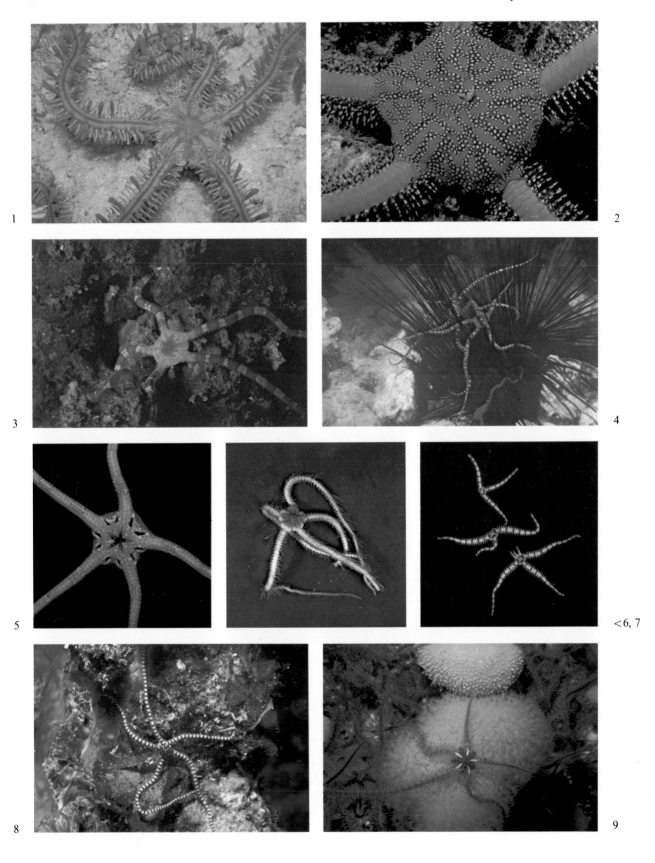

1 *Ophiomastix venosa*
2 *Ophiarachna incrassata*
3 *Ophioderma longicaudum*

4 *Ophiomyxa flaccida*
5 *Pectinura assimilis*
6 Ophiacanthid

7 *Ophiactis savignyi*
8 *Ophiomyxa* sp.
9 *Ophiothrix fragilis*

1 *Ophiothrix purpurea*
2 *Ophiothrix suensoni*
3 *Ophiothrix quinquemaculata*
4 *Ophiothrix quinquemaculata*
5 *Ophiothrix quinquemaculata*
6 *Ophiothrix* sp.
7 *Amphioplus repositus*
8 Ophiotrichid
9 *Placophiothrix* sp.

1

< 2, 3

4

5

6

7

8

9, 10 >

11

1  *Eucidaris tribuloides*
2  *Phyllacanthus parvispinus*
3  *Asthenosoma intermedium*
4  *Eucidaris metularia*

5  *Astropyga magnifica*
6  *Asthenosoma varium*
7  *Astropyga radiata*
8  *Asthenosoma intermedium*

9   *Diadema antillarum*
10  *Diadema savignyi*
11  *Diadema setosum*

**117**

1
2
3, 4 >
5
6
7
8
9, 10 >

1 *Diadema savignyi*
2 *Echinothrix calamaris*
3 *Echinothrix calamaris*
4 *Echinothrix diadema*
5 *Mespilia globulus*
6 *Echinothrix diadema*
7 *Arbacia lixula*
8 *Microcyphus rousseaui*
9 *Sphaerechinus granularis*
10 *Lytechinus variegatus*

1

< 2, 3

4

5

6

7

8

< 9, 10

1 *Pseudoboletia indiana*
2 *Toxopneustes pileolus*
3 *Tripneustes gratilla*
4 *Toxopneustes pileolus*
5 *Tripneustes ventricosus*
6 *Echinus acutus*

7 *Echinus esculentus*
8 *Echinus esculentus*

9 *Paracentrotus lividus*
10 *Sterechinus neumayeri*

1  *Echinometra lucunter*
2  *Echinometra mathaei*

3  *Echinostrephus molaris*
4  *Heterocentrotus trigonarius*
5  *Heterocentrotus mammillatus*
6  *Clypeaster subdepressus*

7  *Echinoneus cyclostomus*
8  *Parasalenia boninensis*
9  *Clypeaster* sp.
10  *Peronella lesueuri*

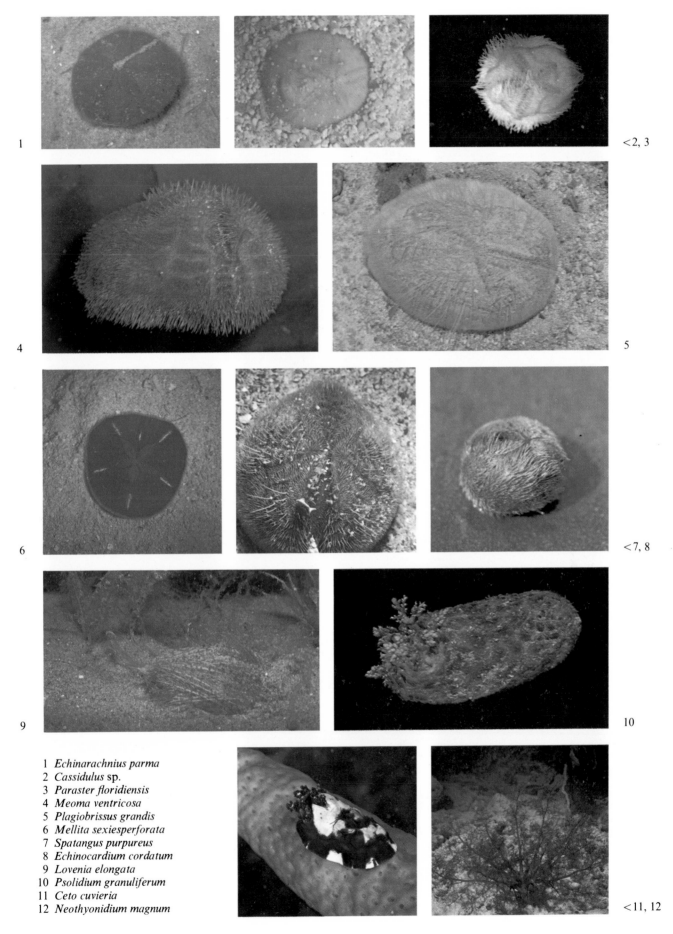

1   *Echinarachnius parma*
2   *Cassidulus* sp.
3   *Paraster floridiensis*
4   *Meoma ventricosa*
5   *Plagiobrissus grandis*
6   *Mellita sexiesperforata*
7   *Spatangus purpureus*
8   *Echinocardium cordatum*
9   *Lovenia elongata*
10  *Psolidium granuliferum*
11  *Ceto cuvieria*
12  *Neothyonidium magnum*

1   *Neothyonidium magnum*
2   *Cucumaria* sp.
3   *Cucumaria miniata*
4   *Ocnus planci*

5   *Pseudocolochirus axiologus*
6   *Actinopyga agassizi/Carapus* sp.
7   *Bohadschia argus*
8   *Actinopyga mauritiana*

9   *Bohadschia graffei*
10  *Holothuria atra*

1, 2>

3

4, 5>

6, 7>

8

9

10

11

1  *Holothuria (Halodeima) edulis*
2  *Holothuria forskali*
3  *Holothuria impatiens*
4  *Holothuria floridana*

5  *Isostichopus badionotus*
6  *Stichopus chloronotus*
7  *Parastichopus californicus*
8  *Astichopus multifidus*

9  *Stichopus regalis*
10  *Stichopus variegatus*
11  *Thelenota ananas*

**123**

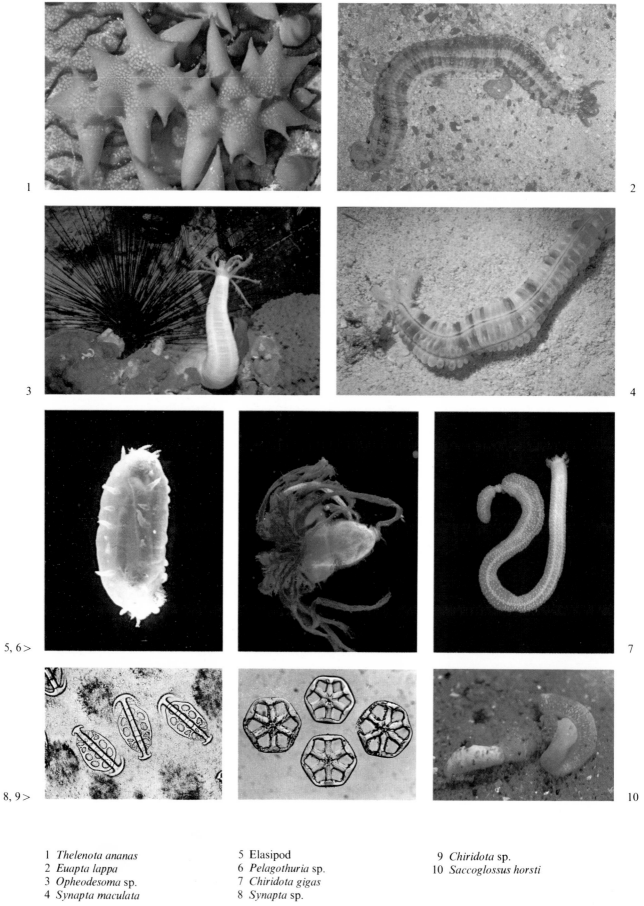

1  *Thelenota ananas*
2  *Euapta lappa*
3  *Opheodesoma* sp.
4  *Synapta maculata*

5  Elasipod
6  *Pelagothuria* sp.
7  *Chiridota gigas*
8  *Synapta* sp.

9  *Chiridota* sp.
10  *Saccoglossus horsti*

**124**

1  *Clavelina lepadiformis*
2  *Archidistoma* sp.
3  *Clavelina lepadiformis*
4  *Clavelina picta*
5  *Archidistoma* sp.
6  *Distoma adriaticum*
7  *Clavelina* sp.
8  *Sycozoa cerebriformis*
9  Clavelinid
10  Clavelinid

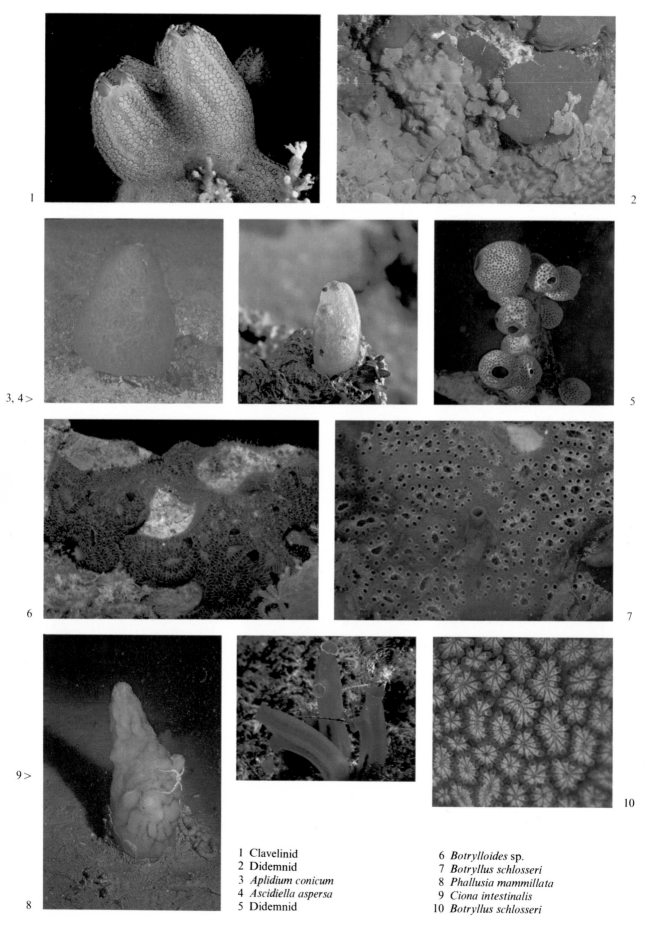

1 Clavelinid
2 Didemnid
3 *Aplidium conicum*
4 *Ascidiella aspersa*
5 Didemnid

6 *Botrylloides* sp.
7 *Botryllus schlosseri*
8 *Phallusia mammillata*
9 *Ciona intestinalis*
10 *Botryllus schlosseri*

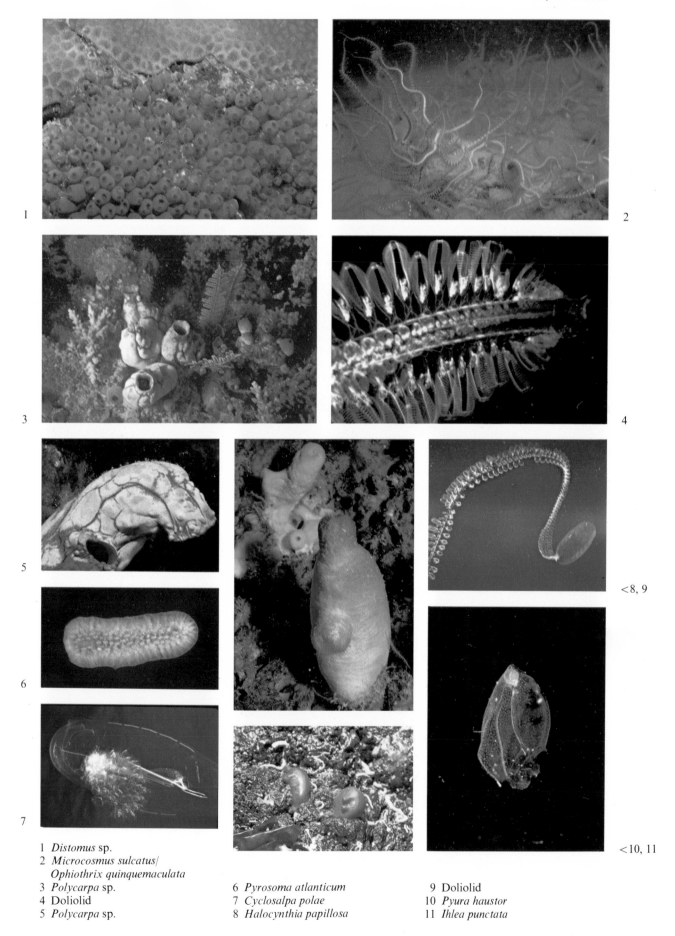

1 *Distomus* sp.
2 *Microcosmus sulcatus*/
 *Ophiothrix quinquemaculata*
3 *Polycarpa* sp.
4 Doliolid
5 *Polycarpa* sp.

6 *Pyrosoma atlanticum*
7 *Cyclosalpa polae*
8 *Halocynthia papillosa*

9 Doliolid
10 *Pyura haustor*
11 *Ihlea punctata*

**127**

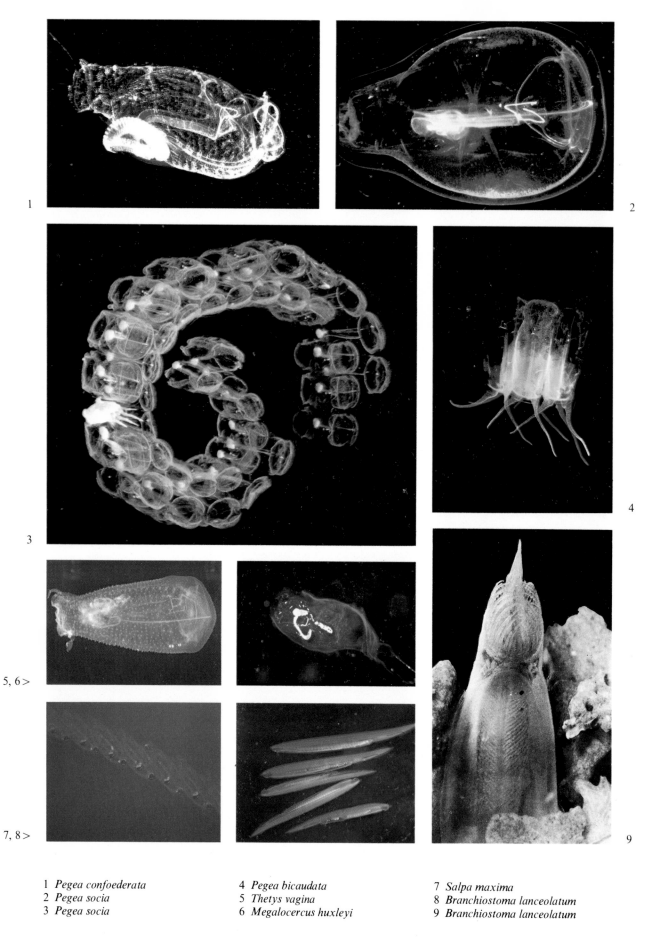

1 *Pegea confoederata*
2 *Pegea socia*
3 *Pegea socia*

4 *Pegea bicaudata*
5 *Thetys vagina*
6 *Megalocercus huxleyi*

7 *Salpa maxima*
8 *Branchiostoma lanceolatum*
9 *Branchiostoma lanceolatum*

Plate 103

# Phylum Brachiopoda

[lamp shells] approx. 300 species

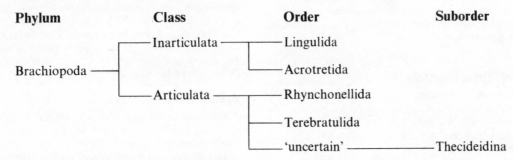

| Phylum | Class | Order | Suborder |
|--------|-------|-------|----------|
| Brachiopoda | Inarticulata | Lingulida | |
| | | Acrotretida | |
| | Articulata | Rhynchonellida | |
| | | Terebratulida | |
| | | 'uncertain' | Thecideidina |

Classification follows Moore (1965) and Rudwick (1970), modified by H. Brunton (personal communication).

An exclusively marine group of bottom-living animals bearing a superficial resemblance to the bivalve molluscs. The body is enclosed within a shell made up of two valves which cover the dorsal and ventral surfaces of the animal, unlike those of the molluscs which are arranged on either side of the body. The valves, which are variable in shape, range from a few millimetres to over 10 cm in length. The ventral valve is typically larger than the dorsal one, and in most species both valves are hinged together with a pair of teeth and sockets. Most brachiopods live permanently attached to hard material on the sea floor either by a fleshy stalk or by cementing the ventral valve directly to the surface. In the few burrowing species the long stalk is situated vertically in the sediment.

The body of the animal occupies only a small portion of the shell cavity at the posterior end, but extensions of the body wall (mantle lobes) line the internal surfaces of the valves forming a large mantle cavity communicating with the exterior when the valves open. The mantle cavity contains a large filament-bearing structure (lophophore) which is used for feeding and respiration. The lophophore may be a simple ring with a few filaments or more frequently it has long coiled lobes fringed with ciliated filaments. In many species the lophophore is partly supported on a calcareous framework arising from the dorsal valve. The beating of the cilia on the lophophore filaments creates a current of water through the mantle cavity bringing a continuous supply of food particles which are directed into the mouth at the base of the lophophore. An anus is lacking in most species and waste matter is expelled through the mouth.

The sexes are normally separate although there are a few hermaphrodite species. Eggs and sperm are usually shed into the water, where fertilisation occurs. A few species, however, brood their eggs. The fertilised egg develops into a free-swimming larva which, after a short planktonic existence, sinks to the bottom and develops into the adult form with the growth of its protective valves.

Brachiopods have a world-wide distribution, occurring in greatest numbers in the shallow seas surrounding the continents where they compete with other attached filter-feeders for food. A few deep-sea species are known.

Modern brachiopods are the descendants of a large and diverse fossil brachiopod record of nearly 600 million years duration. The number of species started to decline from about 250 million years ago.

## Class Inarticulata

approx. 45 species

Brachiopods in which the shell valves, which may be either calcareous (calcium carbonate) or non-calcareous (calcium phosphate) are held together only by muscles. The stalk, when present, either emerges from between the valves or through a hole in the ventral valve. The lophophore has no supporting skeleton and its lobes are normally coiled into spirals. There is an anus present.

## Order Lingulida

Brachiopods with thin, oval-shaped, almost identical, non-calcareous shell valves. There is a long flexible stalk which emerges posteriorly from between the two valves.

Lingulids, unlike any other brachiopod group, live in burrows in soft sediments.

Family: Lingulidae.

### Family Lingulidae

*Lingula* sp. The dozen or so species in this genus live in the Indo-Pacific and North-west Pacific and sometimes occur in such abundance along the Japanese coast that they are collected for food. They normally live buried vertically in soft sediments with only the top of the shell valves protruding slightly through a slit-like opening. The muscular stalk pulls the animal down into the sediment when it is disturbed. Some species are able to survive in foul intertidal muds, although most are distributed subtidally. 103/8

## Order Acrotretida

Brachiopods with circular, often dissimilar, calcareous or non-calcareous shell valves. The stalk, when present, emerges from a slot in the posterior part of the ventral valve. Several species in this group are cemented to the substratum by the ventral valve.

Families: Discinidae, Craniidae.

### Family Craniidae

*Crania* sp. The calcareous shell valves of members of this genus are almost circular and have a diameter of 2–3 cm. The lower valve is cemented to rocks and stones (Plate 103/10). When the animal is feeding the valves gape open revealing the lophophore lobes fringed with ciliated filaments (Plate 103/11). 103/10, 103/11

Plate 103                                                                 Phyla Brachiopoda, Chaetognatha

## Class Articulata
approx. 250 species

Brachiopods with a hinge of an interlocking pair of teeth and sockets between the calcareous valves, which restricts their opening. The stalk, when present, always emerges from a hole, slit, or notch in the ventral valve. The lophophore normally has an internal supporting skeleton. There is no anus.

## Order Rhynchonellida
Brachiopods with strongly convex valves, commonly with radiating ridges. An attachment stalk is present. The lophophore has two spirally coiled lobes supported at their base by a pair of calcareous spikes which project from the dorsal valve.

Families: Cryptoporidae, Basiliolidae, Hemithyrididae, Frieleiidae.

## Order Terebratulida
Brachiopods which usually have smooth pear-shaped convex shells and a stalk. The shell is perforated by minute canals. The lophophore consists of two simple lateral lobes and a coiled central lobe. It is supported by either a short or long calcareous loop extending from the dorsal valve.

Families: Terebratulidae, Dyscoliidae, Cancello-

thyrididae, Megathyrididae, Platidiidae, Kraussinidae, Dallinidae, Laqueidae, Terebratellidae.

## Family Dallinidae
*Terebratalia transversa* (Sowerby). The shell valves of this   103/7
species are very variable in shape but are normally wider than long. The width is usually 2–3 cm. They are attached to hard objects by means of a short stalk. In sheltered areas this species is often found intertidally, but it is found normally in shallow subtidal regions. Occurs in the North-east Pacific.

## Family Terebratellidae
*Terebratella dorsata* Gmelin. The ventral valve of this   103/9
specimen has been removed to reveal the large lophophore lying within the calcareous dorsal valve. The unique hinge arrangement characteristic of the class can also be seen. It is found in the colder waters of the southern oceans.

## Order 'uncertain'
## Suborder Thecideidina
Body with a thick convex ventral valve, commonly cemented to the substratum, and a more or less flat dorsal valve. There is no stalk. The shell may gape widely revealing the lophophore whose lobes fit into grooves in the dorsal valve.

Families: Thecidellinidae, Thecideidae.

# Phylum Chaetognatha

[arrow worms] approx. 50 species

An exclusively marine group of small arrow-shaped animals which live mainly in the upper layers of the sea. The bilaterally symmetrical body, which is often transparent, can reach a length of 10 cm in some species. There is a head, a trunk and a tail region. Fins are present along the sides of the body and on the tail. Large curved spines, which are used for seizing prey, arise from the sides of the head (Figure 39). The spines are covered by a fold of the body wall, the hood, when the animal is not feeding. Rows of small teeth are situated on either side of a ventral depression which leads to the mouth. There are two eyes present on the dorsal surface of the head. Muscles in the trunk and tail region allow the animal to make darting movements through the water, whilst the fins have mainly a stabilising and supportive function.

They are hermaphrodite, and exchange of sperm often occurs between individuals, although some species appear to be self-fertilising. The eggs, which may be released into the water or carried by the parent, hatch out into small larvae, very similar to the adult.

Chaetognaths are some of the most commonly occurring planktonic animals. They are found in all seas from the surface water to depths of over 1000 m. Vertical migrations between upper and lower water layers are common. One genus, *Spadella* is bottom-living. They are voracious carnivores feeding on hydromedusae,

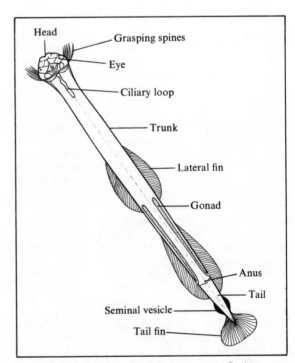

**Figure 39** *The main features of the arrow-worm,* Sagitta.

crustaceans, young fish and even on other chaetognaths. Animals, frequently as large as themselves, are seized by the head spines and swallowed whole.

Most chaetognath species live in well defined regions, e.g. there are cold- and warm-water species, species living in coastal waters, species living in the open ocean. Because of this behaviour they have been used as indicators of the movements of different water masses in the world's oceans.

Seven genera only: *Bathyspadella, Eukrohnia, Heterokrohnia, Krohnitta, Pterosagitta, Sagitta, Spadella.*

104/1   *Eukrohnia fowleri* Ritter-Zahony. This orange-tinted species, which reaches a length of 5 cm, has long narrow lateral fins that are characteristic of the genus as a whole. The gut can be clearly seen in the photograph, extending from the head to the anus which is situated just behind the pair of ovaries. This species has a worldwide distribution, being abundant at depths below 800 m.

104/2   *Sagitta zetesios* Fowler. The genus *Sagitta* is characterised by two pairs of lateral fins, but their translucent nature

makes them difficult to see in the photograph. The head spines, however, are clearly visible. This species, which reaches 5 cm in length, has been recorded most frequently at depths below 200 m in the Atlantic and Pacific Oceans.

*Sagitta* sp. A close-up photograph of a chaetognath head,   104/3
showing the large spines used for seizing and holding the prey.

*Spadella cephaloptera* (Busch). This is a relatively short   104/4
broad species, rarely more than 1 cm in length. It has conspicuous eyes and fins. Developing eggs can be clearly seen through the body wall in the anterior region of the specimen in the photograph. The posterior part of the body is occupied by a large pair of testes. This species is unusual amongst chaetognaths in that it lives on the bottom attached to seaweeds and other objects by adhesive papillae on the underside of its tail. Passing prey is seized by the grasping spines whilst the chaetognath retains its attachment to the bottom. It is found in rock pools low on the shore or on the seabed in shallow waters of the North-east Atlantic and Mediterranean.

# Phylum Echinodermata

approx. 5900 species

The echinoderms (or spiny-skinned animals) are a distinctive group of exclusively marine animals which as adults are mainly bottom-living. They occur from the shore to the deepest parts of the ocean, but because they lack a system for regulating the water and salt balance in their bodies they are excluded from low-salinity areas. Parasitism is little known in the group, whose members range in size from 5 mm to over 1 m. One of their most striking features is the 5-rayed symmetry of the adult body, which can be seen to a greater or lesser degree in all three subphyla which the group comprises. Another universal feature is the presence of a skeleton, lying beneath the skin, composed of calcareous plates or ossicles which in some cases can be fused together to form a rigid shell (test). Spines and knobs (tubercles) often extend from the plates through the skin, which, in some classes, is covered with pincer-like appendages known as pedicellariae. The pedicellariae serve to keep the surface of the body free from debris and are used by some animals for defence and to capture food to supplement their diet. There are a great variety of feeding types in the group, ranging from sediment- and filter-feeders through herbivores to carnivores.

Echinoderms possess a unique internal system known as the water-vascular system which is connected to the exterior in some classes through a sieve-like opening known as the madreporite. The system is continuous with contractile tube-feet projecting through the body wall in radiating lines from the mouth. The regions of the body associated with the radiating lines of tube-feet are often referred to as the ambulacra and the areas between them as the interambulacra. In many species the tube-feet are used for locomotion, but in others they are used exclusively for the collection of food. Although

all echinoderms have a mouth, not all of them have an anus.

Most species have the sexes separate, but a few are hermaphrodite. Eggs and sperms are frequently shed into the seawater, where fertilisation takes place. Planktonic larvae are a consistent feature of echinoderm life-histories, although a few species brood their young. Not all species reproduce exclusively by sexual means; some also reproduce by splitting of the parent body into parts (fission).

## Subphylum Crinozoa

Echinoderms which, at least for some of their life, are attached to the substratum by a stalk. The mouth is directed upwards.

## Class Crinoidea [sea-lilies, feather-stars]
approx. 600 species

Radially symmetrical echinoderms in which the body (=cup) may be either supported on a jointed stalk (sea-lilies) (Figure 40) or on a ring of jointed appendages known as cirri (feather-stars) arising from a single plate on the underside of the cup (Figure 41). The cup extends into five arms which in many species fork repeatedly to produce numerous branches bordered by appendages called pinnules. Finger-like tube-feet on the arms catch plankton and suspended organic matter which is transferred in a mucous string along ciliated grooves to the mouth. The crinoids are unique amongst echinoderms in

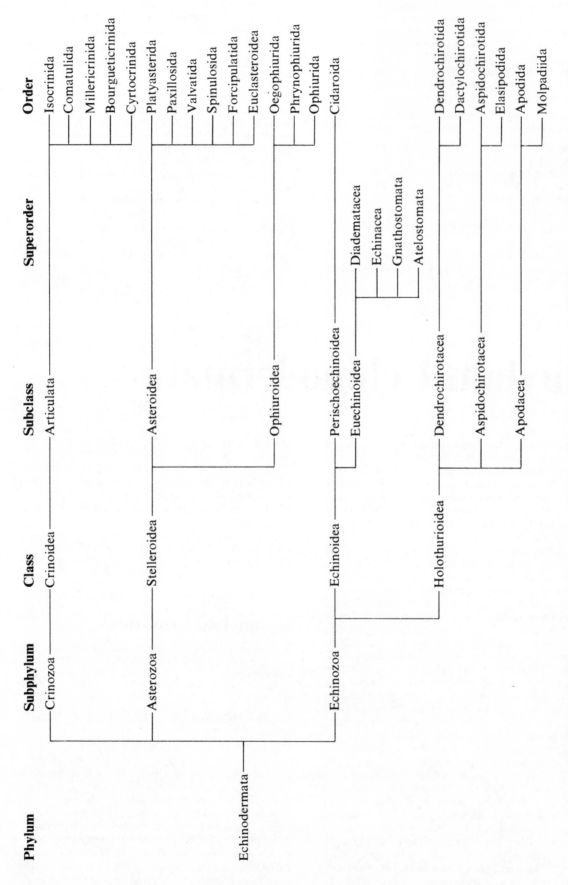

Classification follows Moore (1966, 1967) supplemented by Ubaghs (1953), Clark (1915–1959) and Clark & Clark (1967) for the Crinozoa. The arrangement of the Asterozoa has been modified in the manner suggested by A. M. Clark (personal communication).

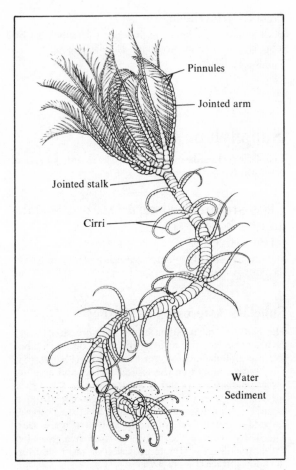

**Figure 40** *The sea-lily,* Endoxocrinus, *with its stalk anchored in sediment.*

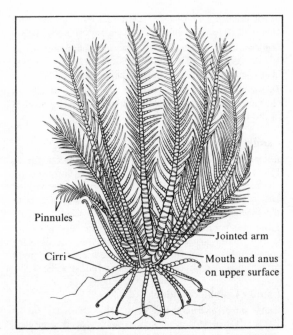

**Figure 41** *Feather-star clinging to a rock with its arms raised in a feeding position.*

that the mouth and the anus are situated on the uppermost surface of the cup. Both the cup and arms of crinoids are heavily armoured with butting calcareous plates. There are no spines or pedicellariae. There is no madreporite.

Arms of feather-stars fragment easily when the animals are attacked by predators but they can be regenerated subsequently. The young stages of feather-stars are attached to the substratum by a stalk, as in the sea-lilies, but later in development break free from the stalk and are capable of swimming with sinuous movements of the arms. They are often found in great numbers in shallow waters clinging with their cirri to any objects on which they can gain a purchase. The stalked sea-lilies, on the other hand, are much rarer and are confined to deeper water.

The class has a long fossil record, being most abundant in the seas over 400 million years ago.

## Subclass Articulata

The only subclass of the crinoids with present-day representatives. Adults may be either stalked or free-living, both kinds with or without cirri.

## Order Isocrinida

The young are attached to the substratum by a stalk. Adults retain the stalk but attach by means of cirri.

Family: Isocrinidae.

## Order Comatulida

They are stalked when young, but break away from the stalk when adult and attach by means of cirri. Within this order the Comasteridae differ from other families by having the mouth towards the edge of the cup and the anus in the centre.

Families: Comasteridae, Zygometridae, Eudiocrinidae, Himerometridae, Mariametridae, Colobometridae, Tropiometridae, Calometridae, Ptilometridae, Asterometridae, Thalassometridae, Charitometridae, Notocrinidae, Aporometridae, Antedonidae, Pentametrocrinidae, Atelecrinidae.

## Family Comasteridae

*Comanthina schlegeli* (Carpenter). Diameter up to 15 cm. The arms repeatedly fork to produce up to 200 branches. The cirri are not readily apparent. There is a wide range of colour variation in this Indo-Pacific species.  104/9

*Comanthus* sp. A form with more than ten arms which occurs on hard bottoms in the Indo-Pacific.  104/8

*Comatula pectinata* (Linnaeus). Diameter up to 20 cm. It has many arms with some pinnules in a characteristic comb-like arrangement. Attachment cirri are visible on the lower surface (Plate 104/5). Found in the Indo-Pacific.  104/5, 104/7

*Comatula* sp. Close-up of cup showing the centrally placed anus on a raised cone. Found in the Indo-Pacific.  104/6, 104/8

*Nemaster rubiginosa* (Pourtalès). Diameter up to 15 cm. Commonly found amongst reef corals with only the arms exposed. It is rare to find the full complement of arms protruding at any one time. Found in the Caribbean.  105/3

## Family Himerometridae

*Amphimetra* sp. Diameter up to 15 cm. Usually with ten arms. The pinnule size varies little along the length of the arm. Found on Indo-Pacific reefs.  105/1

*Heterometra savignii* (Müller). Diameter up to 15 cm. Usually with 20 long arms. The cirri can be seen clinging  105/2, 105/7

245

to the uneven rock surface. It emerges from under coral boulders to feed at night. Found on Indo-Pacific reefs.

105/4 *Himerometra robustipinna* (Carpenter). Diameter up to 15 cm. All arms bear pinnules which are more robust near the cup. Found on Indo-Pacific reefs.

105/6 *Himerometra* sp. Diameter up to 15 cm. It shows characteristic elongated arm pinnules near the cup. Found in the Indo-Pacific.

### Family Mariametridae

105/5, 105/7 *Lamprometra klunzingeri* (Hartlaub). Diameter up to 15 cm. This night-active species, which is common on the fire coral *Millepora*, has many red and white banded arms. Found on Indo-Pacific reefs, sometimes with *Heterometra savignii* (Plate 105/7).

105/9 *Lamprometra palmata* (Müller). Diameter up to 15 cm. It has many arms which in some individuals vary considerably in length. The attachment cirri bear small projections. A night-active species on Indo-Pacific reefs.

### Family Colobometridae

105/8 *Analcidometra armata* (A. H. Clark). Diameter up to 10 cm. It has ten arms with widely spaced pinnules. Occurs on gorgonians in the Caribbean.

106/1 *Oligometra carpenteri* (Bell). Diameter up to 10 cm. Usually with ten arms, whose pinnules give them a rather ragged appearance. The cirri have transverse ridges with spines. Found on corals and gorgonians in the Indo-Pacific.

### Family Ptilometridae

106/6 *Ptilometra macronema* (Müller). Diameter up to 8 cm. It has large pinnules on relatively short arms. Found on hard bottoms in the Indo-Pacific.

### Family Antedonidae

106/2, 106/3, 106/5 *Antedon bifida* (Pennant). Diameter up to 15 cm, with ten arms. The adult attaches by its cirri to any suitable solid object, but is able to swim effectively by flexing the arms (Plate 106/5). There is a wide range of colour variation in this species. Often occurs in very large numbers in the North-east Atlantic.

At an early stage in its development *Antedon* is attached to the substratum by a stalk (Plate 106/3) and after several months of attached life it breaks free from the stalk to lead an independent existence.

106/4 *Florometra seratissima* (Mortensen). Diameter up to 15 cm. It can be seen clinging to a rock pinnacle by its cirri as it feeds at night. Found in the North Pacific.

### Order Millericrinida

Both young and adults are attached to the bottom by a stalk with an expanded base. The wide funnel-like cup bears five unbranched, well separated arms.

Family: Hyocrinidae.

### Order Bourgueticrinida

Both young and adults are permanently attached to the bottom by a stalk with either an expanded base or with branching rootlets at the base. The cup is small and compact, usually inverted-conical or cigar-shaped. It bears five arms which sometimes branch.

Families: Bathycrinidae, Phrynocrinidae.

### Order Cyrtocrinida

Both young and adults live attached to the bottom by a massive, non-segmented stalk. No cirri are present.

Family: Holopidae.

# Subphylum Asterozoa

Free-living echinoderms with arms. The mouth is directed downwards.

## Class Stelleroidea [starfishes, brittle-stars and basket-stars]

approx. 3600 species

Star-shaped radially symmetrical echinoderms in which the body is flattened.

## Subclass Asteroidea [starfishes]

The body is drawn out into five or more stout arms, each arm bearing two or four rows of locomotory tube-feet in an ambulacral groove on its underside (Plate 106/8). They have a mouth on the ventral surface and an anus (occasionally absent) and a madreporite on the dorsal surface. In general the body surface is not strongly armoured but has plates or ossicles (often spiny) embedded in the tissue of the body wall (Figure 42a). The fleshy areas of the body wall are often provided with retractile gills. Minute pedicellariae protrude from the body surface (Figure 42b).

106/8

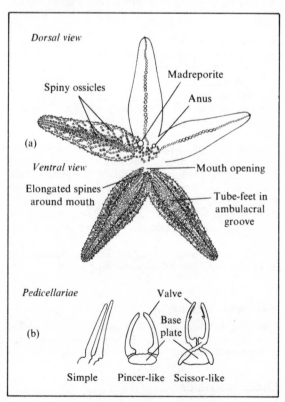

**Figure 42** (a) *Dorsal and ventral views of the starfish,* Asterias, *showing its external features.* (b) *Details of pedicellariae showing the main types that occur in the starfishes.*

Asteroids commonly regenerate lost or damaged parts, and a few species normally reproduce by fission as well as sexually. The majority of starfishes occur on rocky bottoms, although some dwell on sediments. They feed in a variety of ways, some on microscopic organic matter, whilst others eat algae, sponges, corals, polychaetes, crustaceans, bivalves, and even other echinoderms.

## Order Platyasterida

Members of this order have five or more arms. One row of conspicuous plates border the arms. Tube-feet, which occur in two rows, taper and end in a rounded knob. They have simple unstalked pedicellariae which are obviously modified from spines.

Family: Luidiidae.

### Family Luidiidae

106/7 *Luidia ciliaris* (Philippi) [7-rayed starfish]. Diameter up to 50 cm. Very conspicuous plates and spines border the arms. The knobbed tips of the tube-feet are noticeable. A voracious invertebrate predator on sands and gravels of the North-east Atlantic and Mediterranean.

## Order Paxillosida

Normally with five arms. Two well-developed rows of plates border the arms. Tube-feet, which occur in two rows, usually taper to points, but occasionally have suckered tips. Simple unstalked pedicellariae are present.

Families: Astropectinidae, Goniopectinidae, Porcellanasteridae, Benthopectinidae.

### Family Astropectinidae

106/10 *Astropecten aranciacus* (Linnaeus) [red comb star]. Diameter up to 50 cm. A general invertebrate predator on mud and sand in the Mediterranean and the North-east Atlantic.

106/9 *Astropecten irregularis* (Pennant). Diameter up to 15 cm. Spines of the upper plates bordering the arms are smaller and fewer in number than for *A. aranciacus*. It is normally found on sand, but it will migrate on to rock in search of brittle-star prey. Found in the North-east Atlantic and Mediterranean.

106/11 *Astropecten polyacanthus* Müller & Troschel. A night-active species, with large spines, from the Indo-Pacific.

107/1 *Astropecten spinulosus* (Philippi). Diameter up to 10 cm. A night-active Mediterranean species often found in sandy patches between rocks.

## Order Valvatida

Normally with five arms. Two well-developed rows of plates border the arms. Suckered tube-feet occur in two rows. Pedicellariae when present are pincer-like and frequently inset into the underlying ossicles.

Families: Sphaerasteridae, Odontasteridae, Chaetasteridae, Archasteridae, Goniasteridae, Oreasteridae, Ophidiasteridae, Radiasteridae.

### Family Odontasteridae

107/2 *Acodontaster* sp. Characteristically thin plates occur around the edges of the arms in this Antarctic species.

### Family Chaetasteridae

*Chaetaster longipes* (Retzius). Diameter up to 20 cm. A deep-water form from the Mediterranean with cylindrical arms, rarely found shallower than 40 m. — 107/6

### Family Archasteridae

*Archaster laevis* H. L. Clark. Diameter up to 15 cm, with a conspicuous madreporite. Often found partly buried in sand. A general invertebrate predator of the Indo-Pacific. — 107/3

This is the only starfish genus known to pair whilst releasing sexual products.

### Family Goniasteridae

*Ceramaster placenta* (Müller & Troschel). Diameter up to 15 cm. Very large plates border the short arms. The upper surface is covered with a mosaic of small plates. It is normally found on sand in the Mediterranean and North-east Atlantic. — 107/4

*Iconaster longimanus* (Möbius). Diameter up to 10 cm. The upper plates meet on the surface of the narrow arms. The madreporite and the anus are conspicuous on the patterned upper surface. Found on sand and gravel in the Indo-Pacific. — 107/5

*Nectria ocellata* Perrier. Diameter up to 15 cm. Plates on the upper surface have conspicuous raised areas (paxillae) covered with granules. Usually found in rocky areas of the Zealandic region. — 107/8

*Pentagonaster duebeni* Gray. Diameter up to 15 cm. The paxillae are smooth and encircled by small white tubercles. Found in rocky areas of the Zealandic region, where it feeds on encrusting organisms. — 107/9

### Family Oreasteridae

*Choriaster granulatus* Lütken. Diameter up to 30 cm. The upper surface is covered by a smooth skin through which small gills protrude giving it a granular appearance. Occurs on reefs in the Indo-Pacific, where it feeds on coral polyps and on other invertebrates. — 107/7, 107/10

*Culcita novaeguineae* Müller & Troschel [large pin-cushion star]. Diameter up to 20 cm. This animal is much more easily recognised as a starfish when juvenile – as it matures the arms become less obvious. If rolled over, *Culcita* is able to right itself by inflating half of its body until the tube-feet can get a grip (Plate 108/2). Feeds on coral in the Indo-Pacific. — 108/1, 108/2

*Culcita schmideliana* (Retzius) [pin-cushion star]. Diameter up to 10 cm. A deep-bodied starfish with a leathery surface. Found on rocks, sand and gravel in the Indo-Pacific. — 108/4

*Culcita* sp. An unidentified specimen from Indo-Pacific reefs. — 108/5

*Oreaster reticulatus* (Linnaeus). Diameter up to 30 cm. It has a reticular pattern of ossicles bearing spines and tubercles. Occurs commonly on sea-grass beds and rarely on coral reefs, where it feeds on sponges. Found in the Caribbean. — 109/1

*Pentaceraster mammillatus* (Audouin). Diameter up to 15 cm. There are large, characteristically rounded, tubercles on the upper surface. Found on reefs in the Indo-Pacific. — 109/2

*Protoreaster lincki* (de Blainville). Diameter up to 30 cm. The reticular skeleton bears large spines. Found on rock and sand in the Indo-Pacific. — 108/7, 108/8

## Family Ophidiasteridae

108/6, 108/9 *Fromia elegans* H. L. Clark. Diameter up to 8 cm. Two specimens are shown, one a juvenile with conspicuous black tips to the arms, the other an adult moving amongst blue colonial tunicates. Found on coral reefs in the Indo-Pacific.

108/10 *Fromia ghardaqana* Mortensen. Diameter up to 8 cm. There is a great deal of colour variation in this species. A sex change from male to female occurs with increasing age. Found on coral reefs in the Indo-Pacific.

108/3 *Fromia monilis* Perrier. Diameter up to 10 cm. A common species on coral reefs in the Indo-Pacific.

109/3 *Gomophia aegyptiaca* Gray. Diameter up to 15 cm. It has conspicuously conical plates crowned with a spine. Found on coral reefs in the Indo-Pacific.

109/4 *Hacelia attenuata* Gray. Diameter up to 20 cm. The rounded arms taper for most of their length. Its suckered tube-feet are guarded by rows of short spines. Found on rocks in the Mediterranean and adjacent Atlantic.

109/7 *Linckia laevigata* (Linnaeus). Diameter up to 40 cm. The stout finger-like arms have a brilliant blue coloration. Often found amongst coral rubble in the Indo-Pacific.

109/5 *Nardoa novaecaledoniae* (Perrier). A species up to 15 cm in diameter, with variable-length arms. Its plates are covered with small tubercles and pores for gills are visible between the plates. It has a conspicuous madreporite. Found on coral reefs in the Indo-Pacific.

109/6 *Nardoa pauciforis* (von Martens). Diameter up to 20 cm. A conspicuous madreporite occurs between the base of two arms. Sometimes seen amongst corals at low tide on Indo-Pacific reefs.

109/8 *Ophidiaster confertus* H. L. Clark. Diameter up to 25 cm. The long finger-like arms are covered by thick skin. Found on rocks and coral reefs in the Zealandic region.

110/1 *Ophidiaster ophidianus* Lamarck. Diameter up to 20 cm. The long cylindrical arms taper abruptly at the tips. Found on rocky bottoms in warm Mediterranean waters.

109/9 *Pharia pyramidata* (Gray). Diameter up to 30 cm. Regular rows of plates are present on the upper surface. It has a large irregular-shaped madreporite. Found on reefs and rocks in the eastern Pacific.

110/2 *Tamaria megaloplax* (Bell). Diameter up to 30 cm. A specimen with a distinctive block patterning on the upper surface. This 'comet' form has four arms regenerating from one. Some species in this family deliberately break off their arms and reproduce asexually in this way. Found on hard bottoms in the Indo-Pacific.

## Family Radiasteridae

110/6 *Radiaster gracilis* (H. L. Clark). Diameter up to 15 cm. It has regular rows of short conical spines on the upper surface and a prominent madreporite. Found on hard bottoms and sand in the Indo-Pacific.

## Order Spinulosida

Members of this order have five or more arms. Two inconspicuous rows of plates border the arms. Suckered tube-feet occur in two (rarely four) rows. Pedicellariae are rare, but if present consist of grouped spines.

Families: Solasteridae, Korethrasteridae, Python-asteridae, Pterasteridae, Asterinidae, Ganeriidae, Poraniidae, Asteropseidae, Echinasteridae, Valvasteridae, Acanthasteridae, Mithrodiidae, Metrodiridae.

## Family Solasteridae

*Crossaster papposus* (Linnaeus) [sun-star]. Diameter up to 30 cm. There is a large central disc with 8–13 short arms. It has a conspicuous madreporite. Found on rock and sand in the North Atlantic, where it preys particularly on other starfish. 110/7

*Solaster endeca* (Linnaeus) [purple sun-star]. Diameter up to 30 cm with 7–13 arms. The upper surface is less spiny than that of *Crossaster papposus*. Found on rocky bottoms in the North Atlantic. 110/3

*Solaster stimpsoni* Verrill. Diameter up to 30 cm, usually with ten arms. Found on rocks and sand in shallow waters and in the intertidal zone of the North Pacific. 110/4

## Family Pterasteridae

*Euretaster cribrosus* (von Martens). Diameter up to 10 cm. It has a soft membrane supported like a roof on the paxillae of the upper surface. The chamber beneath this roof serves a respiratory function and is also used for brooding the young. The specimen is seen in the act of expelling water from the chamber. Found in the Indo-Pacific. 110/5

## Family Asterinidae

*Anseropoda placenta* (Pennant) [goose-foot starfish]. Diameter up to 20 cm. The body is wafer-thin with short arms, the edges of which often look tattered. The upper and lower surfaces of the body are covered with overlapping 'tiles'. This ventral view of the body shows the suckerless tube-feet in ambulacral grooves. It lives on sand and mud in the North-east Atlantic and Mediterranean, where it feeds on shrimps and small crabs. 110/8

*Asterina gibbosa* (Pennant) [cushion star]. Diameter up to 5 cm. It is more tolerant of low salinity conditions than most starfish. Often found intertidally in rock pools, under rocky overhangs and under stones in the Mediterranean and the North-east Atlantic. 110/9

*Patiriella exigua* (Lamarck). Diameter up to 3 cm. A shallow-water species of the Indo-Pacific. 111/5

## Family Poraniidae

*Porania pulvillus* (Müller). Diameter up to 15 cm. A cushion-like, deep-bodied starfish with prominent gills. It feeds on small organic particles which are caught in strings of mucus and transferred to the mouth by ciliary action. Found on rocks and sand in the North-east Atlantic. 111/1

## Family Asteropseidae

*Dermasterias imbricata* (Grube) [leather star]. Diameter up to 25 cm. The upper surface has the consistency of soft leather. It feeds on a variety of organisms including other echinoderms. Usually found among rocks and weeds in the North Pacific. 111/2

## Family Echinasteridae

*Echinaster sepositus* Gray. Diameter up to 30 cm. The upper surface is covered by soft skin which is pockmarked. Gills are situated in the pits. Plate 111/4 shows a male shedding sperm in a characteristically arched position. 111/4, 111/7

The eggs develop directly into young starfish. Occurs on rocks and sand in the Mediterranean.

111/6 *Henricia oculata* (Pennant) [blood star]. Diameter up to 20 cm. It has stiff arms, circular in cross-section and is very variable in colour. The members of this species usually have a conspicuous madreporite. Normally occurs on rocky bottoms in the North-east Atlantic, where it feeds on sponges.

111/9 *Henricia ornata* (Perrier). Diameter up to 20 cm. A species which is sometimes also placed in the genus *Echinaster*. The posture of the animal is that frequently seen in starfish brooding their young or when feeding. Found in the South Atlantic.

111/8 *Plectaster decanus* Müller & Troschel [mosaic sea-star]. Diameter up to 20 cm. It has a pronounced mosaic pattern on the upper surface. Found in the Zealandic region.

## Family Valvasteridae

111/3 *Valvaster striatus* (Lamarck). Diameter up to 15 cm. The single row of large bivalved pedicellariae occurring on the upper marginal plates of the arms is characteristic of the family. The conical spines around the edges of the arms are a feature of this Indo-Pacific species.

## Family Acanthasteridae

111/10, *Acanthaster planci* (Linnaeus) [crown-of-thorns starfish].
112/1, Diameter up to 60 cm. There are 9–23 arms with large
112/4 spines on the upper side. The tube-feet have strongly developed suckered tips (Plate 112/1). It browses almost exclusively on coral polyps (Plate 112/4). It occasionally reaches plague proportions locally in the Indo-Pacific, where large areas of coral reef are sometimes laid bare by its feeding activities.

## Family Mithrodiidae

112/5 *Mithrodia clavigera* (Lamarck). Diameter up to 25 cm. Widely spaced conspicuous spines are characteristic of this species. A night-active inhabitant of Indo-Pacific reefs.

## Family Metrodiridae

112/2 *Metrodira subulata* Gray. Diameter up to 10 cm. The dorsal surface has a covering of short spines. Often found intertidally on sand and gravel in the Indo-Pacific.

## Order Forcipulatida

Five or more long arms without conspicuous plates bordering the arms. The suckered tube-feet are in four (sometimes two) rows. The pedicellariae are highly specialised, with two valves which are either straight or crossed in a scissor-like fashion.

Families: Heliasteridae, Zoroasteridae, Asteriidae.

## Family Heliasteridae

112/3 *Heliaster microbrachius* Xantus. Diameter up to 15 cm. In life, with its arms under its body, it could be mistaken for a sea urchin by the casual observer, because of the spiny nature of the body surface. It feeds on a variety of mussels and crustaceans in the Panamanian region.

## Family Asteriidae

112/8 *Asterias forbesi* Verrill. Diameter up to 30 cm. The upper body surface has a warty appearance, with a conspicuous madreporite. Found on all types of bottom in the North-west Atlantic, where it feeds on bivalve molluscs.

112/6, *Asterias rubens* Linnaeus. Diameter up to 50 cm. The
112/9 dorsal surface is covered with spiny tubercles, and it has a conspicuous madreporite. It is found on all types of bottom in the North-east Atlantic; there it feeds particularly on mussels and scallops (Plate (112/6). The shell valves of the molluscs are drawn apart by the pull of the suckered tube-feet and the stomach of the starfish is everted between them.

112/10 *Coscinasterias tenuispina* (Lamarck). Diameter up to 15 cm, with six to ten arms, often unequal in length. The spines on the dorsal side are surrounded by a mass of pedicellariae. Occurs on rocks and stones in the Mediterranean and adjacent Atlantic.

112/7 *Evasterias troschelli* Stimpson. Diameter up to 50 cm, but more usually specimens are less than 30 cm. Found in shallow rocky areas of the North Pacific.

112/11 *Marthasterias glacialis* (Linnaeus) [spiny starfish]. Diameter up to 80 cm. It has long flexible arms, the upper surfaces of which are covered by spines surrounded by pedicellariae. There is a conspicuous madreporite. Occurs on rocks and sand in the North-east Atlantic and Mediterranean where it feeds particularly on bivalve molluscs.

113/2 *Pisaster ochraceus* (Brandt) [ochre star]. Diameter up to 30 cm. A common starfish of rocky intertidal areas which feeds primarily on mussels. There is a wide colour variation in this species. Found in the North Pacific.

113/1, *Pycnopodia helianthoides* (Brandt) [sunflower star].
113/3 Diameter up to 1 m. A many-armed species which is found on rocks, sand and gravel. It feeds on a wide range of organisms but most commonly on bivalve molluscs. Plate 113/3 shows it in a characteristic feeding position. Found in the North Pacific.

113/4 *Uniophora granifera* (Lamarck). Diameter up to 15 cm. It has large and small globular tubercles on its upper surface, and a conspicuous madreporite. Found in the Zealandic region.

## Order Euclasteroidea

Species in this order have many long, narrow arms not bordered by plates, and which are sharply distinct from the small central body. The suckered tube-feet are in two rows. There are numerous pedicellariae which always have crossed valves. An exclusively deep-water group.

Family: Brisingidae.

## Subclass Ophiuroidea [brittle-stars and basket-stars]

They have a body (central disc) sharply separated from the five or more long slender arms. There are no ambulacral grooves visible on the underside of the arms, and the papillae-like tube-feet are not locomotory in function but are used for feeding. The toothed mouth and the madreporite are situated on the underside (Figure 43). The oral surface of the disc is invaginated on either side of the arm bases forming ten internal sacs (bursae) connected to the outside through slit-like openings (Plate 115/5). The bursae function as respiratory chambers and as outlets for the sexual products. There is no anus, the waste being expelled from the mouth. Ophiuroids are the most active echinoderms and move by flexing their arms. The upper surface of the central disc is usually covered with plates or shields, sometimes ornamented with spines or granules. The most conspicuous of these are the radial shields situated at the origin of the arms from the disc. Each arm has a segmented appearance with a series of plates frequently completely surrounding it. There are usually two rows of lateral plates, one row of upper plates and one row of lower plates. In some genera, however, only lateral plates are present but these meet on the upper and lower surfaces of the arms. The arrangement of the plates together with the spines is important in the taxonomy of the group. No pedicellariae are found in ophiuroids.

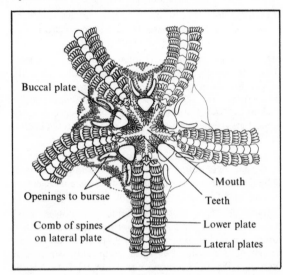

**Figure 43** *The underside of the central disc of a brittle-star showing the main external features.*

Ophiuroids can regenerate lost or damaged arms, and a few species can reproduce asexually. They occur often in dense mats, on rocks, gravel, sand and mud. Many feed on dead organic material and on microscopic organisms suspended in the water. Some of the larger forms prey on worms, crustaceans, and bivalves.

## Order Oegophiurida

The ambulacral grooves are covered by soft skin and not closed over by arm plates. The central disc is covered by skin or by overlapping scales.

Family: Ophiocanopidae.

## Order Phrynophiurida

The lateral arm plates are nearly ventral in position and the dorsal arm plates are absent or rudimentary, allowing the arms to be coiled vertically around objects. The arms may be simple or branched. The plates of both the central disc and the arms are overlain by thick soft skin.

Families: Asteronychidae, Asteroschematidae, Gorgon-ocephalidae, Euryalidae.

### Family Asteroschematidae [serpent-stars]

113/8   *Astrobrachion adhaerens* (Studer). Diameter up to 10 cm. The arms are covered with thick soft skin.

Frequently found entwined amongst the branches of antipatharian corals in the Indo-Pacific.

*Astrobrachion* sp. A serpent-star with its long arms    113/6 coiled around antipatharian coral. Found in the Indo-Pacific.

*Ophiocreas rhabdotum* H. L. Clark. With arms uncoiled,    113/5 may reach a diameter of 15 cm. The arms are very flexible. Members of this species are usually found below 15 m, with their arms tightly coiled around the stems of colonial coelenterates. They are thought to feed on small crustaceans, which are captured by the flexible arms, as well as on organic debris. Found in the Indo-Pacific.

### Family Gorgonocephalidae [basket-stars]

*Astroboa nuda* (Lyman). Diameter up to 75 cm. The    113/7, highly branched arms end in tendril-like tips (Plate 113/7).    113/9 A territorial species, at night they emerge from hiding and spread their arms in an open mesh basket to catch small swimming prey with their tendrils (Plate 113/9). Found in the Indo-Pacific.

*Astrophyton muricatum* (Lamarck). Diameter up to    114/1, 75 cm. During the daytime they hide amongst sponges,    114/2 gorgonians and corals (Plate 114/1). They spread their basket of finely branched arms at night to catch planktonic organisms. Found in the Caribbean.

*Astrospartus mediterraneus* (Risso). Diameter up to    114/3 50 cm. Very similar in habits to the above species. Normally not found shallower than about 50 m in the Mediterranean and adjacent Atlantic.

*Gorgonocephalus* sp. The arms of this genus are less    114/4 finely branched than those of *Astroboa* and *Astrophyton*, although it feeds in a similar manner. A night-active form from the North Pacific.

## Order Ophiurida

The ambulacral grooves are closed by growth of the lateral arm plates. Upper and lower arm plates are nearly always present. The arms are capable of lateral movements and limited vertical movements, but cannot be coiled around objects. The central disc is usually covered by plates (Plate 115/5).

Families: Ophiuridae, Ophioleucidae, Ophiocomidae, Ophionereididae, Ophiodermatidae, Ophiacanthidae, Ophiomyxidae, Hemieuryalidae, Ophiactidae, Amphiuridae, Ophiotrichidae.

### Family Ophiuridae

*Ophiolepis superba* H. L. Clark. Diameter up to 15 cm.    114/5 The arms are conspicuously marked with dark bands. The central disc bears concentric rings of plates. The lateral arm plates bear short spines. Often seen lying inert on reefs during the daytime in the Indo-Pacific.

*Ophiura texturata* Lamarck. Diameter up to 20 cm.    114/7 The stout tapering arms are edged with spines. Small combs of spines occur at the origin of the arms. Usually found on sand and mud in the North-east Atlantic and Mediterranean, where it preys upon small molluscs and crustaceans.

### Family Ophiocomidae

*Ophiocoma pica* Müller & Troschel. Diameter up to    114/8 10 cm. It has a characteristic pattern of radiating lines on the disc and large arm-spines. Found on reefs and sand in the Indo-Pacific.

114/9 *Ophiocoma scolopendrina* (Lamarck). Diameter up to 30 cm, with prominent arm spines. An intertidal territorial species, which specialises in feeding on organic debris in the surface water film. Feeding can be seen occurring as the water rises over the animal on the incoming tide. Found in the Indo-Pacific.

114/6 *Ophiocomina nigra* (Abildgaard). Diameter up to 25 cm. It has a finely granular central disc variously patterned in dark colours and many short, stout arm-spines. It feeds on plankton, occurring in large numbers on hard bottoms in the North-east Atlantic and Mediterranean.

115/1 *Ophiomastix venosa* Peters. Diameter up to 30 cm. The club-like spines situated at intervals along the upper surfaces of the arms, and the black lines on the central disc, are characteristic of this species. Occurs on sand and amongst algae in shallow waters of the Indo-Pacific.

### Family Ophionereididae

114/10 *Ophionereis schayeri* (Müller & Troschel). Diameter up to 15 cm. The central disc has a granular appearance and the short arm-spines project laterally. Found in sand and gravel of the Zealandic region.

### Family Ophiodermatidae

115/2 *Ophiarachna incrassata* (Lamarck). Diameter up to 45 cm. It has a deep-bodied central disc, whose reticulated upper surface is finely granular. The upper arm plates are smooth. This characteristically green species differs from most in the family by having well developed lateral arm spines. Occurs on sand and coral rubble in the Indo-Pacific.

115/3 *Ophioderma longicaudum* (Retzius). Diameter up to 30 cm. It has a leathery central disc and long arms with short lateral spines. A common species on sand and rocks in the Mediterranean.

115/5 *Pectinura assimilis* (Bell). Diameter up to 15 cm. This ventral view shows the toothed mouth and the slit-like bursal openings characteristic of the ophiuroids. The ambulacral grooves are entirely enclosed by the arm plates. The rows of short spines on the lateral plates are closely applied to the sides of the arms. Found on hard-bottoms and sand in the Zealandic region.

### Family Ophiacanthidae

115/6 An unidentified ophiacanthid from abyssal muds. The petaloid central disc is covered with small spines. The arms have characteristic long slender spines. Found in the North-east Atlantic.

### Family Ophiomyxidae

115/4 *Ophiomyxa flaccida* (Say). Diameter up to 15 cm. The thick naked skin is slimy to the touch. Three variously coloured specimens can be seen entwined amongst the spines of the long-spined sea-urchin, *Diadema antillarum*. Found in the Caribbean.

115/8 *Ophiomyxa* sp. Diameter up to 15 cm. Found on Indo-Pacific reefs.

### Family Ophiactidae

115/7 *Ophiactis savignyi* (Müller & Troschel). Diameter up to 5 cm. It is often found with any number of arms up to seven, due to its habit of splitting across the disc and regenerating new arms. Found amongst sponges and coralline algae in the Indo-Pacific.

### Family Amphiuridae

*Amphioplus repositus* (Koehler). Diameter up to 15 cm. Its long slender arms are characteristic of the family. Normally lives buried in sand with only the tips of the arms exposed. Occurs in the Indo-Pacific.   116/7

### Family Ophiotrichidae

*Ophiothrix fragilis* (Abildgaard) [common brittle-star]. Diameter up to 15 cm. It usually has a pentagonal disc with five rows of spines radiating to the points of the pentagon. The long, very fragile, arms are covered with prominent spines. Found on rocks and sand, sometimes in very large aggregations, in the North-east Atlantic and Mediterranean.   115/9

*Ophiothrix purpurea* von Martens. Diameter up to 20 cm. This species usually has a beautifully patterned disc with long spines and a purple line along the top of each arm. Found on Indo-Pacific reefs.   116/1

*Ophiothrix suensoni* Lütken. Diameter up to 20 cm. It is characterised by a darkly pigmented line along the upper surface of the arms, which have very long spines. Found on gorgonians and sponges in the Caribbean.   116/2

*Ophiothrix quinquemaculata* (Delle Chiaje). Diameter up to 15 cm. Plate 116/3 shows a cluster of specimens on the sponge *Suberites domuncula*. Organic particles suspended in the water are trapped on the erect arms of the brittle stars and transferred to the mouth. Plate 116/4 shows arm coiling, which is characteristic at certain current speeds.   116/3, 116/4, 116/5

When they occur on soft mud, two arms only are used for feeding with the other three forming a tripod-like support for the animal (Plate 116/5). Occurs in dense aggregations on soft bottoms in the Mediterranean.

*Ophiothrix* sp. An unknown Mediterranean species which lives buried in the sediment. Only two of its five arms are extended into the water for feeding.   116/6

*Placophiothrix* sp. Diameter up to 15 cm. A long-spined species which may suspend itself across water currents to aid its feeding. Found on Indo-Pacific reefs.   116/9

A beautifully patterned species of ophiotrichid inhabiting Indo-Pacific reefs.   116/8

# Subphylum Echinozoa

Free-living echinoderms without arms. The mouth is directed downwards or forwards.

## Class Echinoidea
### [sea-urchins, heart-urchins, sand-dollars]
approx. 800 species

Radially symmetrical or secondarily bilaterally symmetrical echinoderms without arms, whose bodies are either subspherical, ovate or flattened. The skeletal plates butt together (sometimes overlap) to form a test through which protrude five double rows of tube-feet flanking the ambulacral areas (Figure 44*a*). The tube-feet are almost exclusively locomotory. Movable spines articulating with tubercles on the test are also often used for locomotion. Small spherical tubercles not associated with spines are frequently borne on the ambulacral plates. The mouth is located on the underside of the body and in the majority of species is ringed with five teeth forming part of a complex chewing apparatus known as 'Aristotle's lantern'. The anus is situated either centrally on the upper side or marginally near the junction of the upper and lower surfaces. The mouth and anus are surrounded by flexible membranes bearing dissociated plates. Many different types of stalked pedicellariae (usually three-valved) may occur (Figure 44b).

Echinoids have a certain regenerative ability, but are unable to repair extensive damage to the test. The sexes are separate and, as far as is known, none are able to reproduce asexually. They are exclusively bottom-living, occurring both on hard and soft bottoms. Most species feed on both plants and animals, although a few may be regarded as primarily plant-feeders or debris-feeders.

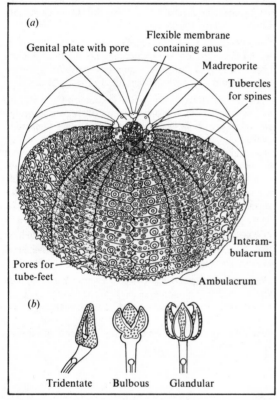

**Figure 44** (a) *Dorso-lateral view of the test of the sea-urchin,* Echinus, *showing the main external features.* (b) *Three types of pedicellariae found in the echinoids.*

## Subclass Perischoechinoidea

Radially symmetrical echinoids with ambulacra consisting of two rows of plates and interambulacra of two rows. There are no gill slits. The ambulacral plates do not bear small spherical tubercules. Only tridentate and glandular pedicellariae are present (Figure 44b).

## Order Cidaroida

Members of this order have a subspherical rigid test. The interambulacral areas are conspicuously wider than the ambulacral ones. The ambulacral plates are simple. Each interambulacral plate bears a massive primary spine whose base is surrounded by small spines. The teeth of Aristotle's lantern are not keeled.

Families: Cidaridae, Psychocidaridae.

### Family Cidaridae

*Eucidaris metularia* (Lamarck). Diameter up to 3 cm. The beautifully patterned upper surface of the test is bare, contrasting with the spines which are often encrusted with organisms. Occurs on rocks and reefs in the Indo-Pacific.   117/4

*Eucidaris tribuloides* (Lamarck). Diameter up to 5 cm. Some of the large blunt-tipped spines have recently been regenerated and are not covered with fouling organisms. It is often found in crevices amongst rocks near the low-tide zone in the Caribbean.   117/1

*Phyllacanthus parvispinus* Tenison-Woods. Diameter up to 15 cm. The primary spines are relatively short being no longer than the horizontal diameter of the test. The older spines are usually covered with encrusting organisms. Found in the Indo-Pacific.   117/2

## Subclass Euechinoidea

Variably shaped echinoids with the test composed of double rows of ambulacral plates and double rows of interambulacral plates. Gill slits are usually present, as are small spherical tubercles on the ambulacral plates. Bulbous pedicellariae frequently occur (Figure 44b).

## Superorder Diadematacea

Members of this superorder have a subspherical rigid or flexible test which is radially symmetrical. Groups of three or more ambulacral plates are frequently fused together to form compound plates. The tubercles are perforated and bear long, brittle, usually hollow, venomous spines. The teeth of the lantern are not keeled.

Familes: Echinothuriidae, Diadematidae, Lissodiadematidae, Micropygidae, Aspidodiadematidae, Pedinidae.

### Family Echinothuriidae

*Asthenosoma intermedium* H. L. Clark. Diameter up to 15 cm. It has fragile primary spines sparsely scattered over the upper surface. The beaded secondary spines occur in comb-like rows over the upper surface. The blue-coloured tips of these spines contain poison sacs (Plate 117/8). A night-active species on Indo-Pacific reefs.   117/3, 117/8

*Asthenosoma varium* Grube. Diameter up to 15 cm. The long primary spines protrude beyond the densely packed secondary spines which bear white poison sacs at their tips. A night-active species on Indo-Pacific reefs.   117/6

### Family Diadematidae

*Astropyga magnifica* A. M. Clark. Diameter up to 20 cm. A close-up photograph showing the large anal cone. The ambulacral areas are spineless and are marked with small iridescent blue spots. In the centre of the animal the bladder that can be seen is used by the animal for storing faeces prior to powerful expulsion. Found in the Caribbean.   117/5

*Astropyga radiata* (Leske). Diameter up to 20 cm. The iridescent blue spots of the spineless ambulacral areas are clearly visible. The specimen is seen amongst seagrass in the Indo-Pacific.   117/7

*Diadema antillarum* (Phillipi) [long-spined sea-urchin]. Diameter up to 30 cm. It has extremely long, brittle spines amongst which small fish, shrimps and plankton   117/9 are often found sheltering. These black (occasionally white) hollow spines are covered with a glandular skin secreting a poisonous fluid. This fluid frequently causes severe irritation when the spines penetrate the skin. These sea-urchins are attacked and eaten by the trigger fish, *Balistes*. Occurs in shallow waters of the Caribbean and the warm Atlantic.

117/10, 118/1   *Diadema savignyi* Michelin. Diameter up to 30 cm. This species is noted for the iridescent star pattern on the top of the test. Plate 118/1 shows a close-up of this species at night, revealing a lightening of the coloration of the spines which occurs at low light-intensities. The banding of the spines is a feature common in juveniles of all species of *Diadema*. A prominent faecal sac can be seen in the centre of the test. Found on hard bottoms in the Indo-Pacific.

117/11   *Diadema setosum* (Leske). Diameter up to 30 cm. Although very similar in appearance to *D. savignyi* it can be distinguished by the presence of a red ring around the anus and by white flecks on the test. Occurs on hard bottoms in the Indo-Pacific.

118/2, 118/3   *Echinothrix calamaris* (Pallas). Diameter up to 20 cm. It has numerous long white barbed spines, often beautifully banded, and smaller brown secondary spines. Plate 118/3 shows an animal with its large primary spines lowered revealing the 5-radiate structure of the phylum and the distended faecal sac. Found in the Indo-Pacific.

118/4, 118/6   *Echinothrix diadema* (Linnaeus). Diameter up to 20 cm. The large, conspicuously banded, primary spines are finely barbed, making removal from flesh very difficult; it has smaller secondary spines. Plate 118/6 shows an *Echinothrix* during the day in shallow water with two specimens of the shrimp *Stegopontonia commensalis* lying along the spines. The shrimp is protected from predators by the spines of the sea-urchin. Found in the Indo-Pacific.

## Superorder Echinacea

Members of this superorder have a subspherical rigid test which is radially symmetrical. The ambulacral plates are compound. The tubercles and spines are solid. The lantern has keeled teeth. Glandular pedicellariae containing poison glands are sometimes present.

Families: Saleniidae, Phymosomatidae, Stomechinidae, Arbaciidae, Temnopleuridae, Toxopneustidae, Echinidae, Echinometridae, Strongylocentrotidae, Parasaleniidae.

### Family Arbaciidae

118/7   *Arbacia lixula* (Linnaeus) [black urchin]. Diameter up to 15 cm. It has solid sharp-tipped spines. The upper surface is never covered with fragments of algae or shells. Occurs commonly in shallow rocky areas of the Mediterranean.

### Family Temnopleuridae

118/5   *Mespilia globulus* (Linnaeus). Diameter up to 5 cm. Ten conspicuous spineless areas on the test are carpeted with small glandular pedicellariae. The upper surface is frequently covered with algal fragments. Occurs on reefs in the South Pacific.

118/8   *Microcyphus rousseaui* Agassiz. Diameter up to 5 cm. It has conspicuous spineless areas on the test. Found on rocks and weeds in the Indo-Pacific.

### Family Toxopneustidae

*Lytechinus variegatus* (Leske). Diameter up to 10 cm. A species that is variable in colour which occurs in shallow waters of the Caribbean on coral rubble and on sand amongst eel-grass and turtle-grass.   118/10

*Pseudoboletia indiana* (Michelin). Diameter up to 10 cm. The white spines usually have violet tips. It often covers itself with shells or pieces of algae, perhaps to protect it from strong sunlight in shallow water. Found in the Indo-Pacific.   119/1

*Sphaerechinus granularis* (Lamarck) [violet urchin]. Diameter up to 15 cm. This Mediterranean species has short densely packed spines.   118/9

*Toxopneustes pileolus* (Lamarck). Diameter up to 15 cm. The spines are short and inconspicuous amongst the covering of large glandular pedicellariae containing poison. Small animal prey are probably captured by these pedicellariae to supplement the urchin's main diet. The pedicellariae, whose jaws may be up to 5 mm across when fully open, can cause intense pain and irritation to man (Plate 119/2). Found on hard bottoms and sand in the Indo-Pacific.   119/2, 119/4

*Tripneustes gratilla* (Linnaeus). Diameter up to 15 cm. The glandular pedicellariae are inconspicuous amongst the spines. It can be seen using its suckered tube-feet to cross a rocky area. Often occurs amongst turtle-grass in the Indo-Pacific. The roes of this species are edible.   119/3

*Tripneustes ventricosus* (Lamarck). Diameter up to 15 cm. Inconspicuous pedicellariae occur amongst the short white spines. Found on reefs and amongst eel-grass and turtle-grass in the Caribbean.   119/5

### Family Echinidae

*Echinus acutus* (Lamarck). Diameter up to 15 cm. Its conical test is sparsely covered with long spines. Found on rocks and soft bottoms in the Mediterranean and North-east Atlantic, rarely shallower than 20 m.   119/6

*Echinus esculentus* Linnaeus [edible sea-urchin]. Diameter up to 20 cm. A large number of small spines cover the test which is somewhat flattened in young specimens but high-domed in adults. The ventral mouth has five prominent jaws (Plate 119/7). Main propulsion is by means of extensible suckered tube-feet. The roes are sometimes eaten. Occurs on rocks in the North-east Atlantic, often in very shallow water.   119/7, 119/8

*Paracentrotus lividus* (Lamarck). Diameter up to 12 cm. The colour varies from brown to green. It is sometimes mistaken for *Arbacia lixula* but the spines are shorter and never black. The suckered tube-feet often hold pieces of algae and shells on the upper surface, whereas *A. lixula* never has such a covering, since the tube-feet have no suckers. It sometimes lives in cavities which it has excavated into soft rock in the Mediterranean and North-east Atlantic.   119/9

*Sterechinus neumayeri* (Meissner). Diameter up to 10 cm. Species of this genus are confined to the Antarctic and Subantarctic regions.   119/10

### Family Echinometridae

*Echinometra lucunter* (Linnaeus). Diameter up to 10 cm. Strong spines covering the oval test enable this species to enlarge holes in soft rock at low tide level. Found in the Caribbean.   120/1

*Echinometra mathaei* (de Blainville). Diameter up to 10 cm. The test is oval and bears strong spines. An   120/2 abundant species in the Indo-Pacific, where it sometimes excavates cavities in coral and rock.

120/3 *Echinostrephus molaris* (de Blainville). Diameter up to 5 cm. It has long thin spines on the upper surfaces. Lives in deep burrows excavated out of coral or rock. Found in the Indo-Pacific.

120/5 *Heterocentrotus mammillatus* (Linnaeus) [slate pencil urchin]. Diameter up to 25 cm. It has a few large heavy spines, which contrast with short flat-tipped secondary spines forming a mosaic over the test. The tridentate pedicellariae have narrow valves, meeting only at their tips. The large spines were in the past used for writing on slates and are now used commonly for wind chimes and jewellery. It is active at night and wedges itself into crevices during the day. An Indo-Pacific species.

120/4 *Heterocentrotus trigonarius* (Lamarck). Diameter up to 25 cm. The tridentate pedicellariae have leaf-shaped valves, closely applied along their lengths. Found in the Indo-Pacific.

### Family Parasaleniidae

120/8 *Parasalenia boninensis* Mortensen. Diameter up to 5 cm. The oval test bears large primary spines with distinct white collars around their bases. Found on hard bottoms in the North-west Pacific.

## Superorder Gnathostomata

Members of this superorder have a bilaterally symmetrical rigid test which is frequently markedly flattened with a pronounced boundary between the upper and lower surfaces. The ambulacral areas are often modified into a petaloid design. The ambulacral plates are not compound. The primary tubercles are usually perforated and the spines hollow. The lantern, when present, has keeled teeth. The anus is posterior in all living species. They feed primarily on organic debris extracted from the soft sediments in which they burrow.

Families: Echinoneidae, Clypeasteridae, Arachnoididae, Fibulariidae, Laganidae, Dendrasteridae, Echinarachniidae, Mellitidae, Astriclypeidae, Rotulidae.

### Family Echinoneidae

120/7 *Echinoneus cyclostomus* (Leske). Diameter up to 5 cm. An oval species covered with short dense erect spines which lives concealed beneath coral rubble and stones in the Caribbean and the Indo-Pacific.

### Family Clypeasteridae

120/6 *Clypeaster subdepressus* (Gray). Diameter up to 15 cm. It is oval in shape, flattened marginally and raised in the central area of the test. The test is covered with small spines. A Caribbean species, normally living buried in coral sand from which it extracts organic particles.

120/9 *Clypeaster* sp. Diameter up to 15 cm. It is oval in shape, but less flattened than *C. subdepressus*. Found in coral sands of the Indo-Pacific.

### Family Laganidae

120/10 *Peronella lesueuri* Agassiz. Diameter up to 15 cm. It has four genital pores, with the madreporite pores scattered and inconspicuous. Lives in sands of the Indo-Pacific.

### Family Echinarachniidae

*Echinarachnius parma* (Lamarck). Diameter up to 10 cm.   121/1 The disc is intact. The petaloid ambulacral areas are open at their outer ends. Abundant on sandy bottoms of the North-west Atlantic, where it is a common food of flatfish and cod.

### Family Mellitidae

*Mellita sexiesperforata* (Leske) [sand-dollar]. Diameter   121/6 up to 10 cm. The test is a flattened disc perforated by six slits. It lives buried in coral sands and muds in the Caribbean; if exposed, it is able to bury itself in two or three minutes.

## Superorder Atelostomata

Members of this superorder have a bilaterally symmetrical rigid test. They are somewhat flattened but are without a pronounced boundary between the upper and lower surfaces. The ambulacral areas are usually modified into a petaloid design, with the anterior ambulacral area often depressed into the test. The ambulacral petals are frequently surrounded by a ciliated band, whose function is to produce water currents. Ciliated bands sometimes occur elsewhere on the test. The ambulacral plates are not compound. When present, the primary tubercles are perforated and the spines hollow. There is no lantern apparatus. They feed on organic particles extracted from soft sediments.

Families: Echinolampadidae, Cassidulidae, Pliolampadidae, Holasteridae, Urechinidae, Calymnidae, Pourtalesiidae, Toxasteridae, Hemiasteridae, Palaeostomatidae, Pericosmidae, Schizasteridae, Aeropsidae, Brissidae, Spatangidae, Loveniidae, Asterostomatidae, Neolampadidae.

### Family Cassidulidae

*Cassidulus* sp. An unidentified specimen with an open   121/2 petaloid design, which burrows into coarse sand in the Caribbean.

### Family Schizasteridae

*Paraster floridiensis* (Kier & Grant). Diameter up to   121/3 10 cm. It has a heart-shaped body with a deep frontal ambulacral depression covered by an arch of spines. The majority of spines lie backwards over the test. Burrows in sand in the Caribbean.

### Family Brissidae

*Meoma ventricosa* (Lamarck). Diameter up to 15 cm.   121/4 The spines are on prominent tubercles in the petaloid region bounded by the ciliated band. There are smaller spines covering the rest of the test. Found in Caribbean sands.

*Plagiobrissus grandis* (Gmelin). Diameter up to 20 cm.   121/5 The anterior ambulacrum is deeply furrowed. Prominent patches of long backward-pointing primary spines occur on the test. The specimen is seen burrowing into sand in the Caribbean.

### Family Spatangidae

*Spatangus purpureus* Müller [purple heart-urchin]. Di-   121/7 ameter up to 15 cm. The anterior ambulacrum is in a shallow furrow. The test is covered with short spines and occasional longer ones. The species is invariably reddish-purple in colour. Occurs in coarse sand and shell-gravel in the Mediterranean and the North-east Atlantic.

### Family Loveniidae

121/8 *Echinocardium cordatum* (Pennant) [sea-potato]. Diameter up to 10 cm. The anterior ambulacrum lies in a deep furrow. The test is covered with a fur of backward-pointing short spines. A group of dorsal spines maintains an opening to the surface of the sand or muddy sand in which it burrows. A cosmopolitan species.

121/9 *Lovenia elongata* (Gray). Diameter up to 10 cm. The test is more elongated and the backward-pointing long spines more numerous than in *E. cordatum*. Burrows in sand in the Indo-Pacific.

## Class Holothurioidea [sea-cucumbers]
approx. 900 species

Cucumber-shaped or worm-like echinoderms without arms, whose secondary bilateral symmetry obscures their basic radial symmetry. They generally lie on one side with the mouth at one end of the body and the anus at the other (Figure 45). The skeleton is frequently reduced to scattered calcareous plates or spicules embedded in the leathery body wall. The spicules are an important taxonomic feature, as they are usually characteristic for each species. In most holothurians there are five bands of tube-feet along the length of the body, those of the three more ventral bands being modified for locomotion or attachment. In some species, however, tube-feet are irregularly distributed over the whole body surface or entirely absent. The mouth is surrounded by a set of food-collecting tentacles which are modified tube-feet. The pharynx is often ringed by calcareous plates, probably homologous with Aristotle's lantern of the echinoids. Most species respire by drawing water into paired respiratory trees situated within the body and opening into the hind gut.

Holothurians have extensive powers of regeneration especially of the gut. Shedding of the gut and respiratory trees is a frequent response to attack by predators. Although mostly confined to the bottom, a few species are pelagic. Many bottom-dwellers feed on plankton and/or organic debris which is ensnared on the tentacles. Others plough through the sediment, ingesting large quantities of bottom-deposit from which they extract nutritive material.

## Subclass Dendrochirotacea

Holothurians with two distinct types of oral tentacle either large and bushy or finger-like. Respiratory trees are present. Ventral tube-feet are often reduced in number, reflecting the inactivity of members of the group. There is a tendency for the mouth to be displaced dorsally.

## Order Dendrochirotida

Members of this order have richly branched bush-shaped tentacles. Some species have an almost complete covering of plates on the upper surface, others have only small ossicles in the body wall. The calcareous ring around the pharynx varies in shape and size.

Families: Placothuriidae, Paracucumidae, Psolidae, Phyllophoridae, Sclerodactylidae, Cucumariidae.

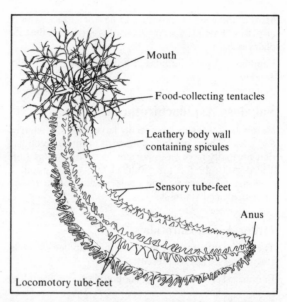

**Figure 45** *Ventro-lateral view of a sea-cucumber showing the main external features.*

### Family Psolidae

*Ceto cuvieria* (Cuvier). Body length up to 5 cm. There is a dorsally displaced mouth (surrounded by tentacles) and anus. A ventrally placed sole is being used by the black and white specimen to creep along a finger of orange sponge. Found in the Zealandic region.   121/11

*Psolidium granuliferum* H. L. Clark. Body length up to 5 cm. This dorsal view of the species shows the mouth surrounded by tree-like tentacles, and the anus on a raised papilla at the rear end. The dorsal surface is covered with calcareous scales, whilst the ventral surface is modified into a sole for clinging to hard surfaces. Found in the Zealandic region.   121/10

### Family Phyllophoridae

*Neothyonidium magnum* (Ludwig). Body length up to 15 cm. The tree-like tentacles are arranged in two rings around the mouth. The tentacles are either held up in the water to catch floating organic matter, or wiped across the surface of the sediment. The tentacles are retracted one at a time and wiped across the mouth to remove their entrapped food (Plate 122/1). The body lies buried in the sand. Found in the Indo-Pacific.   121/12, 122/1

### Family Cucumariidae

*Cucumaria miniata* (Brandt). Body length up to 20 cm. The tentacles trap organic particles suspended in the water. Protrudes from rock crevices in the North Pacific.   122/3

*Cucumaria* sp. A small creeping species from the Indo-Pacific.   122/2

*Ocnus planci* (Brandt). Body length up to 15 cm. A mobile species which is found on a variety of bottoms in the Mediterranean and North-east Atlantic.   122/4

*Pseudocolochirus axiologus* H. L. Clark. Body length up to 10 cm. A brightly coloured species with feathery white tentacles. A reef-dweller in the Indo-Pacific.   122/5

### Order Dactylochirotida

Members of this order have simple finger-like tentacles.

Most of these small holothurians have a surface armour of plates. There is a simple calcareous ring around the pharynx.

Families: Ypsilothuriidae, Vaneyellidae, Rhopalodinidae.

## Subclass Aspidochirotacea

Members of this subclass usually have small leaf-shaped tentacles. They possess numerous ventral tube-feet that normally have an important locomotory function. The dorsal tube-feet are modified into warts, papillae, or elongated processes. There is a tendency for the mouth to be displaced ventrally.

## Order Aspidochirotida

The dorsal tube-feet are modified into papillae or warts. Respiratory trees are present.

Families: Holothuriidae, Stichopodidae, Synallactidae.

### Family Holothuriidae

122/6    *Actinopyga agassizi* (Selenka). Body length up to 30 cm. The creeping sole with numerous tube-feet is clearly delimited from the dorsal warty surface. The commensal pearlfish, *Carapus*, is often seen protruding from its anus. A common species in eel-grass beds and reef areas of the Caribbean.

122/8    *Actinopyga mauritiana* (Quoy & Gaimard). Body length up to 30 cm. A distinctly bicolour species with pointed spines on the upper surface. Inhabits sea-grass beds and reef areas in the Indo-Pacific.

122/7    *Bohadschia argus* Jäger. Body length up to 30 cm. Easily identified by its conspicuous 'eye' markings. Found on coral sands in the Indo-Pacific, where it is often partly covered with sand grains.

122/9    *Bohadschia graffei* Semper. Body length up to 30 cm. It has short white-tipped papillae in the darker coloured areas of the dorsal surface. The specimen can be seen with its short, dark, leaf-like tentacles extended over encrusting reef organisms. Found in the Indo-Pacific.

122/10   *Holothuria atra* Jäger. Body length up to 20 cm. This common species found on the sand in sea-grass beds of the Indo-Pacific is seen extruding sticky white threads.

123/1    *Holothuria (Halodeima) edulis* Lesson. Body length up to 35 cm. The thick soft skin has inconspicuous dorsal warts. There is a distinctly pink coloration on the underside of this species. Occurs on sand and reefs in the Indo-Pacific.

123/4    *Holothuria floridana* (Pourtalès). Body length up to 15 cm. Numerous papillae are present on the dorsal surface. The specimen is extruding sticky white threads from its anus. The threads, which are defensive organs, are normally attached to the respiratory trees. Found in the Caribbean.

123/2    *Holothuria forskali* Delle Chiaje [cotton spinner]. Body length up to 25 cm. Many spiky papillae are present on the dorsal surface. The underside has three bands of suckered locomotory tube feet. Sticky white threads are extruded when the animal is disturbed. Found on muddy sand and rocks in the North-east Atlantic and Mediterranean.

123/3    *Holothuria impatiens* Forskål. Body length up to 25 cm. Large warts are present on the dorsal surface. The skin

is decidedly gritty to the touch. Occurs on coral sands in the Indo-Pacific.

### Family Stichopodidae

*Astichopus multifidus* (Sluiter). Body length up to 40 cm. A conspicuously warty species which extracts organic material from coral sand in the Caribbean.    123/8

*Isostichopus badionotus* Selenka. Body length up to 20 cm. Two of the three ventral bands of tube-feet are visible on this characteristically marked inhabitant of sea-grass beds. Found in the Caribbean.    123/5

*Parastichopus californicus* Stimpson. Body length up to 50 cm. Like most aspidochirotids, it scoops up bottom material containing organic matter with its tentacles and passes it into the mouth. Occurs on soft bottoms and rocks in the North Pacific.    123/7

*Stichopus chloronotus* (Brandt). Body length up to 20 cm. It is always dark green in colour with large, usually red-tipped, warts on the dorsal surface. Often found lying in the open on coral rubble in the Indo-Pacific.    123/6

*Stichopus regalis* Cuvier. Body length up to 30 cm. It has a flattened sole-like ventral surface which is characteristic of this family. The mouth opens on the underside. Occurs on soft bottoms in the Mediterranean and North-east Atlantic.    123/9

*Stichopus variegatus* Semper. Body length up to 20 cm. A yellow or brown species often with a spotted pattern on the dorsal surface. Occurs on Indo-Pacific reefs.    123/10

*Thelenota ananas* (Jäger) [prickly red fish]. Body length may exceed 1 m. A large body with few spicules has led to this species being sought after for the 'beche de mer' fishery. Large teat-like papillae (Plate 124/1) are characteristic of the dorsal surface. Lives on sand and reefs in the Indo-Pacific, where it leaves obvious casts of faecal material.    123/11, 124/1

## Order Elasipodida

Members of this order often have their dorsal tube-feet modified into elongate processes. There are no respiratory trees. The thick, jelly-like body wall has few spicules. Species belonging to this group are mainly bizarre deep-water inhabitants.

Families: Deimatidae, Laetmogonidae, Elpidiidae, Psychropotidae, Pelagothuriidae.

A small unidentified elasipod from the depths of the Atlantic Ocean, with a translucent body wall containing minute spicules. The gut is visible through the body wall.    124/5

### Family Pelagothuriidae

*Pelagothuria* sp. A genus adapted for a swimming existence, lacking spicules in the translucent jelly-like body wall. The body is supported in the water by a web spread between elongated papillae. The large terminal mouth is surrounded by a circlet of tentacles with forked tips. Ranges the equatorial Atlantic, from the ocean surface to considerable depths.    124/6

## Subclass Apodacea

They have simple tentacles with a series of digits branching from a central stalk (pinnate tentacles). The tube-feet are greatly reduced or absent. Most are burrowing forms but some crawl about on the sea-bed.

## Order Apodida

124/8,
124/9
Members of this order are worm-like sea-cucumbers without tube-feet or respiratory trees. Characteristic anchor (Plate 124/8) and/or wheel spicules (Plate 124/9) are embedded in the body wall of species in this order.

Families: Synaptidae, Chiridotidae, Myriotrochidae.

### Family Synaptidae

124/2
*Euapta lappa* (Müller). Body length up to 1 m. The extensible worm-like body has spicules projecting from the body surface making it sticky to the touch. A night-active species from the Caribbean, hiding amongst coral during the day.

124/3
*Opheodesoma* sp. [alabaster sea-cucumber]. Body length up to 20 cm. A white translucent species which crawls amongst the reefs feeding on organic particles with its pinnate tentacles. Found in the Indo-Pacific.

124/4
*Synapta maculata* (Chanisso & Eysenhardt). Body

length up to 2 m. This worm-like species has a highly extensible body. When handled, the animal tends to stick to the fingers due to the presence of spicules in the body wall. Crawls amongst coral rubble of Indo-Pacific reefs.

### Family Chiridotidae

*Chiridota gigas* Dendy & Hindle. Body length up to 20 cm. Stubby pinnate tentacles are present. Large wheel spicules (Plate 124/9) are clearly visible in the translucent body wall. Found in sands of the Zealandic region.    124/7

## Order Molpadiida

Tube-feet are reduced to a few anal papillae. Respiratory trees are present. There is a tendency for the posterior part of the body to taper into a tail region.

Families: Molpadiidae, Caudinidae, Gephyrothuriidae.

# Phylum Hemichordata

approx. 90 species

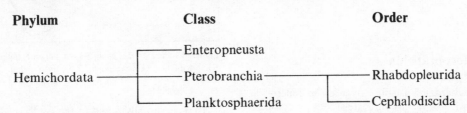

An exclusively marine group of animals, which are either solitary or colonial. Many of them are worm-like. Others are small and provided with tentacles. The body is divided into three regions – the proboscis, the collar and the trunk – each of which contains part of the body cavity. The body surface is ciliated and frequently coated with mucus. The proboscis is round, conical, or shield-shaped and connected to the collar by a short stalk. The short collar bears a mouth on its ventral side. In the tentaculate species, two or more arms with ciliated tentacles arise from the dorsal side of the collar. The trunk region is elongated in the worm-like forms, while in the tentaculate species it is divided into an anterior plump region and a posterior attachment stalk. Frequently, slit-like openings (gill slits) in the anterior part of the gut (pharynx) connect to the body surface through pores in the anterior part of the trunk wall. The beating of cilia on the walls of the slits draws a water current in through the mouth and drives it out through the pores. The through current has a respiratory function. In the worm-like forms, the anus is situated at the end of the trunk, but in the tentaculate species the gut is U-shaped and the anus is just behind the collar.

The sexes are usually separate and fertilisation of the egg occurs in the water. The egg often develops into a free-swimming planktonic larva, although it may develop directly into a juvenile in some species. Asexual reproduction by budding occurs in a few forms.

Hemichordates are found in all seas but many prefer

warm and temperate waters. The majority live in the intertidal zone and shallow waters, although a few occur at great depths. They are mainly ciliary-feeders, trapping small organisms and particles in mucous secretions which are then conveyed by the cilia on the proboscis, or arm-tentacles, to the mouth. Some of the burrowing species appear to extract organic matter from the sand and mud ingested during burrowing.

## Class Enteropneusta [acorn-worms]
approx. 70 species

Relatively large solitary, worm-like, animals whose length normally ranges from 5 to 20 cm although some species can reach a length of over 2 m. The proboscis is usually short and conical, but is elongated with a deep mid-dorsal groove in a few species. The collar does not have tentacle-bearing arms. There is an elongated trunk which has a terminal anus. The numerous gill slits in the pharynx wall either open directly to the exterior or into a pouch-like cavity which itself is open to the exterior (Figure 46).

The fertilised eggs may hatch as planktonic larvae or as young worms.

Acorn-worms live in U-shaped burrows in sand and mud, or amongst rocks and seaweeds, mainly in the intertidal zone and shallow waters. They are slow-moving animals with soft, slimy, bodies which break

Plate 124

Phylum Hemichordata

easily when handled. The mucus of several species has a characteristic smell which may be offensive to possible predators, such as fish.

Families: Harrimaniidae, Spengelidae, Ptychoderidae.

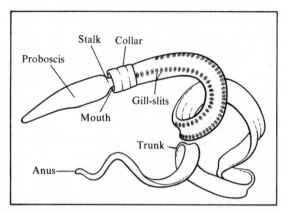

**Figure 46** *External features of the acorn-worm* Dolichoglossus.

### Family Harrimaniidae

124/10 *Saccoglossus horsti* Brambell & Goodhart. The long proboscis is usually a creamy-white colour in this species, which has over 100 pairs of gill openings. It is found in intertidal muddy sand and gravel in the North-east Atlantic.

## Class Pterobranchia

approx. 20 species

Small hemichordates, usually only a few millimetres in length, which live mainly in aggregations or colonies. With one exception, they live in horny secreted interconnected tubes attached to the bottom. Individuals in a colony may be connected to one another by a strand of soft tissue. The proboscis is shield-shaped, but their most characteristic feature is one to several pairs of tentacle-bearing arms on the collar. The trunk has a plump anterior region and a stalk which is either attached to the tube or connected to other stalks. A pair of gill slits may or may not be present, and the anus opens dorsally just behind the collar.

Little is known of the life cycle of these hemichordates, but a larva with a brief swimming phase is present in some species. New individuals are frequently formed by budding from the parent.

Pterobranchs are bottom-dwellers. They are often found attached to other animals such as sponges and bryozoans.

### Order Rhabdopleurida

Colonial tube-dwelling pterobranchs whose individuals are connected to one another. The collar gives rise to one pair of arms. Gill slits are lacking.

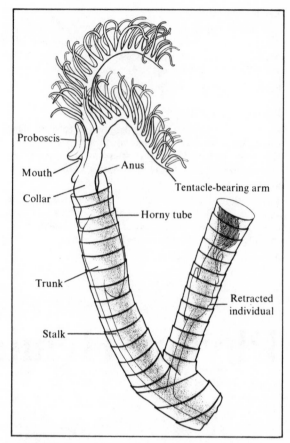

**Figure 47** *Part of the branching colony of the pterobranch,* Rhabdopleura, *showing the main features of individuals.*

They are usually found in the colder regions of the world.

One genus: *Rhabdopleura* (Figure 47).

### Order Cephalodiscida

These pterobranchs usually live in aggregations, although adult individuals are not connected to one another. There are several pairs of arms on the collar and one pair of gill slits on the trunk.

The majority have been found in deep waters of the southern hemisphere, although one non-colonial species without a tube (*Atubaria*) lives on hydroids in northern waters.

Two genera: *Cephalodiscus, Atubaria*.

### Class Planktosphaerida

A group known only from planktonic larvae, which have branching bands of cilia on the body surface. Internal structures clearly indicate that the larvae belong to the hemichordates although the adults are unknown.

# Phylum Chordata

approx. 48 000 species

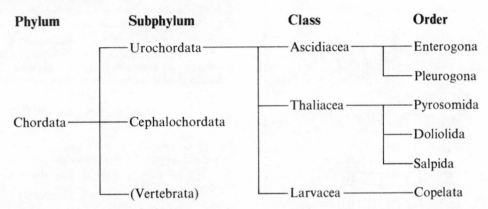

| Phylum | Subphylum | Class | Order |
|--------|-----------|-------|-------|
| Chordata | Urochordata | Ascidiacea | Enterogona |
|  |  |  | Pleurogona |
|  | Cephalochordata | Thaliacea | Pyrosomida |
|  |  |  | Doliolida |
|  |  |  | Salpida |
|  | (Vertebrata) | Larvacea | Copelata |

Classification of the Urochordata follows Berrill (1950).

This phylum consists mainly of animals with backbones (vertebrates), which are not being considered in this book. However members of two of the three subphyla are regarded as invertebrates since they lack a true backbone, although possessing several chordate characteristics.

Chordates are a diverse, highly successful, group of animals that have colonised a wide range of habitats. In at least some stage of their life cycle they possess a supporting skeletal rod (notochord) lying beneath the dorsal tubular nerve cord and a pharynx perforated by gill slits. A post-anal muscular tail, containing extensions of the notochord and nerve cord, is present.

In the majority of chordates the sexes are separate, although hermaphroditism predominates in the Urochordata.

## Subphylum Urochordata
### [tunicates]
approx. 1375 species

An exclusively marine group of animals, in which most of the chordate characteristics are found only in the larval stages. The majority are barrel-shaped, solitary or colonial animals which may be attached to the bottom or be free-swimming. The body is enclosed within a gelatinous or leathery tunic containing cellulose as one of its structural components. The tunic usually has two main openings, an anterior inhalent siphon through which a water current enters the animal, and an exhalent siphon (anteriorly or posteriorly situated) through which water is expelled. The inhalent siphon leads into a large pharynx whose wall is perforated by two or more gill slits. The gill slits usually open into a special exhalent chamber communicating with the exterior through a siphon, but in a few species the gill slits open directly on to the body surface. The pharynx has both a respiratory and a feeding function. A ciliated groove in its floor produces streams of mucus which trap food particles. Both the anus and the reproductive system open into the exhalent chamber.

Most tunicates are hermaphrodite and the eggs are fertilised either in the open water or inside the exhalent chamber. The fertilised egg usually develops into a free-swimming tadpole-like larva with a notochord in the tail. After a short, non-feeding, free existence the larva usually becomes attached by its front end and then develops into the adult. In a few species eggs develop directly without a tadpole stage. Many tunicates reproduce asexually by budding and in the life cycles of colonial forms there is often a complex alternation of sexual and asexual phases.

Tunicates are found in all seas, attached to any suitable surface, or swimming freely in the water. Attached tunicates can be found intertidally, but the majority occur in shallow waters. The free-swimming forms are found mainly in the surface waters of the open ocean.

## Class Ascidiacea [sea-squirts]
approx. 1250 species

Sea-squirts are solitary or colonial tunicates, living attached to the bottom, submerged objects, or other organisms. Solitary forms may reach a length of over 20 cm and colonies can extend for more than 50 cm. The body wall may sometimes be covered by growths of other organisms. Two siphons, often very prominent in solitary species, are situated at the free end of the body. The numerous gill slits in the wall of the pharynx are usually subdivided by horizontal and transverse bars of tissue which give the complete structure a basket-like appearance (Figure 48). Water is drawn into the pharynx by the beating of many cilia situated round the edges of the gill slits. Food particles are trapped on mucus in the pharynx. After the particles have been extracted the water leaves the body through the siphon of the exhalent chamber. In simple colonial species, individuals are connected by branches covered by the tunic. In others, individuals are embedded in a common tunic, whilst in some specialised species individuals discharge their water through a common exhalent chamber and siphon.

There is usually a free-swimming tadpole-like larva in the life cycle. This becomes attached to a surface and changes into an adult. In colonial species asexual reproduction by budding is common.

Sea-squirts are found in all seas from the intertidal zone to deep waters.

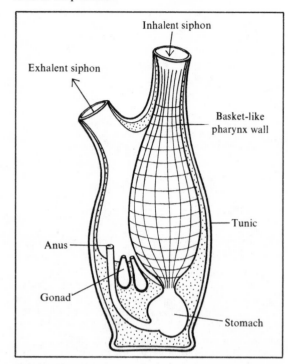

**Figure 48** *Generalised diagram of a sea-squirt showing the main body features.*

## Order Enterogona

Solitary or colonial sea-squirts with the single gonad situated in or near the loop of the gut. The body is often divided into two regions – an anterior region containing the pharynx, and an abdominal region containing the gut and other organs. In the Polyclinidae there is a third distinct posterior region, the post abdomen, containing the reproductive organs and heart.

Families: Clavelinidae, Polyclinidae, Didemnidae, Cionidae, Diazonidae, Perophoridae, Corellidae, Ascidiidae, Hypobythiidae, Agnesiidae.

### Family Clavelinidae

125/2, 125/5    *Archidistoma* sp. A colonial form with an uncanny resemblance to a sponge, due to the fact that the exhalent siphons of individuals do not open directly to the exterior but into a common cloaca which resembles a sponge osculum. Found in the Mediterranean and adjacent Atlantic.

125/1, 125/3    *Clavelina lepadiformis* (Müller). Forms colonies of transparent delicate individuals up to 3 cm high, joined at their bases by a network of stolons. The basket-like pharynx is often delineated by opaque white or yellow markings. It is found in clusters attached to hard objects from low-tide level downwards. Occurs in the North-east Atlantic and Mediterranean.

125/4    *Clavelina picta* (Verrill). Similar to the above species, individuals in the colonies being up to 2 cm high, with parts of the internal anatomy delineated by opaque purple or red markings. Clusters are often found surrounding branches of alcyonarians and corals in the Caribbean.

*Clavelina* sp. A beautifully marked colony of an unknown species from the Indo-Pacific.     125/7

*Distoma adriaticum* Drasche. Forms mushroom-like colonies up to 10 cm high in which the individuals are joined together throughout their length. Found attached to hard objects on open ground, often in areas with a high silt content. A Mediterranean species.     125/6

*Sycozoa cerebriformis* (Quoy & Gaimard). A colonial species with a characteristic convoluted surface with the individuals situated along the edges of the convolutions. Found on Indo-Pacific reefs.     125/8

An unknown species of clavelinid from the Indo-Pacific.     125/9

Two photographs of an unknown clavelinid from the Indo-Pacific, Plate 125/10 showing the growth form, and Plate 126/1 a close-up view of individuals within the colony.     125/10, 126/1

### Family Polyclinidae

*Aplidium conicum* Olivi. The colony of individuals forms a cone-shaped fleshy mound up to 20 cm high. Found on rocks and shells on sand or gravel bottoms in the Mediterranean.     126/3

### Family Didemnidae

Colonies of brilliant blue didemnids found growing amongst sponges in the Caribbean. Several individuals share a common exhalent opening. Members of this family are frequently mistaken for sponges as they often form thin encrusting sheets and contain calcareous spicules.     126/2

These globular didemnid colonies, about 2 cm high, are found on shallow reefs in the Indo-Pacific.     126/5

### Family Cionidae

*Ciona intestinalis* (Linnaeus). A tubular sea-squirt up to 15 cm high with a terminal inhalent siphon and the exhalent siphon situated close to it. The soft transparent body often has yellow or green hues and the openings are frequently ringed with yellow. Common on rocks, piers, piles and other hard objects from the lower shore to depths in excess of 500 m, often in areas with a high silt content. A cosmopolitan species.     126/9

### Family Ascidiidae

*Ascidiella aspersa* (Müller). A solitary sea-squirt with a rough, whitish-brown, oval body up to 10 cm high. Found on rocks and other surfaces in the intertidal zone and to depths of 50 m or more. Found in the north-east Atlantic and Mediterranean.     126/4

*Phallusia mammillata* (Cuvier). A solitary species up to 15 cm high, whose surface is covered with many rounded smooth swellings. The exhalent siphon is situated about halfway down the body. Commonly found attached to stones in silty areas of the North-east Atlantic and Mediterranean.     126/8

## Order Pleurogona

Solitary or colonial sea-squirts in which the body is not divided into distinct regions. The gonads are situated in the walls of the exhalent cavity.

Families: Styelidae, Pyuridae, Molgulidae.

## Family Styelidae

126/6 *Botrylloides* sp. A fleshy encrusting colonial form with the individuals arranged in rows around an elongated common exhalent opening. Found on Indo-Pacific reefs.

126/7, 126/10 *Botryllus schlosseri* (Pallas). Forms flattened fleshy colonies with individuals arranged in a characteristic star-shaped pattern around a common exhalent opening. It is sometimes found encrusting kelp holdfasts, and other tunicates, as well as rocks, in the North Atlantic and Mediterranean.

127/1 *Distomus* sp. A beautifully coloured colony from the Great Barrier Reef.

127/3, 127/5 *Polycarpa* sp. A solitary tunicate up to 15 cm high, with a thick tough brightly-coloured leathery tunic with irregular swellings on its surface. The exhalent siphon is situated halfway down the body. An Indo-Pacific species.

## Family Pyuridae

127/8 *Halocynthia papillosa* Linnaeus. A characteristically coloured solitary ascidian, up to 10 cm high, with a firm leathery tunic. The inhalent and exhalent openings are surrounded by bristles. Found subtidally on open rocks and in caves in the Mediterranean.

127/2 *Microcosmus sulcatus* Coquebert [sea-egg]. An edible ascidian up to 20 cm long, with a tough tunic which is frequently covered with encrusting organisms. The inhalent and exhalent openings are spaced widely apart. The specimen in the photograph is covered by the brittle-star *Ophiothrix quinquemaculata*. Found attached to rocks and shells on sandy and muddy bottoms, usually at depths below 20 m, in the Mediterranean.

127/10 *Pyura haustor* (Stimpson). A typically red, smooth-surfaced, ascidian which reaches a length of 5 cm when fully extended. The inhalent and exhalent openings are situated closely together. Commonly found in the intertidal zone and in shallow waters of the North-east Pacific.

# Class Thaliacea

approx. 65 species

A group of solitary or colonial tunicates which live exclusively in the plankton. The siphons are situated at opposite ends of the body and the water current which passes through is used for locomotion as well as for feeding and respiration. The wall of the pharynx has two or more gill slits opening into an exhalent chamber situated posteriorly.

The life cycles are complex, particularly in the Doliolida, and involve asexual budding. A small tadpole-like larva with a notochord in its tail is sometimes produced.

Thaliaceans mainly occur in tropical and subtropical seas, although several species are found in colder waters.

# Order Pyrosomida

Hollow conical colonies whose length ranges from a few centimetres to over 2 m. Individuals are situated in a common tunic around a central chamber open at the wide end of the cone. Each individual has numerous gill slits in the pharynx wall. Their anterior inhalent siphons are on the external surface of the colony and the posterior exhalent siphons open into the central chamber. The constant flow of water from the opening of the central chamber propels the colony through the water.

Colonies are hermaphrodite and fertilisation is internal. There is no larval stage. The fertilised egg remains in the central chamber of the colony, where it develops into an individual which leaves the parent colony and produces a new colony by further budding.

Pyrosomids are found mainly in the plankton of tropical and subtropical seas. Luminescent organs are situated on the pharynx of each individual within the colony.

Family: Pyrosomidae.

## Family Pyrosomidae

127/6 *Pyrosoma atlanticum* Peron. A colony of individuals can reach a length of 50 cm, although inshore forms are usually smaller. The common tunic is covered with rounded papillae and finger-like tentacles are usually present. Colonies are markedly luminescent. Found in the warmer waters of the Atlantic and in the Mediterranean.

# Order Doliolida

Small transparent thaliaceans with barrel-shaped bodies which may reach a length of 10 mm. The gelatinous tunic is thin, and there are eight or nine distinct muscle bands completely encircling the body. The pharynx, which occupies most of the anterior part of the body, has several ciliated gill slits opening into the posterior exhalent chamber.

The fertilised egg develops into a tailed larva similar to that of ascidians. The larva transforms slowly into an individual with a fleshy process on its posterior end. This 'nurse' individual first produces buds which develop into feeding and 'carrier' individuals which become attached to the fleshy process. Food gathered by the feeding individuals is used to maintain the nurse and carriers. Later the nurse buds off reproductive individuals which attach to the carriers. The carriers detach themselves from the fleshy lobe and transport the reproductive individuals until these mature and themselves break free.

Doliolids occur in the plankton of both tropical and temperate waters.

Family: Doliolidae.

## Family Doliolidae

127/4, 127/9 Plate 127/9 shows a doliolid nurse individual with feeding, carrier and reproductive individuals situated on its long fleshy process. Plate 127/4 is a close-up of the feeding individuals which maintain the entire colony. Found in warm waters of the Atlantic.

# Order Salpida [salps]

Transparent thaliaceans which are either solitary or joined together in chains. Individuals, whose length varies from a few millimetres to several centimetres, have a thick gelatinous tunic and muscle bands which do not form complete rings around the body. The pharynx has only two large gill slits.

Individuals grouped in chains eventually separate to become solitary breeding forms. Fertilisation is internal and the egg develops attached to the wall of the exhalent cavity of the parent. It develops into an individual with a small ventral process and is eventually expelled from the exhalent cavity. The ventral process segments, each segment forming an individual salp. The resulting chain of small salps often remains united for some time and may reach a length of 30 cm or more.

Salps are found in both tropical and temperate waters.

Family: Salpidae.

### Family Salpidae

127/7    *Cyclosalpa polae* Sigl. Solitary individuals can reach a length of 8 cm. The salp in the photograph is carrying a whorl of small individuals which have been budded from the ventral process; such a whorl is characteristic of the cyclosalps. Found in warm waters of the Atlantic.

127/11    *Ihlea punctata* (Forskål). A salp with a characteristically spotted tunic, from the warmer waters of the Atlantic.

128/4    *Pegea bicaudata* (Quoy & Gaimard). A whorl of laterally united sexual individuals. Each individual has two long processes at the posterior end. Small planktonic organisms present in the water passing through the body are caught on a mucous net. Found in warm waters of the Atlantic.

128/1    *Pegea confoederata* Forskål. Solitary individuals may reach a length of 12 cm in this species which is characterised by very short muscle bands barely reaching the side margins of the body. The specimen in the photograph has a chain of sexual individuals budding from the ventral process. Cosmopolitan in warm waters.

128/2,    *Pegea socia* (Bosc). The photograph of a solitary in-
128/3    dividual (Plate 128/2) shows the ciliated groove in the pharynx, the gill and stomach. A chain of developing sexual individuals is seen in Plate 128/3. Found in warm waters of the Atlantic.

128/7    *Salpa maxima* Forskål. The photograph shows part of a long chain of sexual individuals; the chain may reach a length of over 1 m. The stomach and gonad are the only opaque parts of these transparent individuals. Found in the warmer waters of the Atlantic and in the Mediterranean.

128/5    *Thetys vagina* (Tilesius). One of the largest salps – up to 20 cm long. The body is broader anteriorly, and there are a pair of appendages situated at the posterior end. The body surface is covered with small pointed papillae. Cosmopolitan in warm waters.

## Class Larvacea
approx. 60 species

Transparent planktonic animals often only a few millimetres in length. The U-shaped body consists of a small trunk region and a long tail containing a notochord which emerges at right angles to the body axis on the ventral side. The pharynx has two gill slits opening directly on to the body surface. The animal is enclosed within, or attached beneath, a gelatinous 'house' through which water is drawn by movements of the tail (Figure 49). The house, which does not contain cellulose, has its inhalent opening covered by a screen of fibres which only allows the smallest plankton to pass through. Further sieving of food particles occurs within the house and then inside the pharynx. Expulsion of water through an exhalent opening drives the animal through the water. When the filters become clogged with inedible material the animal abandons the house and secretes a new one.

The fertilised egg develops into a free-swimming larva

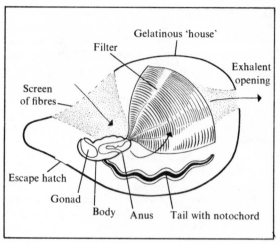

**Figure 49** *A larvacean, filter-feeding within its gelatinous 'house'.*

which assumes adult characters whilst remaining planktonic. Several larval features (e.g. the tail and notochord) are retained throughout adult life.

Larvaceans occur in the plankton of all seas, being particularly common in shallow coastal waters, although a few are found at great depths. These delicate animals are rarely collected intact in plankton samples.

## Order Copelata

Characters identical to those of the class.

Families: Fritillariidae, Oikopleuridae, Kowalevskaiidae.

### Family Oikopleuridae

*Megalocercus huxleyi* (Ritter & Byxbee). The body plus tail of this animal, seen here in its gelatinous house, may reach a length of over 2 cm. A relatively common Indo-Pacific species.

128/6

# Subphylum Cephalochordata
approx. 25 species

Members of this subphylum are frequently called 'amphioxus' or 'lancelets' They are translucent, fish-like invertebrates which may reach a length of 5–6 cm. The laterally compressed body tapers at both ends and has a post-anal tail. A narrow fin runs back from the head on the dorsal side, around the end of the tail and continues forward along the ventral surface as far as the opening of the exhalent chamber, situated mid-ventrally. The animal, which has a supporting notochord extending along its whole length, swims by means of rapid flexures of its body. The ventral mouth is surrounded by small tentacles and leads into a large pharynx with numerous gill slits in its wall. The gill slits open into an exhalent chamber which extends for a considerable distance along the body. A water current, used both for feeding and respiration, flows through the mouth and pharynx into the exhalent chamber and leaves through the small ventral opening. The anus is situated near the posterior end of the animal and opens towards the left-hand side.

The sexes are separate and fertilisation occurs in the water. The egg develops into a transparent planktonic larva which is similar in shape to the adult. Evidence suggests that the larvae migrate to the sea-bottom during the day and move up towards the surface at night.

Adults are found in coarse sands and shell gravels in shallow seas of both tropical and temperate regions. In coarse gravel they are often totally buried, while in finer sediments the anterior end protrudes into the water. They feed, like tunicates, by trapping small organisms on mucus inside the pharynx. Adults tend to congregate in large numbers in specific localities, and as a result are fished commercially in some parts of the world.

Two genera only: *Branchiostoma, Asymmetron.*

*Branchiostoma lanceolatum* (Pallas) [amphioxus or lancelet]. These specimens reveal the essential features of the body as given in the subphylum characteristics. The mouth tentacles, V-shaped muscle blocks, and segmented gonads are particularly obvious in the photograph. Plate 128/9 shows the anterior end of the animal protruding from the sediment.

128/8, 128/9

# Bibliography

## INVERTEBRATE CLASSIFICATION AND BIOLOGY

### General

Barnes, R. D. 1969. *Invertebrate zoology* (2nd edn). 743 pp. Philadelphia: Saunders.

Barrington, E. J. W. 1967. *Invertebrate structure and function.* 549 pp. London: Nelson.

Bayer, F. M., & Owre, H. B. 1968. *The free-living lower invertebrates.* 229 pp. New York: Macmillan.

Buchsbaum, R., & Milne, L. J. 1960. *Living invertebrates of the world.* 303 pp. London: Hamish Hamilton.

Clark, R. B., & Panchen, A. L. 1971. *Synopsis of animal classification.* 126 pp. London: Chapman & Hall.

Faune de France. 1921–66. Vols. 1–68. Paris: Lechevalier.

Fretter, V., & Graham, A. 1976. *A functional anatomy of invertebrates.* 589 pp. New York, London: Academic Press.

Grassé, P.–P. (ed.). 1948– *Traité de Zoologie. Anatomie, Systématique, Biologie.* Paris: Masson.

Grzimek, B. (ed.). 1974. *Grzimek's animal life encyclopedia,* vol 1 (598 pp.), vol. 3 (540 pp.). New York, London: Van Nostrand Reinhold.

Hyman, L. H. 1940–67. *The invertebrates,* vols. I–VI. New York: McGraw-Hill.

Kaestner, A. 1967. *Invertebrate zoology,* vol. I (597 pp.), vol. II (472 pp.), vol. III (523 pp.). New York: Wiley–Interscience.

Kermack, D. M. (ed.). 1970– *Synopsis of the British fauna* (new series), vols. 1– London: Academic Press/Linnean Society.

Kükenthal, W., & Krumbach, T. (eds.). 1923– *Handbuch der Zoologie,* vols. 1– Berlin: de Gruyter.

Marshall, A. J., & Williams, W. D. (eds.). 1972. *Textbook of zoology. Invertebrates.* 874 pp. London: Macmillan.

Meglitsch, P. A. 1967. *Invertebrate zoology* (2nd edn). 961 pp. Oxford University Press.

Moore, R. C. (ed.). 1953– *Treatise on invertebrate paleontology.* Pt A– Geological Society of America and University of Kansas Press.

Rothschild, Lord. 1965. *A classification of living animals* (2nd edn). 134 pp. London: Longmans.

Smith, J. E., Carthy, J. D., Chapman, G., Clark, R. B., & Nichols, D. 1971. *The invertebrate panorama.* 406 pp. London: Weidenfeld & Nicolson.

Webb, J. E., Wallwork, J. A., & Elgood, J. H. 1978. *Guide to invertebrate animals* (2nd edn). 305 pp. London: Macmillan.

### Porifera

Bergquist, P. R. 1978. *Sponges.* 268 pp. London: Hutchinson.

Burton, M. 1963. *A revision of the classification of the calcareous sponges.* 693 pp. London: British Museum (Natural History).

Fry, W. G. (ed.). 1970. The biology of the Porifera. *Symp. zool. Soc. Lond.,* No. 25, 512 pp.

Hartman, W. D., & Goreau, T. F. 1970. Jamaican coralline sponges: their morphology, ecology, and fossil relatives. *Symp. zool. Soc. Lond.,* No. 25, 205–43.

Hartman, W. D., & Goreau, T. F. 1975. A Pacific tabulate sponge, living representative of a new order of sclerosponges. *Postilla,* **167,** 1–14.

Ijima, I. 1927. The Hexactinellida of the Siboga expedition. *Siboga Exped.,* **6,** 1–383.

Lévi, C. 1973. Systématique de la Classe des Demospongiaria (Demosponges), 37–631. In: Grassé P.–P. (ed.). *Traité de Zoologie, III, Spongiaires.* Paris: Masson.

Long, M. E. 1977. Consider the sponge . . . *Natn. geogr. Mag.,* **151,** 392–407.

Minchin, E. A. 1900. Porifera, 1–178. In: Lankester, E. R. (ed.), *A treatise in zoology,* Pt 2. London: Adam & Charles Black.

Vacelet, J. 1970. Les éponges Pharetronides actuelles. *Symp. zool. Soc. Lond.,* No. 25., 189–203.

Wiedenmayer, F. 1977. *Shallow-water sponges of the western Bahamas.* 287 pp. Basel: Birkhäuser.

### Coelenterata

Alvariño, A. 1971. Siphonophores of the Pacific with a review of the world distribution. *Bull. Scripps Instn. Oceanogr.,* **16,** 1–432.

Deas, W., & Domm, S. 1976. *Corals of the Great Barrier Reef.* 127 pp. Sydney: Ure Smith.

Friese, U. E. 1972. *Sea anemones.* 128 pp. Hong Kong: T.F.H.

Kramp, P. L. 1961. Synopsis of the medusae of the World. *J. Mar. biol. Ass. U.K.,* **40,** 1–469.

Millard, N. A. H. 1975. Monograph on the Hydroida of Southern Africa. *Ann. S. Afr. Mus.,* **68,** 1–513.

Moore, R. C. (ed.). 1956. *Treatise on invertebrate paleontology,* Pt F, *Coelenterata.* 498 pp. Geological Society of America and University of Kansas Press.

Petersen, K. W. 1979. On the taxonomy of Athecata (Cnidaria, Hydrozoa) with a discussion of the development of coloniality in the group. In: *Biology and systematics of colonial organisms.* London: Systematics Association.

Rees, W. J. (ed.). 1966. The Cnidaria and their evolution. *Symp. zool. Soc. London.,* No. 16, 449 pp.

Russell, F. S. 1953. *The medusae of the British Isles,* vol. I. *Anthomedusae, Leptomedusae, Limnomedusae, Trachymedusae and Narcomedusae.* 530 pp. Cambridge University Press.

Russell, F. S. 1970. *The medusae of the British Isles,* vol. II. *Pelagic Scyphozoa with a supplement to the first volume on Hydromedusae.* 284 pp. Cambridge University Press.

Stoddart, D. R., & Yonge, Sir Maurice. 1971. Regional variation in Indian Ocean coral reefs. *Symp. zool. Soc. Lond.,* No. 28, 584 pp.

Totton, A. K. 1965. *A synopsis of the Siphonophora.* 230 pp. London: British Museum (Natural History).

Walton-Smith, F. G. 1972. *Atlantic reef corals.* 164 pp. University of Miami Press.

### Ctenophora

Greve, W. 1975. Ctenophora. *Fich. Ident. Zooplancton,* **146,** 1–6.

## Platyhelminthes

Ax, P. 1963. Relationships and phylogeny of the Turbellaria, 191–224. In Dougherty (ed.) *The lower Metazoa*. University of California Press.

Jennings, J. B. 1974. Symbioses in the Turbellaria and their implications in studies on the evolution of parasitism, 127–160. In Vernberg, W. B. (ed.) *Symbiosis in the sea*. University of South Carolina Press.

Riser, N. W., & Morse, M. P. (eds.) 1974. *Biology of the Turbellaria*. 530 pp. New York: McGraw Hill.

## Nemertea

Gibson, R. 1972. *Nemerteans*. 224 pp. London: Hutchinson.

## Rotifera

Donner, J. 1966. *Rotifers*. 80 pp. London: Warne.

## Nematoda

Andrassy, I. 1976. *Evolution as a basis for the systematization of nematodes*. 288 pp. London: Pitman.

Chitwood, B. G., & Chitwood, M. B. 1974. *Introduction to nematology*. 334 pp. Baltimore: University Park Press.

Crofton, H. D. 1966. *Nematodes*. 160 pp. London: Hutchinson.

Croll, N. A. (ed.). 1976. *The organisation of nematodes*. 439 pp. New York, London: Academic Press.

Croll, N. A., & Matthews, B. E. 1977. *Biology of nematodes*. 201 pp. Glasgow: Blackie.

Gerlach, S. A., & Riemann, F. 1973, 1974. The Bremerhaven checklist of aquatic nematodes. Pts 1 & 2. *Veröff. Inst. Meeresforsch. Bremerh*, Suppl. 4, 734 pp.

## Entoprocta

Nielsen, C. 1971. Entoproct life-cycles and the entoproct/ ectoproct relationship. *Ophelia*, **9**, 209–341.

## Sipuncula and Echiura

Gibbs, P. E. 1977. British sipunculans. *Synopses Br. Fauna* (N.S.), No. 12, 35 pp.

Stephen, A. C., & Edmonds, S. J. 1972. *The phyla Sipuncula and Echiura*. 528 pp. London: British Museum (Natural History).

Rice, M. E., & Todorović, M. (eds.). 1975. *Proceedings of the international symposium on the biology of the Sipuncula and Echiura*. Belgrade: Institute for Biological Research. Washington: National Museum of Natural History.

## Pogonophora

Ivanov, A. C. 1963. *Pogonophora*. 479 pp. New York, London: Academic Press.

Southward, E. C. 1963. Pogonophora. *Oceanogr. mar. Biol.*, **1**, 405–28.

Southward, E. C. 1971. Recent researches on the Pogonophora. *Oceanogr. mar. Biol.*, **9**, 193–220.

Webb, M. 1969. *Lamellibrachia barhami*, gen. nov., sp. nov. (Pogonophora) from the north-east Pacific. *Bull. mar. Sci.*, **19**, 18–47.

## Annelida

Brinkhurst, R. O., & Jamieson, B. G. M. 1971. *Aquatic Oligochaeta of the world*. 860 pp. Edinburgh: Oliver & Boyd.

Clark, R. B. 1969. Systematics & phylogeny: Annelida, Echiura, Sipuncula, 1–68. In: Florkin, M., & Scheer, B. (eds.) *Chemical zoology*, vol. 4.

Dales, R. P. 1962. The polychaete stomodeum and the interrelationships of the families of Polychaeta. *Proc. zool. Soc. Lond.*, **139**, 389–428.

Dales, R. P. 1963. *Annelids*. 200 pp. London: Hutchinson.

Day, J. H. 1967. *A monograph of the Polychaeta of Southern Africa*. Pt 1 *Errantia*. Pt 2. *Sedentaria*. 878 pp. London: British Museum (Natural History).

Fauchald, K. 1977. The polychaete worms. Definitions and keys to the orders, families and genera. *Nat. Hist. Mus. Los Ang. Cty*, Sci. Ser., **28**, 1–190.

Hartman, O. 1959 & 1965. Catalogue of the polychaetous annelids of the World. Pts 1 & 2. *Occ. Pap. Allan Hancock Fdn.*, **23**, 628 pp. Supplement 1960–1965 & index, 197 pp.

Storch, V. 1968. Zur vergleichenden Anatomie der segmentalen Muskelsysteme und zur Verwandschaft der Polychaeten-Familien. *Z. Morph. Tiere*, **63**, 251–342.

## Crustacea

Barnard, J. L. 1969. The families and genera of marine gammaridean Amphipoda. *Bull. U.S. natn. Mus.*, **271**, 535 pp.

Birstein, Ya. A. [1973]. *Deepwater isopods (Crustacea, Isopoda) of the North-western part of the Pacific Ocean*. (Translated from Russian.) 316 pp. New Delhi: Indian National Scientific Documentation Centre.

Delamare, C., & Chappuis, P. A. 1954. Morphologie des Mystacocarides. *Arch. Zool. Exp. Gén.*, **91**, 7–24.

Healy, A., & Yaldwyn, J. 1970. *Australian crustaceans in colour*. 112 pp. Sydney: Reed.

Kaestner, A. 1970. *Invertebrate zoology*. vol. III. *Crustacea*. 523 pp. New York: Wiley-Interscience.

Manton, S. M. 1973. Arthropod phylogeny – a modern synthesis. *J. Zool. Lond.*, **171**, 111–30.

Manton, S. M. 1977: *The Arthropoda. Habits, functional morphology, and evolution*. 527 pp. Oxford: Clarendon.

Moore, R. C. (ed.). 1961. *Treatise on invertebrate paleontology*. Pt Q. *Arthropoda*. 3, *Crustacea, Ostracoda*. 442 pp. Geological Society of America and University of Kansas Press.

Naylor, E. 1972. British marine isopods. *Synopses Br. Fauna* (N.S.), No. 3, 86 pp.

Newman, W. A., & Ross, A. 1976. Revision of the balanomorph barnacles; including a catalogue of the species. *Mem. S. Diego Soc. nat. Hist.*, **9**, 108 pp.

Schultz, G. A. 1969. *How to know the marine isopod crustaceans*. 359 pp. Iowa: Brown.

Southward, A. J. 1976. On the taxonomic status and distribution of *Chthamalus stellatus* (Cirripedia) in the north-east Atlantic region with a key to the common intertidal barnacles of Britain. *J. mar. biol. Ass. U.K.*, **56**, 1007–28.

Tinker, S. W. 1965. *Pacific Crustacea*. 134 pp. Rutland, Vermont: Tuttle.

Waterman, T. H., & Chace, F. A. 1960. General crustacean biology, pp. 1–33. In Waterman, T. H. (ed.) *The physiology of Crustacea*. New York and London: Academic Press.

## Chelicerata

Hedgpeth, J. W. 1954. On the phylogeny of the Pycnogonida. *Acta. zool. Stockh.*, **35**, 1–21.

King, P. E. 1973. *Pycnogonids*. 144 pp. London: Hutchinson.

Newell, I. M. 1947. A systematic and ecological study of the Halacaridae of Eastern North America. *Bull. Bingham oceanogr. Coll.*, **10**, (3), 1–232.

Webb, J. E., Wallwork, J. A., & Elgood, J. H. 1978. *Guide to invertebrate animals* (2nd edn). 305 pp. London: Macmillan.

## Tardigrada

Morgan, C. I., & King, P. E. 1976. British tardigrades. *Synopses Br. Fauna* (N.S.) No. 9, 133 pp.

## Mollusca

Abbott, R. T. 1962. *Seashells of the world.* 160 pp. New York: Golden Press.

Abbott, R. T. 1968. *Seashells of North America.* 280 pp. New York: Golden Press.

Abbott, R. T. 1976. *Seashells.* 159 pp. New York: Bantam Books.

Boss, K. J. 1971. Critical estimate of the number of recent Mollusca. *Occ. Pap. Mollusks, Harv.*, **3**, 81–135.

Coleman, N. 1975. *What shell is that?* 308 pp. Sydney: Hamlyn.

Chun, C. 1975. *The Cephalopoda.* Pt 1 *Oegopsida.* Pt 2 *Myopsida, Octopoda.* 436 pp. Jerusalem: Keter Publishing House.

Clarke, M. R. 1966. A review of the systematics and ecology of oceanic squids. *Adv. mar. Biol.*, **4**, 91–300.

Graham, A. 1971. British prosobranchs. *Synopses Br. Fauna* (N.S.), No. 2, 112 pp.

Humfrey, M. 1975. *Seashells of the West Indies.* 351 pp. London: Collins.

Keen, A. M. 1971. *Sea shells of tropical west America* (2nd edn). 1064 pp. Stanford University Press.

Lemche, H., & Wingstrand, K. G. 1959. The anatomy of *Neopilina galathea* Lemche, 1957 (Mollusca Tryblidiacea). *Galathea Rep.*, **3**, 9–71.

Moore, R. C. (ed.). 1960. *Treatise on invertebrate paleontology.* Pt I *Mollusca 1.* 351 pp. Geological Society of America and University of Kansas Press.

Moore, R. C. (ed.) 1969. *Treatise on invertebrate paleontology.* Pt N. *Mollusca 6. Bivalvia.* 951 pp. Geological Society of America and University of Kansas Press.

Morris, P. A. 1973. *A field guide to shells of the Atlantic and Gulf coasts and the West Indies.* (3rd edn). 330 pp. Boston: Houghton Mifflin.

Morton, J. E. 1967. *Molluscs.* (4th edn). 244 pp. London: Hutchinson.

Nixon, M., & Messenger, J. B. (eds.). 1977. The biology of cephalopods. *Symp. zool. Soc. Lond.*, **38**, 615 pp.

Oliver, A. P. H. 1975. *The Hamlyn guide to shells of the world.* 320 pp. London: Hamlyn.

Palmer, C. P. 1974. A supraspecific classification of the scaphopod Mollusca. *Veliger*, **17**, 115–25.

Penniket, J. R., & Moon, G. J. H. 1970. *New Zealand seashells in colour.* 112 pp. Wellington: Reed.

Russell, H. D. 1971. *Index Nudibranchia. A catalogue of the literature 1554–1965.* 141 pp. Greenville: Delaware Museum of Natural History.

Taylor, D. W., & Sohl, N. F. 1962. An outline of gastropod classification. *Malacologia*, **1**, 7–32.

Tebble, N. 1976. *British bivalve seashells.* 212 pp. London: British Museum (Natural History).

Thompson, T. E. 1976. *Biology of opisthobranch molluscs. 1.* 207 pp. London: Ray Society.

Thompson, T. E., & Brown, G. H. 1976. British opisthobranch molluscs. *Synopses Br. Fauna* (N.S.), No. 8, 201 pp.

Voss, G. L. 1977. A classification of recent cephalopods. *Symp. zool. Soc. Lond.*, No. 38, 575–9.

Voss, G. L., & Williamson, G. R. 1971. *Cephalopods of Hong Kong.* 138 pp. Hong Kong Government Press.

Warmke, G. L., & Abbott, R. T. 1961. *Caribbean seashells.* 346 pp. Narbeth, Pennsylvania: Livingston.

Wilson, B. R., & Gillett, K. 1971. *Australian shells.* 168 pp. Sydney: Reed.

## Bryozoa

Hayward, P. J. 1976. The marine fauna of Lundy. Bryozoa. *Rep. Lundy Fld. Soc.*, **27**, 16–34.

Ryland, J. S. 1970. *Bryozoans.* 175 pp. London: Hutchinson.

Ryland, J. S. 1976. Physiology and ecology of bryozoans. *Adv. mar. Biol.*, **14**, 285–443.

Ryland, J. S., & Hayward, P. J. 1977. British anascan bryozoans. *Synopses Br. Fauna* (N.S.), No. 10, 188 pp.

## Phoronida

Emig, C. C. 1971. Taxonomie et systématique des phoronidiens. *Bull. Mus. natn. Hist. nat. Paris*, Ser. 3, **8**, (Zool. 8), 474–95.

## Brachiopoda

Moore, R. C. (ed.). 1965. *Treatise on invertebrate paleontology*, Pt H. *Brachiopoda.* vol. 1 (521 pp.) vol. 2 (404 pp.). Geological Society of America and University of Kansas Press.

Rudwick, M. J. S. 1970. *Living and fossil brachiopods.* 199 pp. London: Hutchinson.

## Chaetognatha

Alvariño, A. 1965. Chaetognaths. *Oceanogr. mar. Biol.*, **3**, 115–94.

Tokioka, T. 1965. The taxonomical outline of Chaetognatha. *Publ. Seto mar. biol. Lab.*, **12**, 335–57.

## Echinodermata

Bakus, G. J. 1973. Biology and ecology of tropical holothurians, 325–67. In: Jones, O. A., & Endean, R. (eds.). *Biology and geology of coral reefs*, vol. 2. Biol. 1. New York, London: Academic Press.

Boolootian, R. A. (ed.). 1966. *Physiology of Echinodermata.* 822 pp. New York: Wiley–Interscience.

Clark, A. H. 1915–1959. A monograph of the existing crinoids I. The comatulids. *Bull. U.S. natn. Mus.*, No. 82, Pt 1–4.

Clark, A. H., & Clark, A. M. 1967. A monograph of the existing crinoids. I. The comatulids. *Bull. U.S. natn. Mus.*, No. 82, Pt 5.

Clark, A. M. 1977. *Starfishes and related echinoderms.* (3rd edn), 160 pp. London: British Museum (Natural History) & T.F.H.

Clark, A. M., & Courtman-Stock, J. 1976. *The echinoderms of southern Africa.* 277 pp. London: British Museum (Natural History).

Clark, A. M., & Rowe, F. W. E. 1971. *Monograph of shallow-water Indo-West Pacific echinoderms.* 238 pp. London: British Museum (Natural History).

Millott, N. (ed.). 1967. Echinoderm biology. *Symp. zool. Soc. Lond.*, No. 20, 240 pp.

Bibliography

Moore, R. C. (ed.). 1966, 1967. *Treatise on invertebrate paleontology*, Pts S & U. *Echinodermata*. vol. 1 (650 pp.), vol. 3 (695 pp.). Geological Society of America and University of Kansas Press.

Nichols, D. 1969. *Echinoderms* (4th edn). 192 pp. London: Hutchinson.

Ubaghs, G. 1953. Classe des Crinoides. In: Priveteau, J. (ed.). *Traité de paléontologie*, vol. 3.

## Hemichordata and Chordata

Barrington, E. J. W. 1965. *The biology of Hemichordata and Protochordata*. 176 pp. Edinburgh: Oliver & Boyd.

Berrill, N. J. 1950. *The Tunicata, with an account of the British species*. 354 pp. London: Ray Society.

Hubbs, C. L. 1922. A list of the lancelets of the world with diagnoses of five new species of *Branchiostoma*. *Occ. Pap. Mus. Zool. Univ. Mich.*, **105**, 1–16.

Metcalf, M. M., & Bell, M. 1918. The Salpidae: a taxonomic study. *Bull. U.S. natn. Mus.*, **100** (2), 5–189.

Name, Van, W. G. 1945. The North & South American ascidians. *Bull. Amer. Mus. nat. Hist.*, **84**, 1–476.

Thompson, H. 1948. *Pelagic tunicates of Australia*. 196 pp. Melbourne: Commonwealth Council for Scientific and Industrial Research.

# IDENTIFICATION GUIDES

## General

Angel, M., & Angel, H. 1974. *Ocean life*. 72 pp. London: Octopus.

Faulkner, D., & Smith, C. L. 1970. *The hidden sea*. 149 pp. London: Macdonald.

Hardy, A. C. 1956. *The open sea. Its natural history: the world of plankton*. 335 pp. London: Collins.

Heezen, B. C., & Hollister, C. D. 1971. *The face of the deep*. 659 pp. New York: Oxford University Press.

Newell, G. E., & Newell, R. C. 1973. *Marine plankton. A practical guide* (Revised edn). 244 pp. London: Hutchinson.

Petron, C., & Lozet, J.–B. 1975. *The Guinness guide to underwater life*. 218 pp. Enfield, England: Guinness Superlatives.

Riefenstahl, L. 1978. *Coral gardens*. 223 pp. London: Collins.

Roessler, C. 1978. *The underwater wilderness – life around the great reefs*. 319 pp. New York: Chanticleer Press.

Wickstead, J. H. 1965. *An introduction to the study of tropical plankton*. 160 pp. London: Hutchinson.

## North Atlantic & Mediterranean

Barrett, J. H., & Yonge, C. M. 1958. *Collins pocket guide to the seashore*. 272 pp. London: Collins.

Campbell, A. C. 1976. *The Hamlyn guide to the seashore and shallow seas of Britain and Europe*. 320 pp. London: Hamlyn.

Eales, N. B. 1967. *The littoral fauna of the British Isles* (4th edn). 306 pp. Cambridge University Press.

Gosner, K. L. 1971. *Guide to identification of marine and estuarine invertebrates*. 693 pp. New York: Wiley-Interscience.

Hamner, W. M. 1974. Ghosts of the Gulf Stream. Blue-water plankton. *Natn. geogr. Mag.*, **146**, 530–45.

Hass, W. de., & Knorr, F. 1966. *The young specialist's look at marine life*. 356 pp. London: Burke.

Luther, W., & Fieldler, K. 1976. *A field guide to the Mediterranean seashore*. 272 pp. London: Collins.

Miner, R. W. 1950. *Field book of seashore life*. 888 pp. New York: Putnams.

Riedl, R. 1966. *Biologie der Meereshöhlen*. 636 pp. Hamburg: Verlag Paul Parey.

Riedl, R. 1970. *Fauna und flora der Adria*. 702 pp. Hamburg: Verlag Paul Parey.

Sisson, R. F. 1976. Adrift on a raft of Sargassum. *Natn. geogr. Mag.*, **149**, 188–99.

Yonge, C. M. 1966. *The seashore*. (Revised edn). 311 pp. London: Collins.

## Caribbean

Andrews, J. 1977. *Shells and shores of Texas*. 365 pp. Austin: University of Texas Press.

Colin, P. 1978. *Caribbean reef invertebrates and plants*. 512 pp. Hong Kong: T.F.H.

Fotheringham, N., & Brunenmeister, S. 1975. *Common marine invertebrates of the northwestern Gulf coast*. 197 pp. Houston: Gulf Publishing Co.

Greenberg, J., & Greenberg, I. 1972. *The living reef*. 110 pp. Miami: Sea Hawk Press.

Starck, W. A., & Starck, J. D. 1972. Probing the deep reefs' hidden realm. *Natn. geogr. Mag.*, **142**, 867–86.

Voss, G. L. 1976. *Seashore life of Florida and the Caribbean*. 168 pp. Miami: Seeman.

Zeiller, W. 1974. *Tropical marine invertebrates of southern Florida and the Bahama Islands*. 132 pp. New York: Wiley–Interscience.

## South Atlantic

Day, J. H. 1969. *A guide to marine life on South African shores*. 300 pp. Cape Town: Balkema.

## Indo-Pacific

Basson, P. W., Burchard, J. E., Hardy, J. T., & Price, R. G. 1977. *Biotopes of the Western Arabian Gulf. Marine life and environments of Saudi Arabia*. 284 pp. Dhahran, Saudi Arabia: ARAMCO.

Bennett, I. 1971. *The Great Barrier Reef*. 183 pp. Melbourne: Lansdowne.

Clark, E., & Doubilet, D. 1975. Strange world of the Red Sea reefs. *Natn. geogr. Mag.*, **148**, 338–65.

Dakin, W. J. 1960. *Australian seashores*. 372 pp. Sydney: Angus & Robertson.

Darom, D. 1976. *The Red Sea*. 106 pp. Tel. Aviv: Sadan.

Earle, S. A., & Giddings, A. 1976. Life springs from death in Truk Lagoon. *Natn. geogr. Mag.*, **149**, 578–613.

Edmonds, C. 1974. *Dangerous marine animals of the Indo-Pacific region.* 235 pp. Newport, Australia: Wedneil.

Faulkner, D. 1974. *This living reef.* 184 pp. New York: Quadrangle/The New York Times Book Co.

Faulkner, D., & Fell, B. 1976. *Dwellers in the sea.* 194 pp. New York: Reader's Digest.

Fricke, H. W. 1972. *The coral seas.* 224 pp. London: Thames & Hudson.

Gillett, K., & McNeill, F. 1959. *The Great Barrier Reef and adjacent isles.* 194 pp. Sydney: Coral Press.

Hobson, E., & Chave, E. H. 1972. *Hawaiian reef animals.* 135 pp. Hawaii University Press.

Mayland, H. J. 1975. *Korallenfische und Nierclere Tiere.* 295 pp. Hannover: Landbuch-Verlag.

Parish, S. 1974. *Australia's ocean of life.* 128 pp. Newport, Australia: Wedneil.

Schuhmacher, H. 1976. *Korallenriffe.* 275 pp. Munich: BLV.

Taylor, V., Macleish, K., & Taylor, R. 1973. Australia's Great Barrier Reef. *Natn. geogr. Mag.*, **143**, 728–79.

Vine, P. 1972. *Life on coral reefs in the Seychelles.* 56 pp. London: Phillips.

### Zealandic

Doak, W. 1971. *Beneath New Zealand seas.* 113 pp. Wellington: Reed.

Heath, E., & Dell, R. K. 1971. *Seashore life of New Zealand.* 72 pp. Wellington: Reed.

Miller, M., & Batt, G. 1973. *Reef and beach life of New Zealand.* 141 pp. Auckland: Collins.

Morton, J., & Miller, M. 1968. *The New Zealand seashore.* 638 pp. Auckland: Collins.

### North Pacific & Panamanian

Brusca, R. C. 1973. *A handbook to the common intertidal invertebrates of the Gulf of California.* 427 pp. Tucson: University of Arizona Press.

Carefoot, T. 1977. *Pacific seashores. A guide to intertidal ecology.* 208 pp. Seattle: University of Washington Press.

Kozloff, E. N. 1974. *Keys to the marine invertebrates of Puget Sound, the San Juan Archipelago, and adjacent regions.* 226 pp. Seattle: University of Washington Press.

Kozloff, E. N. 1976. *Seashore life of Puget Sound, the Strait of Georgia, and the San Juan Archipelago.* 282 pp. Vancouver: Douglas.

Ricketts, E. F., & Calvin, J. 1962. *Between Pacific tides.* (3rd edn revised by J. W. Hedgpeth). 516 pp. Stanford University Press.

Smith, R. I., & Carlton, J. T. 1975. *Lights Manual: Intertidal invertebrates of the central Californian coast.* (3rd edn). 716 pp. Berkeley: University of California Press.

Steinbeck, J., & Ricketts, E. F. 1941. *Sea of Cortez.* 598 pp. New York: Viking Press.

# Glossary

*Adductor muscle*. Muscle which closes the shell valves in bivalve molluscs and the carapace valves in barnacles.
*Ambulacrum*. Region of echinoderm body which bears the tube-feet.
*Amphid*. Anterior sense-organ of nematodes.
*Antenna*. Sensory appendage on head of annelids and crustaceans.
*Aristotle's lantern*. Complex scraping and chewing apparatus of a sea-urchin.

*Bilateral symmetry*. Symmetry which divides a body along a single plane into identical halves.
*Biramous*. Two-branched.
*Bridle*. Raised ridge of tissue more or less encircling the body behind the tentacles of many pogonophores.
*Bursa*. Internal pouch found in brittle-stars.
*Byssal threads*. Attachment threads of bivalve molluscs.

*Calcareous*. Composed of calcium carbonate or chalk.
*Carapace*. Chitinous and/or calcareous skin fold enclosing part of or whole body of crustaceans and some arachnids.
*Caudal furca*. Pair of processes found on end of tail of certain crustaceans.
*Cerata*. Processes on the backs of some shell-less molluscs.
*Chelicerae*. First pair of appendages of arachnids that are often used for seizing prey.
*Cilia*. Short hair-like processes that are often used in locomotion and food-gathering.
*Cirrus*. Small flexible tentacle-like sensory appendage.
*Clitellum*. Swollen glandular portion of skin of some annelids.
*Comb rows*. Short transverse rows of fused cilia found in ctenophores.
*Commensal*. A term applied to two organisms living in close association with one another.
*Compound eye*. Eye with many similar facets found in arthropods.

*Dorsal*. The back (usually upper surface) of an animal.

*Fission*. Splitting of a body into two or more parts.
*Flagellum*. Whip-like hair projecting from a cell.

*Girdle*. A ring of setae on the body of a pogonophore or that part of the mantle which encircles the shell plates of chitons.
*Gonad*. Organ which produces eggs or sperm.

*Hectocotylus*. Modified tentacle of male cephalopod mollusc used to transfer sperm to the female.
*Hermaphrodite*. An animal which possesses both male and female reproductive organs.

*Interambulacrum*. Area of echinoderm body between rows of tube-feet.
*Introvert*. Extensible anterior region of sipunculan worms.

*Larva*. A pre-adult form which hatches from the egg and which often leads a different life from that of the adult.
*Lophophore*. Food-gathering or respiratory organ with tentacles or filaments.

*Madreporite*. Sieve-like plate connecting the water-vascular system of echinoderms with the exterior.
*Mantle*. That part of the body wall of molluscs which secretes the shell, covers the visceral mass and encloses the mantle cavity.
*Megasclere*. Large supporting spicule of a sponge.
*Mesogloea*. Middle jelly-like layer in body wall of coelenterates.
*Metamorphosis*. Transformation from the larva to the adult form.
*Microsclere*. Minute sponge spicule.

*Nematocyst*. Stinging cell of coelenterates.
*Neuropodium*. Lower lobe of polychaete parapodium.
*Notochord*. Supporting skeletal rod present in some stage of development of all chordate animals.
*Notopodium*. Upper lobe of polychaete parapodium.

*Operculum*. A plug closing the aperture of a tube or shell.
*Opisthosoma*. Posterior part of the body of chelicerates.
*Osculum*. Large opening in a sponge through which the water leaves the body.
*Ossicles*. Small calcareous plates.

*Palp*. Sensory appendage near the mouth.
*Parapodial lobe*. Large lateral lobe found on the foot of certain molluscs.
*Parapodium*. Segmental lateral appendage on the body of a polychaete worm.
*Parasite*. An organism living on or in the body of another organism (the host) from which it obtains food.
*Paxillae*. Raised areas of skin plates on certain starfish.
*Pedipalp*. Second appendage of an arachnid.
*Pedicellariae*. Pincer-like appendages on body surface of echinoderms.
*Peristomium*. Second head zone of a polychaete which bears the mouth, palps and tentacles.
*Pharynx*. Anterior region of the gut which is sometimes eversible.
*Phasmid*. Posterior sense organ found in some nematodes.
*Photosynthesis*. Process in which green plants convert carbon dioxide and water into carbohydrates with the aid of energy from sunlight.
*Phylogeny*. Evolutionary history.
*Pinnule*. Sub-branch of a feather-like appendage, particularly in crinoid echinoderms.
*Plankton*. The drifting life of the sea.
*Proboscis*. Projection, often tubular and protrusible, in the head region.
*Prosoma*. Anterior part of the body of chelicerates.
*Prostomium*. First head zone of a polychaete, which may bear eyes, antennae, and palps.

*Radial canals*. Canals passing from the centre to the edge of the bell in coelenterate medusae.
*Radial shield*. Conspicuous plate on central disc of brittle stars.
*Radial symmetry*. Symmetry in which the body parts are arranged around a central axis.
*Radula*. Toothed tongue-like organ found in molluscs.
*Respiratory trees*. Respiratory processes opening into hind gut of sea cucumbers.
*Rhabdite*. Small rod-like structure present in the skin of flatworms.
*Rostrum*. Anterior extension of the carapace in Crustacea.

*Seta (Chaeta)*. Chitinous bristle.

*Siliceous*. Composed of silica, a glass-like material.

*Siphon*. Tubular projection of the body common in molluscs and sea-squirts, through which water is drawn, or expelled.

*Siphonal canal*. Elongated groove in shell which carries the mantle siphon of gastropod molluscs.

*Spicule*. Skeletal element.

*Spongin*. Fibrous skeletal material of some sponges.

*Stolons*. Creeping branches which often connect individuals in a colony.

*Stylet*. Small pointed bristle.

*Symbiosis*. An association between organisms for their mutual benefit.

*Telson*. Tail piece of Crustacea.

*Test*. Rigid shell of sea-urchins.

*Torsion*. Twisting of the body during the development of gastropod molluscs which results in the mantle cavity, with gills and anus, being situated anteriorly.

*Tubercle*. Small conical projection on the body surface.

*Tube-foot*. Appendage of echinoderms, connected to the water vascular system, which often has a locomotory function.

*Tunic*. Gelatinous or leathery coat surrounding the body of tunicates.

*Uncini*. Short, toothed setae.

*Uniramous*. With one branch only.

*Velum*. Inward projection of the edge of the bell in some coelenterate medusae.

*Ventral*. Under-surface of an animal. Opposite to dorsal.

*Visceral mass*. That part of the molluscan body containing the gut and other organs.

*Viviparous*. Giving birth to live young.

*Wheel organ*. Ciliated structure on the head of rotifers, used in locomotion and feeding.

*Zooid*. An individual in a colony, particularly in a coelenterate or bryozoan colony.

*Zooxanthella*. A single-celled plant (alga) living symbiotically in the tissues of another organism.

# Photographic credits

The copyright © of the photographs used in this book is vested in the individual photographers (1979).

**Anthony Baverstock**
58/7

**Peter Bertorelli**
79/2

**British Museum (Natural History)**
103/9

**British Museum (Natural History): David George**
2/1; 2/2; 2/3; 2/4; 2/5; 53/5; 57/9

**Dick Clarke**
4/10; 9/2; 9/6; 9/7; 10/5; 11/7; 11/8; 24/7; 25/3; 26/2; 26/8; 30/2; 34/5; 35/1; 35/8; 35/10; 36/2; 36/5; 36/12; 39/6; 41/1; 41/2; 42/8; 55/4; 58/1; 70/4; 74/3; 105/8; 114/1; 123/8; 125/4

**Richard Chesher**
36/9; 39/8; 40/8; 40/9; 41/10; 42/4; 43/10; 44/1; 44/3; 44/9; 45/8; 77/1; 83/4; 89/3; 98/2; 104/9; 107/7; 112/8; 115/4; 118/5; 119/5; 120/6; 120/7; 120/8; 121/1; 121/3; 121/4; 122/2; 122/6; 123/4

**Peri Coelho**
1/2; 5/9; 9/10; 72/2; 94/4; 119/7

**Neville Coleman**
15/4; 19/5; 19/7; 20/1; 26/12; 28/3; 31/7; 71/7; 72/6; 76/6; 76/7; 80/10; 87/14; 93/3; 106/6; 107/3; 110/6; 112/2; 113/5; 114/10; 115/5; 115/7; 116/7; 120/10; 121/11; 124/7

**Mike Coltman**
9/1; 111/10

**William Cross**
17/2

**Peter David**
10/9; 11/3; 16/4; 16/5; 16/6; 17/1; 18/1; 18/3; 29/5; 51/2; 51/3; 59/8; 59/11; 59/12; 60/1–60/4; 60/9; 61/3; 61/10; 62/1; 62/3; 63/6; 64/8; 64/9; 65/1; 65/4; 65/5; 65/6; 65/8; 66/1–66/9; 67/1–67/6; 68/5; 69/7–69/11; 70/10; 72/4; 77/5; 80/2; 83/7; 87/8–88/2; 89/10; 92/11; 93/2; 98/4; 98/6; 99/1–99/8; 100/1–100/4; 100/6; 101/1; 101/2; 101/3; 104/1; 104/2; 124/5; 124/6; 127/6; 128/5

**Jean Deas**
45/9; 46/11; 50/5; 90/1

**Walter Deas**
1/8; 5/8; 6/2; 6/10; 8/4; 8/5; 8/6; 12/4; 12/6; 12/8; 13/3; 14/9; 15/8; 20/4; 21/6; 21/10; 22/4; 22/8; 22/9; 22/10; 23/10; 24/2; 24/5; 26/5; 26/10; 26/11; 27/3; 27/8; 31/4; 31/5; 33/1; 33/9; 33/10; 35/3; 35/4; 35/6; 36/7; 37/2; 37/3; 37/4; 37/5; 37/8; 38/2; 38/3; 39/1; 39/10; 40/2; 40/5; 41/5; 41/7; 42/2; 43/4; 43/6–43/9; 44/5; 44/6; 44/11; 46/3; 46/4; 46/9; 47/1; 47/6; 47/7; 49/1; 49/2; 49/7; 49/10; 50/1; 50/2; 50/6; 58/6; 61/1; 67/7; 67/11; 70/3; 70/7; 72/9; 74/9; 75/2; 77/7; 82/8; 83/1; 84/1; 84/2; 85/3; 85/6; 85/7; 88/8; 90/4; 90/6; 90/7; 90/9; 91/3; 91/5; 91/6; 92/3; 92/5; 94/2; 96/4; 96/7; 98/1; 98/5; 100/7; 100/8; 104/5; 104/7; 104/8; 105/1; 105/4;
107/8; 107/9; 108/9; 109/5; 109/7; 109/8; 111/8; 112/1; 113/4; 113/6; 113/8; 114/5; 115/2; 115/8; 116/8; 117/2; 117/3; 117/8; 118/4; 119/1; 119/2; 119/3; 120/2; 121/12; 122/1; 122/5; 123/1; 123/3; 123/10; 124/1; 125/7; 125/8; 125/10; 126/1

**David George**
1/1; 1/6; 2/7; 2/10; 3/2; 4/1; 4/9; 5/2; 5/5; 6/4; 6/8; 9/4; 11/1; 13/1; 13/4; 13/5; 13/6; 13/7; 18/7; 18/10; 21/9; 26/7; 29/1; 29/3; 29/6; 29/8; 29/9; 29/11; 30/7; 33/2; 33/3; 33/7; 33/8; 34/9; 38/9; 39/7; 42/6; 45/3; 46/7; 47/2; 49/9; 51/1; 51/6; 51/8; 52/3–52/6; 53/1–53/4; 53/6; 53/7; 53/9; 54/1; 54/2; 54/4–54/7; 54/9; 55/1; 55/5–55/10; 56/1; 56/2; 56/5–56/8; 57/1; 57/7; 57/8; 57/10; 57/12; 58/8; 58/9; 58/10; 59/3; 59/4; 59/5; 59/7; 59/10; 60/6; 60/8; 60/10; 61/4; 61/5; 61/8; 62/5; 62/7; 62/8; 63/7–63/13; 64/1–64/7; 65/2; 69/4; 71/8; 72/8; 73/9; 74/8; 75/7; 76/1; 78/6; 78/8; 78/9; 78/10; 79/6–79/9; 80/7; 81/6; 81/8; 83/5; 83/10; 84/12; 88/4; 89/5; 91/1; 91/9; 93/1; 93/10; 93/11; 94/1; 94/3; 94/6; 94/8; 95/9; 96/5; 96/6; 96/9; 96/10; 97/1; 97/3; 97/4; 98/7; 100/5; 101/4; 102/1; 104/4; 106/3; 106/7; 106/9; 107/4; 107/5; 109/1; 109/3; 109/4; 110/7; 110/8; 111/1; 114/6; 114/7; 115/6; 115/9; 117/6; 117/9; 119/8; 121/8; 125/2; 126/4; 126/10

**William Gladfelter**
2/6; 3/5; 3/8; 4/11; 7/7; 9/9; 9/11; 10/7; 10/8; 11/2; 15/6; 20/8; 21/2; 23/8; 25/5; 25/6; 26/3; 28/7; 30/1; 32/2; 32/3–32/5; 32/7; 32/10; 34/2; 34/4; 35/2; 35/7; 38/1; 39/3; 39/4; 39/9; 42/10; 43/1; 43/2; 43/3; 43/5; 45/10; 49/4; 52/7; 57/21; 68/10; 69/3; 70/8; 71/2; 72/5; 72/7; 73/2; 73/10; 74/7; 75/1; 76/5; 76/9; 84/11; 89/2; 95/4; 105/3; 120/1; 121/2; 121/5; 124/2; 126/2

**David Guiterman**
61/7; 62/2; 63/4; 97/7; 97/8; 104/3

**John Harvey**
74/4; 84/3

**Keith Hiscock**
2/8; 15/2; 28/9; 101/5; 101/6; 102/6

**Karl Kleemann**
1/4; 1/9; 8/10; 10/6; 12/10; 14/11; 20/3; 21/4; 21/5; 22/6; 24/1; 25/1; 25/4; 25/7; 26/1; 27/2; 28/5; 36/1 36/6; 36/11; 37/1; 37/6; 37/7; 39/5; 40/1; 40/3; 40/4; 40/6; 41/3; 41/4; 41/6; 41/8; 41/9; 42/3; 42/5; 44/4; 44/7; 44/8; 45/1; 45/4; 45/5; 45/7; 49/5; 50/10; 68/3; 88/1; 89/8; 91/2; 92/4; 93/7; 93/8; 95/6; 95/10; 103/1; 103/6; 104/6; 105/6; 108/1; 108/2; 109/6; 111/9; 122/7; 124/4; 127/1; 127/5

**Gertraud Krapp-Schickel**
63/2; 63/5; 64/10; 64/11

**Michael Laverack**
17/8; 47/11; 48/7; 49/8; 50/11; 52/9; 53/10; 56/3; 56/4; 77/9; 79/1; 84/9; 92/7; 92/10; 96/8; 97/6; 102/3; 103/7; 103/8; 103/10; 103/11; 114/4; 124/4; 124/8; 124/9; 124/10; 127/10; 128/8; 128/9

**Jan Lenart**
9/3; 46/10; 91/7; 103/2

**Roger Lincoln**
62/6; 62/9; 63/3; 79/5; 84/10

**John Lythgoe**
5/7; 36/3; 55/2; 59/2; 81/1; 81/2; 87/7

Larry Madin
12/2; 12/3; 12/5; 15/5; 15/7; 15/9; 16/2; 16/3; 16/7;
17/3–17/7; 17/9; 18/2; 18/6; 47/5; 47/9; 47/10; 48/2;
48/6; 54/3; 65/3; 65/7; 65/9; 65/10; 87/5; 87/6; 87/9;
127/4; 127/7; 127/9; 127/11; 128/1–128/4; 128/6

Michael Mastaller
49/6; 52/12; 58/5; 69/2; 71/1; 76/3; 79/4; 80/8; 85/1;
93/9; 94/5; 96/1; 114/8; 116/1; 117/4; 120/5; 121/9

Maria Mizzaro
50/4; 50/9; 51/5; 52/11; 61/6; 62/4; 73/5; 77/2; 77/10;
78/2; 80/3; 81/7; 84/5; 85/8; 88/3; 89/1; 89/4; 89/6;
89/7; 89/9; 90/3; 92/9; 97/5

Pat Morgan
126/9

Christian Petron
1/7; 2/11; 3/1; 3/3; 3/4; 3/6; 4/2–4/5; 5/1; 5/3; 5/4;
5/6; 5/10; 6/3; 6/6; 6/9; 7/1; 7/4; 7/5; 7/6; 7/8; 8/1;
8/2; 8/3; 8/8; 8/9; 8/11; 9/8; 11/4; 11/6; 11/9; 12/1; 19/6;
19/8; 20/10; 22/3; 22/5; 23/1; 23/2; 23/5; 26/6; 27/1;
30/4; 30/6; 31/6; 32/1; 35/5; 35/9; 36/4; 36/8; 37/11;
38/4; 38/5; 38/6; 38/7; 38/8; 40/7; 44/2; 44/10; 45/6;
46/8; 47/3; 49/3; 50/3; 50/7; 54/8; 55/3; 56/9; 57/4; 59/6;
68/8; 70/2; 70/5; 70/6; 70/9; 71/5; 71/6; 73/5; 75/5; 75/8;
77/6; 78/5; 80/4; 80/6; 80/9; 81/4; 81/5; 81/9; 81/10;
82/1; 82/3–82/7; 82/9; 83/2; 83/3; 83/6; 84/6; 84/7;
84/8; 85/2; 85/4; 85/5; 85/9; 85/10; 86/1–86/10; 87/1;
87/2; 87/3; 88/5; 90/5; 90/8; 90/10; 91/8; 92/1; 92/2;
94/9; 95/1; 95/2; 95/3; 95/5; 95/8; 96/2; 96/3; 97/9;
98/8; 100/9; 100/11; 105/9; 106/1; 106/8; 107/10;
108/3; 108/5; 108/6; 108/10; 109/9; 110/2; 112/4; 114/2;
116/2; 117/1; 117/5; 117/10; 118/1; 118/9; 118/10; 119/4;
120/4; 120/9; 121/6; 121/7; 122/8; 122/9; 122/10; 123/5;
123/6; 123/11; 125/9; 126/5

John Place
90/2

Howard Platt
51/7; 51/9; 51/10; 52/1; 52/2

Rod Salm
30/9; 36/10; 76/2; 107/2; 112/3; 119/10; 127/3

Howard Sanders
77/3

Helmut Schuhmacher
3/7; 52/8; 77/4; 97/2

Peter Scoones
18/4; 18/5; 18/8; 19/9; 21/11; 23/7; 25/2; 25/10; 29/4;
30/3; 32/9; 32/11; 32/12; 33/6; 45/2; 50/8; 52/10; 57/3;
61/2; 61/9; 67/12; 71/3; 72/3; 74/2; 75/6; 78/11; 91/10;
108/7; 110/9; 111/6; 111/7; 112/6; 112/9; 112/10; 118/7;
119/9; 123/2; 125/3; 126/7

Heinz Splechtna
51/4; 75/3; 75/4; 78/1; 78/3; 78/4; 79/10; 80/1; 81/3;
83/8; 86/1

Alan Southward
9/5

Armin Svoboda
1/3; 1/5; 2/9; 2/12; 4/6; 4/7; 4/8; 6/1; 6/5; 7/2; 7/3;
7/9; 8/7; 10/1; 10/2; 10/3; 10/4; 11/5; 12/7; 13/2; 13/8;
13/9; 13/10; 14/1–14/8; 14/10; 15/1; 15/3; 16/1; 18/9;
19/1–19/4; 20/2; 20/5; 20/6; 20/7; 20/9; 21/1; 21/3; 21/7;
21/8; 22/1; 22/2; 22/7; 23/3; 23/4; 23/6; 23/9; 24/3;
24/4; 24/6; 24/8; 24/9; 24/10; 25/8; 25/9; 26/4; 26/9;
27/4; 27/5; 27/6; 27/7; 27/9; 28/1; 28/2; 28/4; 28/6;
28/8; 29/2; 29/7; 29/10; 30/5; 30/8; 31/1; 31/2; 31/3;
31/8; 31/9; 32/6; 32/8; 33/4; 33/5; 34/1; 34/3; 34/6;
34/7; 34/8; 34/10; 34/11; 37/9; 37/10; 39/2; 42/1; 42/7;
42/9; 46/1; 46/2; 46/5; 46/6; 47/4; 47/8; 48/3; 48/4;
48/5; 48/8; 53/8; 57/5; 57/6; 57/11; 58/2; 58/3; 58/4;
59/1; 60/5; 60/7; 63/1; 67/8; 67/9; 67/10; 68/1; 68/2;
68/4; 68/6; 68/7; 68/9; 68/11; 69/1; 69/5; 69/6; 70/1;
71/4; 72/1; 72/10; 73/1; 73/3; 73/4; 73/6; 73/8; 74/1;
74/5; 74/6; 76/4; 76/8; 77/8; 78/7; 79/3; 80/5; 82/2;
83/9; 83/11; 84/4; 88/6; 88/7; 88/9; 91/4; 92/6; 92/8;
93/4; 93/5; 93/6; 94/7; 95/7; 98/3; 98/9; 100/10; 101/7;
101/8; 102/2; 102/4; 102/5; 102/7; 102/8; 102/9; 103/3;
103/4; 103/5; 105/2; 105/5; 105/7; 106/2; 106/4; 106/10;
106/11; 107/1; 107/6; 108/4; 108/8; 109/2; 110/1; 110/3;
110/4; 110/5; 111/2; 111/3; 111/4; 111/5; 112/5; 112/7;
112/11; 113/1; 113/2; 113/3; 113/7; 113/9; 114/3; 114/9;
115/1; 115/3; 116/3; 116/4; 116/5; 116/6; 116/9; 117/7;
117/11; 118/2; 118/3; 118/6; 118/8; 119/6; 120/3; 122/3;
122/4; 123/7; 123/9; 124/3; 125/1; 125/5; 125/6; 126/3;
126/6; 126/8; 127/2; 127/8; 128/7

Neil Swanberg
48/1

Peter Vine
106/5

Roman Vishniac
59/9

# Index